P9-EJO-885

# JOHN F. KENNEDY AND THE RACE TO THE MOON

## Palgrave Studies in the History of Science and Technology

James Rodger Fleming (Colby College) and Roger D. Launius (National Air and Space Museum), Series Editors

This series presents original, high-quality, and accessible works at the cutting edge of scholarship within the history of science and technology. Books in the series aim to disseminate new knowledge and new perspectives about the history of science and technology, enhance and extend education, foster public understanding, and enrich cultural life. Collectively, these books will break down conventional lines of demarcation by incorporating historical perspectives into issues of current and ongoing concern, offering international and global perspectives on a variety of issues, and bridging the gap between historians and practicing scientists. In this way they advance scholarly conversation within and across traditional disciplines but also to help define new areas of intellectual endeavor.

Published by Palgrave Macmillan:

*Continental Defense in the Eisenhower Era: Nuclear Antiaircraft Arms and the Cold War*
By Christopher J. Bright

*Confronting the Climate: British Airs and the Making of Environmental Medicine*
By Vladimir Janković

*Globalizing Polar Science: Reconsidering the International Polar and Geophysical Years*
Edited by Roger D. Launius, James Rodger Fleming, and David H. DeVorkin

*Eugenics and the Nature-Nurture Debate in the Twentieth Century*
By Aaron Gillette

*John F. Kennedy and the Race to the Moon*
By John M. Logsdon

# John F. Kennedy and the Race to the Moon

John M. Logsdon

JOHN F. KENNEDY AND THE RACE TO THE MOON
Copyright © John M. Logsdon, 2010.

All rights reserved.

First published in 2010 by
PALGRAVE MACMILLAN®
in the United States—a division of St. Martin's Press LLC,
175 Fifth Avenue, New York, NY 10010.

Where this book is distributed in the UK, Europe and the rest of the world, this is by Palgrave Macmillan, a division of Macmillan Publishers Limited, registered in England, company number 785998, of Houndmills, Basingstoke, Hampshire RG21 6XS.

Palgrave Macmillan is the global academic imprint of the above companies and has companies and representatives throughout the world.

Palgrave® and Macmillan® are registered trademarks in the United States, the United Kingdom, Europe and other countries.

ISBN: 978–0–230–11010–6

Library of Congress Cataloging-in-Publication Data

Logsdon, John M., 1937–
John F. Kennedy and the race to the moon / John M. Logsdon.
p. cm.—(Palgrave studies in the history of science and technology)
Includes bibliographical references.
ISBN 978–0–230–11010–6 (hardback)
1. Space race—United States. 2. Astronautics and state—United States. 3. Kennedy, John F. (John Fitzgerald), 1917–1963—Influence. 4. United States. President (1961–1963 : Kennedy) 5. United States—Politics and government—1961–1963. I. Title.

TL789.8.U5L64 2011
629.45′40973—dc22 2010025629

A catalogue record of the book is available from the British Library.

Design by Newgen Imaging Systems (P) Ltd., Chennai, India.

First edition: December 2010

10 9 8 7 6 5 4 3

Printed in the United States of America.

629.454

*To my students, past and present, at The George Washington University, International Space University, and The Catholic University of America. Thank you for listening, and then going on to make me proud of your accomplishments.*

# Contents

# Preface

This book has had an extremely long gestation period. Understanding its evolution is important to an appreciation of its character and intent.

My involvement with President John F. Kennedy's role in the race to the Moon began as I prepared my doctoral dissertation in political science at New York University in the late 1960s, even as I began my academic career in Washington, DC, at The Catholic University of America in September 1966. I actually signed a contract to publish the dissertation before I defended it. As I moved from Catholic to The George Washington University in 1970, the MIT Press brought out a hardcover edition of the dissertation as *The Decision to Go to the Moon: Project Apollo and the National Interest.* The University of Chicago Press published a paperback edition in 1976. (As a side note, working on the study of the Apollo decision provided an opportunity to be present at the July 16, 1969, launch of the Apollo 11 mission, and also the Apollo 14 and Apollo 17 launches. Those experiences alone were worth the effort that went into research and writing the dissertation and subsequent book.)

My detailed study of the decision-making process by which President Kennedy became convinced that it was in the national interest for the United States to enter, with the intent of winning, the space race with the Soviet Union has been described as "classic" and "powerful and seminal" by leading space historian Roger Launius.[1] Such an assessment is, of course, very gratifying. But as the years passed, I became increasingly dissatisfied with the completeness of the study's narrative elements. The basic story stood the test of time, but because my research for the book was carried out even before Neil Armstrong and Buzz Aldrin reached the Moon, only a very limited base of primary documents on which to base the study was available. The John F. Kennedy Presidential Library had not yet opened, and Lyndon B. Johnson was still president. That meant the narrative lacked the fullness made possible only by using the documentary record; also, many oral histories discussing the Kennedy presidency were not yet available. The flip side of this situation was that the events and considerations that led to the decision to go to the Moon were still fresh in the minds of the key participants in that decision, and I was fortunate enough to be able to interview most of them. Of those involved with the decision to go to the Moon, only Robert McNamara and President Johnson declined interview requests; of course, by that time both John and Robert Kennedy had been assassinated. Early on,

NASA chief historian Gene Emme and through him NASA administrator James Webb became convinced that I was trying to prepare an unbiased account of the decision process, and their support greatly facilitated my research. Thus the 1970 book was based primarily on my interviews with participants in the decision process and the secondary literature, although I was able to gain access to a few key documents. That meant that the story of JFK's lunar landing decision was not complete.

I also came to realize that I had totally missed an important theme in President Kennedy's thinking in the January—May 1961 period. His first instinct on coming to the White House had been to seek cooperation in space with the Soviet Union, not competition. Even after he announced his decision to send Americans to the Moon on May 25, 1961, Kennedy had suggested to Soviet leader Nikita Khrushchev, as they met face-to-face for the only time ten days later in Vienna, that the United States and the Soviet Union should go to the Moon together. Khrushchev responded negatively, and at least for the time being, the cooperative alternative was foreclosed. There was no mention of this alternative path in the 1970 book.

I also came to realize that I had told only one part of the story of John F. Kennedy and the lunar landing program. Achieving large-scale objectives through government action has two requirements. One is a well-crafted decision on what objective to pursue. I believe that JFK's lunar landing decision was indeed an example of choosing a course of action only after careful thought and examination of possible alternatives. But turning a decision into action, and carrying that action through to completion, is also needed for success. While there have been a number of studies of Project Apollo that examined its technical and management elements, surprisingly I found that there had been no focused attention paid to the actions and decisions of President Kennedy and his White House associates from May 1961 through November 1963 that generated the political will needed to mobilize the financial and human resources which made the lunar landing program possible.

This recognition led me in 1998 to propose to the NASA Headquarters History Office a comprehensive study of John F. Kennedy and the U.S. space program. Then NASA administrator Dan Goldin and his associate administrator for policy and plans Lori Garver (now NASA deputy administrator) gave top-level support to my request, and NASA's chief historian Roger Launius approved a modest contract to help me get started. Over the next several years, I carried out a first round of gathering primary documents and other material from the Kennedy Library and the NASA Historical Reference Collection at NASA's headquarters, and drafted a few parts of the book. But my duties as professor and director of the Space Policy Institute of George Washington University's Elliott School of International Affairs, plus a seemingly unquenchable appetite for international travel, took me away from sustained writing.

I never lost my interest in finishing the study, however. As I prepared to leave GW's active teaching faculty in June 2008, once again it was Roger

Launius, by now senior curator in the Space History Department at the Smithsonian National Air and Space Museum, who suggested that I apply for the museum's most senior fellowship, the Charles A. Lindbergh Chair in Aerospace History. I was awarded that position, and from September 2008 through August 2009, I was in residence at the museum, finishing another round of research in the Kennedy Library and the NASA archives and getting most of the writing of a first draft completed. As I finished a chapter draft, both Mike Neufeld, chair of the museum's space history department and Wernher von Braun biographer, and Roger Launius provided very useful comments. I returned to GW's Space Policy Institute as professor emeritus in September 2009, and finished my research and drafting of the manuscript there.

The current study is thus much more than a warmed-over version of my 1970 book. It adds a great deal of new material to the account of the initial decision-making process in that study, providing a fuller understanding of the factors at play as Kennedy made his choice. In addition, the MIT Press graciously provided its permission to incorporate as much of the contents of the earlier book into the new manuscript as I wished, and I have drawn upon many text passages and used almost all of the earlier interview material in crafting this narrative. In doing so, I have tried not lose any of the qualities that have made *The Decision to Go to the Moon* the standard account of that decision. The earlier book ended with President Kennedy's speech to a joint session of Congress on May 25, 1961, in which he announced: "We should go to the Moon." This new study carries the story until the tragic day in Dallas when Kennedy's presidency was so abruptly terminated. Even on November 22, 1963, John Kennedy was intending to speak in positive terms about the future of the U.S. space program.

I have attempted to maintain throughout this study a focus on the decisions and actions of President John F. Kennedy, his inner circle of advisers who made decisions and took actions on behalf of the president, the career executive office staff who supported the Kennedy presidency, and the agency heads with whom the president interacted. Kennedy before he was inaugurated assigned a lead role in space policy to his vice president-elect, Lyndon B. Johnson, and I have also characterized Johnson's role with respect to space decisions during the Kennedy administration. What I have not done, except when it was necessary to understand deliberations at the White House level, is give much attention to the technical and management aspects of Project Apollo itself. There is a very large literature on those topics.[2]

This study is not a complete account of John F. Kennedy and the American space program. Providing such an account was my original aspiration, but the realities of time and page count led to a decision to focus only on Kennedy and the race to the Moon, since that is the singular space achievement with which Kennedy will forever be associated. John Kennedy personally had only limited involvement in the steps taken during his administration to bring communication satellites into early use. But he was deeply involved in making sure that there were no international restrictions placed on the ability of the United States to operate photoreconnaissance satellites, in limiting the spread of military conflict into the new environment of outer space, and in banning tests of nuclear weapons beyond the atmosphere. In his annual address to the General Assembly of the United Nations in 1961 and 1962, he laid out the principles that became the basis of the 1967 Outer Space Treaty. Kennedy also became personally involved with all seven Mercury astronauts and particularly friendly with the first American to orbit the Earth, John Glenn. The astronauts represented a personality type quite attractive to Kennedy and about which he had written about in his book *Profiles in Courage*—individuals who had responded successfully to challenging circumstances. So there is more to be written about Kennedy and space than is contained in this study.

While this narrative draws on what is available in the documentary record and has the benefit of interviews and oral histories that took place close to the events being discussed, it can never be *really* complete. One cannot know which of the many memoranda addressed to President Kennedy he actually read, and of those he read, to what issues and views he gave most attention. Kennedy enjoyed discussing policy issues with his advisers and associates; few of those conversations can be re-created. (A fascinating exception is the tape recording of a November 21, 1962, cabinet room meeting during which Kennedy and James Webb debated Apollo's priority. What Kennedy said in this private meeting, including the phrase "I'm not that interested in space," is rather different than his public rhetoric.) John Kennedy's brother Robert was his closest confidant, but there is a very limited record of their discussions about the U.S. space program. So inevitably this study is a reconstruction of history based on extensive, but still partial, evidence.

Given the more than a decade over which I have been working on this book, there are many people to thank, and I am bound to have forgotten to mention some who deserve recognition. It is obvious that I owe multiple expressions of gratitude to Roger Launius, and it is only fitting that this book is part of the Palgrave Studies in the History of Science and Technology series of which Roger is co-editor. At NASA, in addition to the original support provided by Lori Garver and Dan Goldin, I want to thank archivist Jane Odom and Colin Fries and Liz Suchow of the NASA History Office for their responsiveness in helping me locate key documents and other research material. The research staff of the John F. Kennedy Presidential Library has been supportive during my many visits to the library; Maryrose Grossman was particularly helpful in digging through the photo archives to locate several of the images included in this book. (I must add my frustration in not being able to access a still unreleased audio tape of a meeting between President Kennedy and NASA administrator James Webb on September 18, 1963, during which Kennedy told Webb of his plans to invite the Soviet Union to join the United States in sending people to the Moon as he addressed the United Nations General Assembly two days later.) I had a very productive visit to the Lyndon B. Johnson Presidential Library in March 2010; the staff there was also very helpful.

I, of course, have to express thanks to the Smithsonian Institution for the offer of the Lindbergh Chair; without that year to focus on my writing, I might still be procrastinating. Mike Neufeld and his colleagues in the Department of Space History at the National Air and Space Museum were welcoming; I felt quite comfortable working in their midst.

George Washington students Krystal Brun and Megan Ansdell and MIT student Teasel Muir-Harmony provided occasional but valuable research assistance. My colleague Dwayne Day read the draft manuscript and provided useful comments while also catching my many typos. My successor as director of the Space Policy Institute, Professor Scott Pace, welcomed me back to GW after my year at the museum. I am proud of having created the Space Policy Institute in 1987 and of the accomplishments of the many students who have learned about space policy there, my first students during my years at Catholic University four decades ago, and the young people from many countries I encounter during my continuing involvement with International Space University; these students, together with what I have written through the years, are my lasting heritage, and it is to them that this book is dedicated.

At Palgrave Macmillan, Chris Chappell has been enthusiastic about this study, and I have appreciated his guidance in getting the manuscript into print. Sarah Whalen and Heather Faulls have been very helpful in shepherding the manuscript through the production process. I also welcomed the very useful comments by several anonymous reviewers of my book proposal.

It goes without saying that I am responsible for any errors in this account and for the interpretations of all the actions and decisions detailed herein.

In 1970, I dedicated *The Decision to Go to the Moon* to my wife Roslyn. Forty-one years later, she is still my wife and still a constant source of support and encouragement. She deserves in gratitude much more than a book dedication.

# Prologue

# "We Should Go to the Moon"

On May 25, 1961, President John F. Kennedy, then just over four months in the White House, addressed a joint session of Congress to deliver what was billed as a second State of the Union address on "Urgent National Needs." Before the assembled senators and representatives and a national television audience, Kennedy declared: "I believe that this nation should commit itself to achieving the goal, before this decade is out, of landing a man on the moon and returning him safely to earth." Later in his speech, he reiterated: "I believe that we should go to the moon." Sixteen months later, in his most memorable space speech, made before a crowd of 40,000 at Rice University in Houston, Texas, Kennedy gave this reason for undertaking the lunar journey: "We choose to go to the moon in this decade and do the other things, not because they are easy, but because they are hard, because that goal will serve to organize and measure the best of our energies and skills, because that challenge is one that we are willing to accept, one we are unwilling to postpone, and one which we intend to win."[1]

John Kennedy was a very unlikely candidate to decide to send Americans to the Moon. He had shown little interest in space issues in his time as a senator or during his presidential campaign. According to one journalist who had close ties with Kennedy, "Of all the major problems facing Kennedy when he came into office, he probably knew and understood least about space."[2]

Yet just three months after his inauguration, in the aftermath of the Soviet Union on April 12, 1961, sending the first human, cosmonaut Yuri Gagarin, into orbit, Kennedy asked his advisers to find him "a space program which promises dramatic results in which we could win." The answer came back less than three weeks later—sending astronauts to the surface of Earth's nearest neighbor gave the best chance of besting the Soviet Union in a dramatic space achievement. The resulting prestige from winning a race to the Moon, Kennedy was told, would give the United States a major victory "in the battle along the fluid front of the Cold War." Kennedy accepted this

President John F. Kennedy as he addressed a joint session of the Congress on May 25, 1961, and declared: "We should go to the Moon." Others in the image are Vice President Lyndon B. Johnson (left) and Speaker of the House Sam Rayburn (right) (NASA photograph).

advice, and soon after announced his decision to begin what he characterized as "a great new American enterprise."[3]

President Kennedy's involvement with the lunar landing undertaking was much more intimate and continuing than is usually acknowledged. Kennedy not only decided to go to the Moon; over the remaining thirty months of his tragically shortened presidency, he stayed closely engaged in the effort and in making sure the benefits of Project Apollo would outweigh its burgeoning cost. Convinced that this was indeed the case, he pushed hard to make sure that Apollo was carried out in a manner that best served both the country's interests and his own as president. As the authors of the recent study *If We Can Put a Man on the Moon*...comment, "Democratic governments can achieve great things only if they meet two requirements: wisely choosing which policies to pursue and then executing those policies." Many presidents since John Kennedy have announced bold decisions, but few have followed those decisions with the budgetary and political commitments needed to ensure success.[4] This study details the full range of JFK's actions that carried Americans to the Moon.

Kennedy's commitment to the race to the Moon initiated the largest peacetime government-directed engineering project in U.S. history. Project Apollo by the time it was completed cost U.S. taxpayers $25.4 billion, which would be equal to some $151 billion in 2010 dollars. Apollo is frequently compared to the construction of the Panama Canal as an expensive, long-term, government-funded undertaking. By the time the Canal was completed in 1914, the cost of its construction was $375 million, equivalent to $8.1 billion in 2010 dollars, much less than Apollo. Another comparison might be with the multidecade construction of the Interstate Highway System, which began in the mid-1950s and for which the federal government paid $114 billion out of a total cost of $128 billion. By any measure available, Apollo required a historically massive commitment of public funds over a relatively brief period of time.[5]

This study is the first comprehensive account of the impact of John F. Kennedy on the race to the Moon; others have written extensively about the managerial and technical aspects of the Apollo achievement, but none have portrayed JFK's perspective as he continued to push, in the face of growing criticism and concern about increasing costs, for moving ahead with the lunar landing program. The book contains a detailed narrative of the decisions and actions of President Kennedy, his inner circle of advisers who made decisions and took actions on his behalf, the career White House staff who supported the Kennedy presidency, and the agency heads with whom Kennedy interacted. Kennedy before he was inaugurated assigned his vice president, Lyndon B. Johnson, a lead role in space policy; the study also characterizes Johnson's role with respect to space decisions during the Kennedy administration. Except when necessary to understand deliberations at the White House level, the book does not give much attention to the specific details of Project Apollo itself.

John Kennedy saw his choice to go to the Moon "as among the most important decisions that will be made during my incumbency in the Office of the Presidency." Yet most general accounts of JFK's time as president give only passing attention to his involvement with the lunar landing program. One goal of this study is to create as historically complete a record as possible of that involvement. Doing so fills an empty niche in the record of the Kennedy administration. It also provides a detailed case study of how Kennedy went about conducting his presidency, assessing what actions were needed in the national interest, continuously seeking information from multiple sources, but deferring to his agency heads to carry out the programs he set in motion. Readers of this account can decide for themselves what insights Kennedy's space-related efforts provide about his personality and the way he carried out his presidency. The book's concluding chapter, however, reflects on the character and quality of JFK's space decisions, asks whether the way Apollo was conceived and carried out can serve as a model for other large-scale government efforts, and provides a perspective on the impact of Kennedy's commitment to a lunar landing "before this decade is out" on both the evolution of the U.S. space program and the U.S. position in the world of the 1960s and later.

The image of John Kennedy that emerges from this study is at variance from how he is often regarded with respect to space. Rather than a visionary who steered the U.S. space program toward a focus on exploring beyond Earth orbit, he emerges as a pragmatic political leader who soon after entering office came to see the U.S. civilian space program as an important tool to advance U.S. foreign policy and national security goals. He was flexible in his approach to space activities, willing to compete if necessary but preferring to cooperate if possible.

John F. Kennedy with his actions in the spring of 1961 and in the following months took the first steps toward the Moon. Eight years later, on July 20, 1969, Neil Armstrong would take another "small step for a man, but a giant leap for mankind." Historian Arthur Schlesinger, Jr. has suggested that "The 20th Century will be remembered, when all else is forgotten, as the century when man burst his terrestrial bounds."[6] In undertaking the lunar landing program, John Kennedy linked the politics of the moment with the dreams of centuries and the aspirations of the nation. Unfortunately, Kennedy did not live to see the first footprints on the lunar surface, but in the long sweep of history, it is one of the ways in which he will be most remembered.

# Chapter 1

# Before the White House

Public life was not the first choice among possible futures for John F. Kennedy as he returned from World War II. Kennedy in principle could have chosen among many career paths. Kennedy's own inclination seems to have leaned in the direction of becoming a journalist, a writer of nonfiction books, or even an academic. Kennedy's father, Joseph, however, was determined that his sons not enter the business world; he had amassed sufficient wealth to allow his sons to choose a future that did not have to lead to significant additional income. The reality was that if Kennedy had chosen a career other than politics, it would have meant going against the wishes of his strong-willed father. John Kennedy, from the time his older brother, Joseph Jr., was killed in action during World War II, became his father's designated aspirant to high political office; JFK's father, after the end of the war, planned to build "the greatest political dynasty of the age...one remaining son at a time." Kennedy was easily elected to the U.S. House of Representatives in 1948 and to the Senate in 1952 and again in 1958. But his time in Congress was not fulfilling for either his or his father's ambitions. During his fourteen years in Congress, Kennedy "failed to penetrate the inner circle." The conservative Southern senators who controlled the Senate, in particular, "viewed him as too detached, independent, overrated, and overly ambitious." Beginning in 1956, when he was almost selected as Adlai Stevenson's vice presidential running mate, Kennedy set his eyes on being elected president of the United States in November 1960. As he pursued that objective, the candidate was described by one acute observer as a "charming, handsome, rich, young aristocrat."[1]

In his years as a senator, Kennedy said little about space issues except in the context of the linkage between space launch vehicles and strategic missile capabilities. That changed once he became the Democratic nominee for president in July 1960. The growing disparity in global prestige between the United States and the Soviet Union under the Eisenhower administration became a central theme of JFK's campaign, and the fact that the United

States was trailing the Soviet Union in space achievement was frequently cited by Kennedy as very visible evidence of this disparity. Kennedy offered no specific views on future space activities during the campaign, however, and once he was declared the president-elect, he spent little time on space issues prior to his inauguration. This meant that Kennedy's personal views and interests with respect to space, as differentiated from his campaign rhetoric, remained largely unknown as he entered the White House.

## Kennedy and Space before His Presidential Campaign

Even though the significance of the Soviet launches of spacecraft beginning with Sputnik 1 in October 1957 and the dog-carrying Sputnik 2 the following month and the appropriate U.S. response to Soviet space achievements were major issues before the Congress between 1958 and 1960, Senator Kennedy said little about space issues during those years. Kennedy also showed little personal interest in U.S. and Soviet space efforts. He was a member of the Visiting Committee of the Harvard College Observatory, but any curiosity he may have had regarding astronomy apparently did not carry over into the space realm. The head of the Instrumentation Laboratory at the Massachusetts Institute of Technology, Charles Stark Draper, recalled a discussion with John Kennedy and his brother Robert at a Boston restaurant during the post-Sputnik period. Draper tried to get the Kennedy brothers excited about the promise of space flight. According to one account of the evening, the brothers treated Draper and his ideas "with good-natured scorn"; Draper noted that they "could not be convinced that all rockets were not a waste of money, and space navigation even worse."[2] Kennedy did vote in February 1958 to create a Senate Special Committee on Space and Aeronautics and voted in favor of the National Aeronautics and Space Administration (NASA) funding bills in 1959 and 1960; other actions related to NASA during those years were approved by voice votes, without the positions of individual senators being recorded.[3]

Kennedy's primary focus in his years in the Senate was defense and foreign policy, with particular attention after 1957 to what he and many others called an emerging U.S.-Soviet "missile gap." The growing disparity between U.S. and Soviet missile capability, Kennedy argued, would soon put the United States at a significant strategic disadvantage vis-à-vis the Soviet Union. In a speech on the Senate floor on August 14, 1958, Kennedy expressed worry that the Soviet Union's superiority in nuclear-tipped missiles would allow it to use "sputnik diplomacy" and other nonmilitary means to shift the balance of global power against the United States.[4] Kennedy, when he did speak about space issues in the post-Sputnik period, most often linked them to the missile gap issue, frequently using the term "missile-space problem."

Kennedy did gain a seat on the prestigious Committee on Foreign Relations in the mid-1950s. He was one of the first senators to recognize the significance for U.S. interests of the decolonization movement of the late 1950s; he thought it essential that the United States be an appealing ally to Asians and Africans as they chose which social system to pursue as independent nations. Kennedy attracted international attention by denouncing French suppression of Algeria's move toward independence.[5] He was an ardent anticommunist at this point in his life, and the possibility of newly independent nations "going communist" was troubling to him. His interest in presenting a positive image of the United States to the countries of the third world was a factor in his later judgment that the United States could not by default cede space leadership to the Soviet Union.

An intriguing insight into what may have been JFK's views on space in the prepresidential period emerges from his February 16, 1960, response to a handwritten letter from a Princeton University freshman "with a strong Republican background" who told Senator Kennedy that he should "put more money in the space program." In his response, Kennedy noted that the letter had reached him because of the author's "undeviating Republicanism, Princetonian self-assurance and uncomplicated handwriting." Kennedy suggested that "whatever the scale and pace of the American space effort, it should be a scientific program . . . In this interval when we lack adequate propulsion units, we should not attempt to cover this weakness with stunts." He added, "When this weakness is overcome, our ventures should remain seriously scientific in their purpose." Kennedy felt that "with respect to the competitive and psychological aspects of the space program, it is evident that we have suffered damage to American prestige and will continue to suffer for some time." However, he pointed out that "our recent loss of international prestige results from an accumulation of real or believed deficiencies in the American performance on the world scene: military, diplomatic, and economic. It is not simply a consequence of our lag in the exploration of space *vis-à-vis* the Soviet Union."

Kennedy listed these deficiencies in the order in which he thought they should be addressed: "the missile gap," "inadequate or misdirected policies in the underdeveloped areas," "the disarray of NATO," "the weakening of the international position of the dollar," and, finally, "our inferior position in space exploration." He thought that the United States "should accept the costs now of achieving the powerful propulsion unit a few years earlier than now may be the case" and looked forward to "the internationalization of space exploration, first on a Free World basis, then with the USSR." This response suggests that Senator Kennedy had indeed given some thought to space issues before February 1960, and that in his judgment they had a lower priority in comparison to other threats to U.S. leadership; this may well be an accurate reflection of his views as opposed to the strident attitude toward space leadership that characterized much of his campaign rhetoric later in the year. Notable is JFK's interest at this early date in both increasing the lifting power of U.S. rockets and in internationalizing space activity. The letter also suggests, as in the use of the expression "propulsion unit" rather than launch vehicle, that Kennedy or whoever composed the response was not very familiar with the specifics of the U.S. space effort.[6]

According to his closest policy adviser, Theodore C. Sorensen, by mid-1960 Kennedy thought of space "primarily in symbolic terms... Our lagging space effort was symbolic, he thought, of everything of which he complained in the Eisenhower Administration: the lack of effort, the lack of initiative, the lack of imagination, vitality, and vision; and the more the Russians gained in space during the last few years in the fifties, the more he thought it showed up the Eisenhower Administration's lag in this area and damaged the prestige of the United States abroad."[7]

## Space and the 1960 Presidential Campaign

Space issues did not play a major role in John Kennedy's campaign for the Democratic nomination for president. Once the nomination was secured, Kennedy in his July 15 acceptance speech first used the term "New Frontier," saying "we stand today on the edge of a New Frontier—the frontier of the 1960's—a frontier of unknown opportunities and perils—a frontier of unfulfilled hopes and threats" and noting that "beyond that frontier are the uncharted areas of science and space." Foreshadowing the most famous line in his inaugural address, Kennedy said that the New Frontier of which he was speaking "is a set of challenges. It sums up not what I intend to offer the American people, but what I intend to ask of them."[8]

The Democratic platform on which Kennedy would run stated:

> The Republican Administration has remained incredibly blind to the prospects of space exploration. It has failed to pursue space programs with a sense of urgency at all close to their importance to the future of the world.
>
> The new Democratic Administration will press forward with our national space program in full realization of the importance of space accomplishments to our national security and our international prestige. We shall reorganize the program to achieve both efficiency and speedy execution. We shall bring top scientists into positions of responsibility. We shall undertake long-term basic research in space science and propulsion.[9]

### *Campaign Advice on Space*

To develop background material on the various issues he would have to address during his presidential campaign, Senator Kennedy in December 1958 established a "brain trust" drawn primarily from the faculties of Harvard and the Massachusetts Institute of Technology. Even before the presidential campaign began, Harvard law professor Archibald Cox began collecting research memoranda and reports from experts at both universities and from across the country. According to the authoritative account of the Kennedy campaign, "The professors were to think, winnow, analyze and prepare data on the substance of national policy, to channel from university to speech writers to Cox to Sorensen—and thus to the candidate." This process failed in its execution. While an impressive amount of material was generated, little of it was read by Kennedy or used during the campaign. Regarding the products of Cox's efforts, Theodore Sorensen comments that "not all of their material was usable and even less was actually used. But it provided a fresh and reassuring reservoir of expert intellect."[10] The different perspectives of those caught up in the frenzy of Kennedy's presidential campaign, such as Sorensen, and those with the time to reflect on issues that Kennedy would have to address if he was elected were a continuing source of campaign tensions.

Kennedy himself on September 2, 1960, asked Cox to contact Trevor Gardner, former assistant secretary of the U.S. Air Force for research and development, and a man to whom Kennedy looked for advice on space and missile issues. Kennedy wanted from Gardner "an account of the Administration's failures in missiles, 1953 to today" and his "judgment on the significance of our being in a secondary position in space in the sixties." Kennedy also asked, "Will the Soviet Union have a reconnaissance satellite before we do, and what will it mean?"[11]

Another source of largely unused but remarkably prescient input into Kennedy's campaign was the Advisory Committee on Science and Technology of the Democratic Advisory Council, which in turn reported to the Democratic National Committee. Among its inputs was a September 7 "Position Paper on Space Research." Leading the preparation of this paper was physicist Ralph Lapp, who, after working in the Manhattan Project to develop the atomic bomb, spent most of the rest of his career warning about the dangers of nuclear war. The position paper pointed out that "the United States has failed to define its real objective in space. If purely scientific, this should be so stated so that the American people and others understand our objective. If aimed at 'winning the space race' then this must also be stated and the U.S. program must be directed toward this goal." The paper went on to discuss landing a man on the Moon as a possible objective of a competitive space effort, asking, "Can the United States afford to allow the Russians to land on the moon first?" and noting that the answer to this question was "more political" than technical, since "there is no great scientific urgency" in a manned lunar landing. It noted that "in the psycho-political space race the rewards for being first are exceedingly great; there is little pay-off for second place."

The paper outlined two alternative space programs. One of the programs was "an imaginative and vigorous program of research in space science and technology and to exploit useful applications of this new technology . . . in collaboration with other nations." The other suggested program aimed at "American supremacy in the exploration of space," including "early attainment of a thrust capability consistent with manned flights to the Moon." The paper noted that "Senator Kennedy must make the decision, essentially political in character," between the two programs. The costs of the politically driven second program were estimated to be $26 billion from 1960 to 1970, compared to the then planned expenditures during that period of $12 to 13 billion. The scientifically oriented but faster-paced program was estimated to cost $19 billion. While this paper was unlikely to have been read by Kennedy or his top advisers, it was a quite insightful statement of the central space issue that would occupy Kennedy once he entered the White House, and its cost estimates were surprisingly close to the actual costs of the program that President Kennedy in 1961 chose to pursue .[12]

Yet another input into Kennedy's position on space during the campaign was a briefing paper prepared for the candidate's "Position and Briefing Book"; this was a resource that traveled with the campaign team as a ready source of speech material and responses to media questions. The briefing paper suggested "eliminating the unrealistic distinctions between civilian and defense space projects" and said that there should be "one coordinated space program with joint civilian and military space uses." The paper proposed that Kennedy should "place one man in charge of all space activities, reporting directly to the President."[13]

### *Space Statements during the Campaign*

By 1960, it had become customary for specialized publications to ask presidential candidates to state their positions on issues of interest to their readers. Thus the trade magazine *Missiles and Rockets* on October 3, 1960, published an "open letter to Richard Nixon and John Kennedy," proposing a nine-point "defense and space platform" and asking the candidates to reply, "stating your views and making your stand quite clear on these two closely related problems." Kennedy's response, which appeared in the October 10 issue of the magazine, was drafted by Dr. Edward C. Welsh, at that time working for Senator Stuart Symington; Symington had competed with Kennedy for the Democratic presidential nomination and reflected the views of the more military-oriented elements of the Democratic Party. Both Symington and Welsh were vigorous champions of a strong U.S. space effort; the statement was "full of the clash and clamor of the space race."[14] The Kennedy statement said:

> We are in a strategic space race with the Russians, and we are losing. . . . Control of space will be decided in the next decade. If the Soviets control space they can control earth, as in past centuries the nation that controlled the seas has dominated the continents . . . We cannot run second in this vital race. To insure peace and freedom, we must be first.
>
> The target dates for a manned space platform, U.S. citizen on the moon, nuclear power for space exploration, and a true manned spaceship should be elastic. All these things and more we should accomplish as swiftly as possible. This is the new age of exploration; space is our great New Frontier.[15]

How accurately this statement reflected John Kennedy's actual thinking as of October 1960 with regard to the strategic and military importance of space is questionable; the fact that it was prepared by someone without a central role in Kennedy's campaign suggests that neither Kennedy nor his close policy advisers had much involvement in its content. Many in the space community, however, took the statement at face value and anticipated that if elected Kennedy would favor an accelerated space effort and would put additional emphasis on the military dimensions of the U.S. space program.

### *The Symbolic Role of Space*

John Kennedy laid out his basic argument for his candidacy in one of his early campaign speeches. He told an audience in Portland, Oregon, that

> Other countries of the free world—troubled and restless—are looking for new leadership from the United States, and I believe they are willing to accept and respect the leadership of an administration that will move vigorously on these five fronts:
>
> 1. An administration that moves rapidly to rebuild our defenses, until America is once again first in military power across the board;
> 2. An administration that moves rapidly to revamp our goals in education and research, until American science and learning are once again preeminent;
> 3. An administration that moves rapidly to reshape our image here at home, until it is clear to all the world that the revolution for equal rights is still the American revolution;
> 4. An administration that moves rapidly to renew our leadership for peace, until we have brought to that universal pursuit the same concentration of resources and efforts that we have brought to the preparation of war; and
> 5. Finally, an administration that moves rapidly to remold our attitudes toward the aspirations of other nations, until we have a fuller understanding of their problems, their requirements, and their fundamental values.[16]

Theodore Sorensen notes that there was a single theme that Kennedy stressed throughout the campaign: "the challenge of the sixties to America's security, America's prestige, America's progress." Kennedy on the campaign trail proclaimed over and over again that "it is time to get this country moving again." Eventually that phrase or a variation of it appeared in every campaign speech. By the end of October, "the issue of slipping prestige had become the dominant one of the campaign"; according to the polls, Kennedy had a substantial lead over his Republican opponent, Richard M. Nixon, on this issue.[17]

It was in this context that Kennedy made frequent references to the space program in his campaign appearances. For example:

> If the Soviet Union was first in outer space, that is the most serious defeat the United States has suffered in many, many years... Because we failed to recognize the impact that being first in outer space would have, the impression began to move around the world that the Soviet Union was on the march, that it had definite goals, that it knew how to accomplish them, that it was moving and we were standing still. This is what we have to overcome, that

> psychological feeling in the world that the United States has reached maturity, that maybe our high noon has passed...and that now we are going into the long, slow afternoon.[18]

Although a speech devoted solely to space issues was drafted for Kennedy's campaign use, it was never delivered.[19]

Kennedy's campaign rhetoric with respect to the loss of U.S. prestige because of the Soviet space successes was reinforced by a classified U.S. Information Agency report that was leaked to *The Washington Post*. The title of the October 10 report was "The World Reaction to the United States and Soviet Space Programs—A Summary Assessment." The report was based on polls taken in Great Britain, France, West Germany, Italy, and Norway. On the basis of the results of these surveys, the report concluded that "in anticipation of future U.S.-U.S.S.R. standing, foreign public opinion...appears to have declining confidence in the U.S. as the 'wave of the future' in a number of critical areas."[20]

Vice presidential candidate Lyndon B. Johnson in his campaign appearances did not stress the space issue as strongly or as frequently as did Kennedy, even though from the launch of Sputnik 1 on October 4, 1957 on, Johnson had taken the lead in the Senate on space issues. In late October 1960, in response to Richard Nixon's defense of the space record of the Eisenhower administration, Johnson released a "white paper" prepared by the staff of the Senate space committee, which he chaired. Johnson criticized the administration's space policy but stressed that both Democrats and Republicans in Congress had recognized the need for a strong space effort. The paper contended that "The sad truth is that U.S. progress in space has been continually hampered by the Republican administration's blind refusal to recognize that we have been engaged in a space and missile race with the Soviet Union and to act accordingly." In a statement released with his white paper, Johnson echoed the sentiments of Kennedy's October 10 statement regarding the strategic significance of space: "It is a fact that if any nation succeeds in securing control of outer space, it will have the capability of controlling the earth itself."[21]

Throughout the campaign, Kennedy frequently linked the Eisenhower administration's failures in space to its allowing the Soviet Union to achieve a significant advantage vis-à-vis the United States with respect to the development and deployment of ballistic missiles—the so-called "missile gap." On July 23, after the Democratic convention, candidate Kennedy had a highly classified briefing from the director of the Central Intelligence Agency (CIA), Allen Dulles. A wide variety of topics were covered, including "an analysis of Soviet strategic attack capabilities in missiles." Kennedy asked Dulles "how we ourselves stood in the missile race." Dulles told him that "the Defense Department was the competent authority on this question." After subsequent meetings with defense officials, Kennedy told Sorensen that the briefings "were largely superficial" and "contained little he had not read in *The New York Times*."[22]

The reality was that at the time of the Dulles briefing, there was limited information available to the U.S. leadership on Soviet deployment of intercontinental ballistic missiles (ICBMs). The missions flown by the U-2 spy plane prior to the May 1, 1960, downing of a flight over Russia piloted by Francis Gary Powers had suggested that there had been very limited deployment of the initial Soviet ICBM, and thus that a prospective "missile gap" was not likely to emerge. The first successful U.S. spy satellite mission was not launched until August 18, 1960, after Kennedy's CIA briefing, and it took several additional missions later in 1960 and in early 1961 to confirm that the indications from the U-2 flights were correct. (In fact, only four of the original ICBMs were ever deployed; it took some twenty hours to prepare the rocket for launch, making it an unwieldy military weapon. Its main role turned out to be as the workhorse launch vehicle for early Soviet space missions.[23])

According to Sorensen, the U-2 evidence was not made available to Kennedy in the various intelligence briefings he received during the campaign. Also, from Kennedy's September 1960 question to Trevor Gardner about whether the United States or the Soviet Union would be first to have a reconnaissance satellite, it appears Kennedy was not briefed on the CORONA intelligence satellite program that Eisenhower had approved in February 1958; its existence was known to very few people within the Congress. Even so, Eisenhower was "reportedly furious" that Kennedy continued to raise the missile gap issue throughout the campaign, while his opponent, Richard Nixon, could not provide information that would counter Kennedy's claims because of its highly classified nature.[24]

## Kennedy Elected

On November 8, 1960, John F. Kennedy was elected by a very narrow margin as the thirty-fifth President of the United States; his victory was confirmed only shortly after noon on the next day. In the following ten weeks before he took the oath of office, president-elect Kennedy moved forward briskly on many of the issues that he had highlighted in the campaign. However, he paid very limited attention to space topics during his transition activities. While the perceived lack of urgency in the Eisenhower administration's space efforts may have been a useful issue to stress in the campaign, the reality was that the president-elect and his advisers did not give high priority to addressing either immediate or longer-term space questions during the post-election transition period. There was no contact made with the new National Aeronautics and Space Administration (NASA), which had begun operations two years earlier, on October 1, 1958. Kennedy prior to his inauguration nominated no one to replace Eisenhower appointee T. Keith Glennan as NASA administrator. As John F. Kennedy took the oath of office on January 20, 1961, there was thus significant uncertainty about the future of the U.S. space effort.

# Chapter 2

# Making the Transition

As John F. Kennedy's election as president was finally confirmed shortly after noon on November 9, there were seventy-two days before his inauguration—"seventy-two days in which to form an administration, staff the White House, fill some seventy-five key Cabinet and policy posts, name six hundred other major nominees" and "to formulate concrete policies and plans for all the problems of the nation, foreign and domestic, for which he soon would be responsible as President."[1]

Kennedy did not begin this crucial transition period from scratch. After the party conventions, the Brookings Institution, a Washington think tank, had urged both candidates to begin to prepare for the transition, should they be elected. In September, Kennedy had asked high-powered Washington lawyer Clark Clifford, who had been a senior adviser to President Harry Truman and who had successfully represented Kennedy against allegations that he was not the author of the Pulitzer-prize winning *Profiles in Courage*, to be his primary transition adviser. Kennedy also had asked Columbia University professor Richard Neustadt, a leading scholar of the American presidency and author of the recently published *Presidential Power*,[2] to provide his views on how best to organize the presidency.

Neustadt earlier in 1960 had begun to work on transition issues at the request of Senator Henry "Scoop" Jackson (D-WA). Jackson and Kennedy were similar in their strongly anticommunist views and on giving priority to military strength, and Kennedy had included Jackson on his short list as a running mate. Neustadt had prepared a twenty-two page memorandum by September 15 on "Organizing the Transition." In this memo, Neustadt made a prescient observation: "The Vice President-elect will be looking for work." Jackson took Neustadt to meet Kennedy on September 18, and the candidate was immediately impressed by Neustadt's memo. Kennedy asked Neustadt to elaborate his arguments in additional memoranda, saying someone in the campaign would get back to him in due time. As was typical of Kennedy's style, he asked Clifford and Neustadt to work without consulting one another, ensuring that he would have two independent sources of transition advice.[3]

Neustadt heard no more from the Kennedy campaign until late October, when he was contacted by one of JFK's associates, asking how he was progressing. On November 4, Neustadt joined Kennedy on his campaign airplane to hand the candidate several of the memorandums he had prepared. Within thirty minutes, Kennedy came to Neustadt, saying that he found the material "fascinating." What he may have found most interesting in Neustadt's analysis was the recommendation that he adopt a staffing pattern in the White House that was much closer to that used by Franklin D. Roosevelt than the military-like arrangements that had been set up by Dwight Eisenhower. "You would be your own 'chief of staff,'" suggested Neustadt. "Your chief assistants would have to work collegially, in constant touch with one another and with you.... There is room here for a *primus inter pares* to emerge, but no room for a staff *director or arbiter*, short of you... You would oversee, coordinate, and interfere with virtually everything your staff was doing." Neustadt noted that "no one has yet improved on Roosevelt's relative success in getting information in his mind and key decisions in his hands reliably enough and soon enough to give him room for maneuver."[4] This was a pattern that Kennedy as president would adopt for his space decisions, among many others.

Clifford delivered his transition memorandum and back-up notebooks to Kennedy on November 9, the day on which Kennedy's election victory was confirmed. The president-elect then asked Clifford to serve as his liaison with the outgoing Eisenhower administration during the transition. Kennedy's biographer Arthur Schlesinger, Jr., who lunched with the president-elect that day, reports that Clifford's memorandum was "shorter and less detailed than the Neustadt series," but that "in the main the two advisers reinforced each other all along the line." Both Neustadt, who stayed on as a consultant during the transition and in the early months of the Kennedy administration, and Clifford emphasized that Kennedy should organize the White House to serve his needs as president, with no chief of staff to control who had access to the president and thus whose advice he received.[5]

On November 10, John Kennedy met with his key campaign advisers before heading off for a much needed vacation. After the meeting, he announced three key appointments to his personal White House staff: Pierre Salinger as his press secretary, Kenneth O'Donnell as his special assistant and appointments secretary, and Theodore Sorensen as his principal adviser on domestic policy and programs, with the title of special counsel to the president. Salinger was a former journalist who had served as Kennedy's press secretary during the campaign. O'Donnell had served as the organizer and scheduler of Kennedy's senatorial and presidential campaigns, and was part of the protective Massachusetts "Irish mafia" that had emerged around Kennedy during his political career. Sorensen was from Nebraska, the son of a Unitarian minister, and very different in style from Kennedy's Massachusetts associates. He had joined JFK's Senate staff in 1953 and had become his chief speechwriter and domestic policy adviser and in many ways his alter ego on policy matters. Arthur Schlesinger Jr., the Harvard historian who was Kennedy's link to the liberal wing of the Democratic Party, notes that with respect to Sorensen, "it was hard to know him well. Self-sufficient, taut and purposeful, he was a man of brilliant intellectual gifts, jealously devoted to the President and rather indifferent to personal relations beyond his own family."[6] All three of these men were a generation younger than their equivalents in the Eisenhower administration; as they entered the White House in January 1961, Salinger would be 35, O'Donnell, 36, and Sorensen, 32. Kennedy was 43, the second youngest man ever to become U.S. president.

Salinger reports that during the 1960 campaign, O'Donnell "was constantly at JFK's side. He took his orders directly from the candidate and saw to it that the rest of us carried them out—and to the letter."[7] Once Kennedy was in the White House, O'Donnell as appointments secretary controlled most access to the President; he also acted as Kennedy's chief "enforcer" within the executive branch, making sure that senior agency officials acted in accordance with White House priorities. However, it appears that O'Donnell had relatively limited involvement in space-related issues other than those associated with the political ramifications of NASA's facility location and contract award decisions.

Sorensen saw himself serving "as an honest broker, determining which decisions could be made by me and which could only be made by the president...In meetings where the president was not present, I often did not distinguish between my views and his." Sorensen also notes that he and Kennedy were "close in a peculiarly impersonal way"; there was little social contact between him and Kennedy either as senator or president. There was no love lost between O'Donnell and his staff and Sorenson and his deputies during the time they served President Kennedy. Although O'Donnell's hostility toward Sorensen and his staff was well known to most in the White House, Sorensen professes to have been unaware of the antagonism until years later.[8]

On space issues, it clearly was Sorensen who had more influence on President Kennedy's thinking than O'Donnell, particularly during their early months in the White House. Sorensen, in fact, appears to have been something of a latent space enthusiast. As he speculated during the campaign about what job he might want if Kennedy were elected president, he wondered whether he "might fit in at the National Aeronautics and Space Administration, using my ability to translate scientific and technological terms into layman's language." He also comments that "I had no real aptitude for science and astronomy...Space, like so many other issues, was one I learned on the job." Sorensen later suggested that this "was just free-wheeling speculation as to whether I would have a post and what would be the most suitable post for me. But it didn't take me too long to realize that that would have been totally inappropriate for me and inconsistent with the President's wishes."[9]

After this November 10 meeting, the exhausted president-elect did not return to active engagement with the transition process until the end of the month, although Clifford, Sorensen, and others were meeting during this period with their counterparts in the Eisenhower administration. Once he did become fully engaged, Kennedy first focused on selecting the ten members of his Cabinet; his first nominee was announced on December 1 and the final one on December 17. On that latter date, sixty additional key policy posts and several hundred more other key positions remained to be filled. As he met with Neustadt and Clifford while the search for the people to fill administration positions continued, Kennedy complained about the difficulty of finding the best individuals to staff his "ministry of talent," saying: "People, people, people! I don't know any people. I only know voters."[10]

To help him understand the issues he would confront in the White House, the president-elect had during the campaign commissioned seven task forces; in December, an additional nineteen more were added, and three more in January. One of the December groups, discussed in more detail below, was on outer space; it worked under Sorensen's guidance. The *Congressional Quarterly Weekly Report* called these task forces an "important innovation" in the presidential transition process, noting that there was "no precedent for the large number of task forces...with wide memberships."[11] As he received the reports of his twenty-nine teams, which were composed of the best available individuals and who worked without compensation, Kennedy's reactions ranged from "helpful" to "terrific." Twenty-four of the twenty-nine teams, including the outer space group, turned in their reports before the January 20 inauguration.

## Space Issues Requiring Early Attention by the Kennedy Administration

There were several significant space issues that were unresolved in the final months of the Eisenhower administration.[12] Dealing with these issues was to occupy president-elect Kennedy's attention only briefly during the postelection transition period. They included the relationship between civilian and military space efforts; a proposal to abolish the National Aeronautics and Space Council, the White House body for developing national space policy and strategy; the pace at which larger rocket boosters ought to be developed in order to close the weight-lifting gap with the Soviet Union; and NASA's plans for a human space flight program to follow the initial U.S. effort to send an astronaut into orbit, Project Mercury.

### *Civil–Military Relations*

The most fundamental policy question to be addressed in the months following the 1957 launches of Sputniks 1 and 2 was whether a new organization for space was needed, or whether all U.S. government space activities, including those with primarily civilian objectives, should be managed by the Department of Defense. In the weeks following the Sputnik launches, both the Army and the Air Force put forward ambitious space plans, and campaigned vigorously for primacy in the U.S. space effort. The Army claim was based in large part on the fact that German émigré Wernher von Braun and his rocket team worked at the Army Ballistic Missile Agency (ABMA) and that they constituted the country's top reservoir of launch vehicle–related technical talent; in addition, the Army's Jet Propulsion Laboratory had managed the development of the first U.S. satellite, Explorer 1, and could serve as the organization developing future satellites. AMBA in its struggle to gain an important space role developed ambitious plans in 1957 and 1958, beginning with a suborbital launch of a human and extending to establishing outposts on the Moon. The Army campaign was not successful, and by mid-1960, both the Jet Propulsion Laboratory and von Braun's rocket team had been transferred to the new civilian space agency NASA.[13]

The Air Force claim took a different approach. Its chief of staff, General Thomas White, argued that "there is no division, per se, between air and space. Air and space are an indivisible field of operations."[14] To make this point, the Air Force in early 1958 coined the word "aerospace." The implication was that the service was the natural choice for the space role. The Air Force also rapidly developed ambitious plans for its space efforts, including putting a man into orbit as soon as possible and eventually sending humans to the Moon.[15]

President Eisenhower, upset by the competition between the two military services, established a Defense Advanced Research Agency (DARPA) in the Office of the Secretary of Defense as an interim step. DARPA, rather than one of the armed services, was to manage all U.S. space efforts, civilian and military, but it was not successful in establishing itself as the lead U.S. space agency. After several months of discussion within the Eisenhower administration, a March 1958 memorandum to the president argued that "because of the importance of the civil interest in space exploration, the long term organization of Federal programs in this area should be under civilian control." The memorandum recommended that "leadership of the civil space effort be lodged in a strengthened and redesignated National Advisory Committee for Aeronautics."[16] That organization, known by most as NACA, had been the government's primary aeronautical research and development organization since its creation in 1915. President Eisenhower accepted this recommendation, and on April 2, 1958, proposed the establishment of a National Aeronautics and Space Agency; his message to Congress included a draft bill "to provide for research into the problems of flight within and outside the earth's atmosphere and for other purposes."[17] After four months of debate by the Congress, the National Aeronautics and Space Act of 1958, establishing the National Aeronautics and Space Administration, was signed by President Eisenhower on July 29, 1958. The shift in organizational designation from "agency" to "administration" upgraded the status of the new space agency within the executive branch hierarchy.[18]

Eisenhower selected T. Keth Glennan as the first NASA administrator. Glennan was "a Republican with a fiscally conservative bent, an aggressive businessman with a keen sense of public duty and an opposition to government intrusion into the lives of Americans, and an administrator and an educator with a rich appreciation of the role of science and technology in an international setting."[19] He was an engineer by training and had had a wide-ranging career, including a stint as a member of the Atomic Energy Commission. But he knew little about space science or technology, and had not followed the post-Sputnik debates leading to the creation of NASA; in his own words, "I didn't have any idea what astronomy or geodesy or any of those things would mean in this strange world. I literally knew nothing." Glennan agreed to take the position, but only on the condition that Hugh Dryden, the executive director of NACA, would be nominated as deputy administrator.

Many in Washington had expected Dryden to get the top NASA job; trained in aerodynamics, he had been the top official handling day-to-day affairs at NACA since 1949, and was a widely known and respected individual in the U.S. and international aeronautical research communities. However, some in Congress deemed his approach to space to be too cautious for the leader of the organization they had in mind, and thus had indicated to the White House that Dryden would not be acceptable as the first head of the space agency.[20] This outcome was deeply disappointing for Dryden, but he overcame that disappointment to serve as NASA's deputy administrator until his death in 1965. According to Keith Glennan, Dryden was "held in esteem by people all over the world" and was "gentle, quiet, wise, very wise, and an astute politician without being a politician."[21] Dryden stayed on in his position during the Kennedy administration and played a key role in the deliberations leading to the decision to go to the Moon.

NASA inherited three programs from the Department of Defense (DOD) that would later play important roles in the space efforts of the Kennedy administration. In August 1958, DARPA transferred its human space flight project (which in fact was originally an Air Force initiative) to NASA; the effort was soon named Project Mercury. According to the NASA history of the project, NASA "received authorization to carry out this primitive manned venture into lower space mainly because Eisenhower was wedded to a 'space for peace' policy…In 1958 there simply was no clear military justification for putting a man in orbit."[22] Even so, Eisenhower decided that the first U.S. people to fly in space, to be known as "astronauts," should be drawn from the ranks of military test pilots.

Another DOD program inherited by NASA was a much more powerful launch vehicle than any rocket available to NASA in its early years. Known as Saturn C-1 and conceived and managed by the von Braun team, the lift-off thrust of the Saturn C-1 vehicle was to be 1.5 million pounds, four times more powerful than the 360,000 pounds thrust of the Atlas ICBM that NASA was planning to use for orbital flights in Project Mercury. The von Braun team also proposed developing a second version of the Saturn launch vehicle, known as the Saturn C-2. This version would have an upgraded version of the first stage of the Saturn C-1 and upper stages that would use very energetic but difficult to handle liquid hydrogen as its fuel; because of its very low temperature in a liquid state, this fuel was called "cryogenic."

The third inherited DOD project was the very powerful F-1 rocket motor, designed to provide 1.5 million pounds of thrust, equivalent to the eight engines that were to power the first stage of the Saturn C-1. The F-1 was originally an Air Force project. During 1959–1960, the von Braun team developed the concept of an extremely powerful heavy lift launch vehicle called Nova, which was based on the use of up to eight F-1 engines in its first stage.[23]

By the end of 1959, just over a year after it began operation, NASA had developed a ten-year plan that identified various mission milestones and estimated the costs of achieving them. Among the highlights of the plan were the following:

| | |
|---|---|
| 1961–1962 | Attainment of manned space flight, Project Mercury |
| 1965–1967 | First launchings in a program leading to manned circumlunar flight and to a permanent near-earth space station |
| Beyond 1970 | Manned flight to the moon |

The cost of this ten-year program was estimated to be between $12 billion and $13 billion.[24]

While the Army reconciled itself to a minor role in space with the loss of the von Braun team, the Air Force never accepted its loss of space leadership to NASA. Air Force leaders and supporters were encouraged by the tone of candidate Kennedy's public statements on space, especially his strident October 1960 response to *Missiles and Rockets.* The service by 1960 had gained the lead role for space within the Department of Defense from DARPA; then it turned its ambitions to recapturing from NASA the primary role overall in U.S. space activities. After the 1960 election, the Air Force launched "an intense public and internal information campaign to express Air Force views on space to congressmen, journalists, businessmen, and other influential people."[25] For example, *The New York Times* reported that "The Air Force has drafted a publicity offensive to stake out a major role for itself in the nation's space program" and that "this offensive is clearly keyed to the change in administrations. It is the openly expressed belief of the Air Force that the Kennedy administration will look more favorably on military operations in space than does the Eisenhower administration."[26]

To help in its campaign, the Air Force asked Trevor Gardner, former Air Force assistant secretary for research and development and a prime mover in the Atlas missile program, to chair a committee to recommend a more dynamic Air Force space program. As noted in the previous chapter, Gardner had been one of Kennedy's advisers on missile and space issues during the presidential campaign. In December 1960 Gardner also became a member of the group President-elect Kennedy chartered to advise him on space matters during the postelection transition; this was interpreted as another suggestion that Kennedy favored a larger role for the military in space.

### *The Future of the Space Council*

A contentious issue in developing the U.S. organizational approach to space during 1958 had been how best to coordinate the activities of the new space civilian space agency NASA, the Department of Defense and the military services, the Atomic Energy Commission, and other government agencies that might become involved in space activities. The Senate as it considered space legislation hoped to create a single integrated national space program, with civilian and military elements, rather than separate programs carried out by different agencies, and thought that some kind of formal policy-making and coordinating body was needed to achieve this objective. The White House did not want to interpose such a body between the operating space agencies and the president, and thus was resistant to the Senate proposal. After a July 7, 1958, one-on-one meeting at the White House between Dwight Eisenhower and Senate majority leader Lyndon Johnson, a primary advocate of the integrated approach to space, an agreement was reached on creating a policy-level body, to be chaired by the president. According to some accounts, at their meeting Johnson convinced President Eisenhower to chair the board. Other accounts suggest that it was Eisenhower who suggested this approach.[27] Immediately after Johnson left the White House, Eisenhower phoned his deputy chief of staff, Wilton Persons, and told him that he and Johnson had "specifically agreed upon the President's proposal of modeling the advisory group along the lines of the National Security Council: that the authority would be placed with the President."[28]

Following the White House meeting, the policy board was named the National Aeronautics and Space Council. It would have eight members in addition to the president as chair. Other members would include the secretaries of state and defense, the NASA administrator, the chairman of the Atomic Energy Commission, one additional government member, and three individuals from outside the government. Although Eisenhower had accepted the creation of such a Space Council, he made it clear to his associates at the White House that

> he had no intention of convening this body regularly, nor of acting as its presiding officer. He refused to use his discretionary power to appoint an executive secretary or create a separate staff for the Space Council, and he asked [his science adviser James] Killian to preside at its infrequent meetings. Lyndon Johnson may have forced Eisenhower to accept the Space Council as his price for creating NASA, but the president would make sure it would remain a minor body that would never threaten his full control over the nation's space policy.[29]

The Space Council indeed met infrequently and had little influence on space issues in the months following its creation. In a memorandum dated November 16, 1959, NASA administrator Glennan told President Eisenhower that the National Aeronautics and Space Council had not been "particularly useful or effective" and he doubted whether it could "usefully be employed in the management of the nation's space program." He recommended that the president propose amendments to the 1958 Space Act to abolish the Council. However, when Glennan met with Eisenhower on January 8, 1960, he found that the president "seemed to have forgotten our earlier discussion," even though in the interim the proposed changes in the Act had been drafted and were ready to be sent to the Congress. After this meeting, it was clear to Glennan that "discussions between the president and the legislative leaders (especially Lyndon Johnson) as well as with the members of the Space Council would be necessary before the amendment could be proposed to Congress." Eisenhower and Glennan met with Johnson and senior Republican senator Styles Bridges at the White House on January 13 to let them know about the proposed changes in the Space Act; Glennan quotes Johnson as saying "Well, Mr. President, you will remember that you were the one who really wanted this Space Council, and if you want to do away with it now, I'm certain it will be all right with me."[30]

The proposed amendment to the Space Act was sent to Congress in January 1960 and approved by the House of Representatives five months later, on June 9. But the Senate refused to act, primarily because Lyndon Johnson, majority leader and, at the time, still candidate for the Democratic presidential nomination, opposed the changes, despite what he had told Eisenhower in January. Glennan met again with Johnson on June 23, but was unsuccessful in convincing him to bring the proposed changes before the Senate for approval. Glennan reports Johnson as saying "I don't see any reason for giving you a new law at the present time. If I am elected president, you will get a changed law without delay."[31] On August 31, Johnson, speaking not only as Senate majority leader but also by then as the Democratic candidate for vice president, justified his opposition to changing the Space Act in a memorandum inserted in the *Congressional Record*: "One fact is of overriding importance. A new President will take office on January 20, 1961—less than five months from now. The next President could well have different views as to the organization and function of the military and civilian space programs. Any changes in the Space Act at this session will have little or no effect on the space program during the next few months, but could restrict the freedom of action of the next president."[32]

It is unlikely that at this point Johnson envisioned a scenario in which John Kennedy, if elected president, would decide to revitalize the Space Council and turn over its chairmanship to his vice president, i.e., Lyndon Johnson. But four months later, that is precisely what happened.

### *Increasing U.S. Rocket Lifting Power*

Linking Soviet space achievements to the Russian ballistic missile program, as John Kennedy had done during the presidential campaign, was a reasonable thing to do, since even in the 1957–1960 period it was well known in U.S. intelligence and technical circles that the Soviet Union had used its initial R-7 Intercontinental Ballistic Missile as its space launch vehicle. Soviet engineers had been developing this missile since the early 1950s, giving them a several-year head start on the United States. The launch of Sputnik 1 on October 4, 1957, was thus not only a propaganda loss for the United States; it was also a very visible demonstration that the Soviet Union possessed the capability to launch a nuclear warhead across intercontinental distances, and that the United States could be vulnerable to a Soviet nuclear attack.[33]

Among their other impacts, the launches of Sputnik 1 and Sputnik 2 demonstrated that the Soviet Union possessed the capability to lift much heavier payloads into space than did the United States. Sputnik 1 weighed 184.3 pounds, and Sputnik 2 weighed 1,120 pounds. Moreover, the second stage of the R-7 booster also went into orbit on each launch, so in reality the Soviet Union had placed some 12,000–13,000 pounds into space; it was the rocket's upper stage, not the satellite itself, which was visible to the naked eye of observers around the world. By contrast, the first U.S. satellite, Explorer 1, which was launched by the Army team led by Wernher von Braun on January 31, 1958, weighed only 30.8 pounds, with half of that weight being the satellite's last-stage booster rocket.[34]

This disparity in satellite-lifting capability was the by-product of the difficulty the Soviet Union had several years earlier in designing a warhead for an ICBM launch. The three megaton nuclear warhead which was the payload for the R-7 ICBM weighed approximately 11,000 pounds, thus requiring the development of a powerful booster to send it on its intercontinental trajectory. By contrast, the United States a few years later was able to develop a thermonuclear warhead weighing only around 1,600 pounds; this meant that U.S. Atlas and Titan ICBMs did not have to be nearly as powerful as the Soviet R-7 in order to accomplish their military mission. This was acceptable in terms of strategic rocket relationships, but meant that the United States was at a severe disadvantage in sending heavy payloads into space. The United States might be able to launch scientifically sophisticated satellites, but it would not be able to match the Soviet Union in publicly visible space achievements using a converted ICBM as a launch vehicle.

There were two approaches taken during the Eisenhower administration to closing the U.S.-USSR gap in rocket-lifting power. One was to develop the Saturn C-1 launcher, with its first stage having 1.5 million pounds of lift-off thrust. The other was to develop the large F-1 rocket engine, which at some future time could be used to power a much larger launch vehicle. President Eisenhower on January 12, 1960, had indicated his strong support for the Saturn program, and on January 14 told Glennan that "it is essential to push forward vigorously to increase our capability in high thrust space vehicles." Four days later, the Saturn project received the highest national priority, DX, authorizing the use of overtime work and giving it precedence for scarce materials and other program requirements.[35]

However, the Bureau of the Budget (BOB) was determined that the budget to be submitted by Dwight Eisenhower a few days before he left office in January 1961 would be balanced, and this determination took priority over Eisenhower's support for accelerating the Saturn program. NASA had hoped to get a FY1962 budget of $1.4 billion approved; such a budget would have enabled NASA to accelerate its booster and rocket engine development efforts. After tough negotiations with BOB, NASA was held to a $1.1 billion total.[36] At that budget level, there would necessarily be a delay in closing the weight-lifting gap with the Soviet Union. It would be up to the new president to decide whether this was an acceptable situation.

### *The Future of Human Space Flight*

The initial U.S. human space flight effort, Project Mercury, was a very basic undertaking, intended primarily to investigate the human ability to survive being launched into space and returning to Earth and to perform various simple tasks in the weightless environment while in orbit. Planning for a human space flight program to follow Mercury began within NASA in 1959. The first task was to select the objective for that program. To do this, NASA formed a Research Steering Committee on Manned Space Flight in April 1959. At the committee's first meeting the next month, George Low, who was responsible for human space flight at NASA headquarters, suggested that the committee "adopt the lunar landing mission as its present long range objective with proper emphasis on intermediate steps, because this approach will be easier to sell." The committee concluded its meeting without agreeing whether such a mission, or a less ambitious mission to fly around the Moon without landing, was the better choice of a long-range objective.[37] By the time the group met the following month, Low had convinced his colleagues that the lunar landing mission was indeed the appropriate long-term goal for NASA's human space flight program. Operating without top-level political guidance and basing their choice on what constituted from a space program point of view a rational program of human space flight development, NASA planners thus chose the lunar landing objective almost two years before John F. Kennedy made his decision to send Americans to the Moon.

In late July 1960, Administrator Glennan approved the suggestion of the NASA director of space flight programs, Abe Silverstein, that the post-Mercury program be named Project Apollo. Silverstein had picked the name out of a Greek mythology book because he thought that the image of the god Apollo riding his chariot across the Sun gave the best representation of the grand scale of the proposed project. Project Apollo faced a high degree of political uncertainty, however. President Eisenhower and many of his advisers were skeptical of the long-term value of human space flight, and were not inclined to approve an undertaking as ambitious as sending U.S. astronauts to the Moon. As NASA and the BOB developed a space budget for fiscal year 1962, no funds were included for the Apollo spacecraft.

The leadership of the U.S. scientific community reinforced skepticism from White House regarding the human space flight. In December 1960, eighteen months after leaving his position as Eisenhower's science adviser, James Killian suggested that "many thoughtful citizens are convinced that the really exciting discoveries in space can be realized better by instruments than by man." Killian was aware of NASA's ambitious plans, and cautioned that "unless decisions result in containing our development of man-in-space systems and big rocket boosters, we will soon have committed ourselves to a multi-billion dollar space program." He asked, rather rhetorically, "Will several billion dollars a year additional for enhancing the quality of education do more for the future of the United States and its position in the world than several billion dollars a year additional for man-in-space?"[38]

Killian's successor as science adviser, Harvard professor George Kistiakowsky, also was skeptical of the concept of a fast-paced human space flight program intended to win a "space race" with the Soviet Union. In a 1959 discussion paper on "To Race or Not to Race?" Kistiakowsky noted that "there is a well-established military maxim that advises against engaging in battle on the field of the enemy's choosing, but that is precisely what we have committed ourselves to, publicly interpreting Soviet achievements as a challenge for a contest based on that unique and narrow technological specialty—rocketry—in which they excel." He asked "to what extent and how?" should the United States disengage from such a space race and noted that "our strength is that our satellites and space probes have provided us with more scientific information than the Soviets' did." He suggested that "we should hammer this home" and "dismiss the current weight superiority of Soviet payloads as unessential."[39] Kistiakowsky formed an ad hoc panel of the President's Science Advisory Committee (PSAC) in September 1960 "to present the costs of our [human] space activities to the President and to the attention of the next administration." In an early meeting of the panel, Kistiakowsky asked its members to spell out in its report "what cannot be done in space without man." His view that there were relatively few things that met this criterion, and thus building the rockets necessary for ambitious human missions "has to be thought of as mainly a political rather than a scientific enterprise."[40]

The six-person Ad Hoc Panel on Man-in-Space completed its report in December. The panel was chaired by PSAC member and Princeton University chemist Donald Hornig. The panel worked closely with NASA during its investigation. The panel's report noted that "the most impelling reason for our [space] effort has been the international political situation which demands that we demonstrate our technological capabilities if we are to maintain our position of leadership." The report called Project Mercury "a somewhat marginal effort, limited by the thrust of the Atlas booster." As a result of this limitation, there was not "a high probability of a successful flight while also providing adequate safety for the Astronaut." The report noted that "a difficult decision will soon be necessary as to when or whether a manned flight should be launched" and that " the chief justification for pushing Project

Mercury on the present schedule lies in the political desire either to be the first nation to send a man into orbit, or at least to be a close second."[41]

With respect to future missions, the panel noted that they depended on the availability of the Saturn launch vehicle, and that a circumlunar mission could only be attempted only when the Saturn C-2 with a hydrogen-fueled upper stage became available. The panel estimated that this would be in 1968 or 1969, with an initial attempt at a circumlunar mission around 1970. This was several years later than what NASA had indicated was possible. The Saturn C-2 would also be needed to launch the three-person Apollo spacecraft into Earth orbit to function as a space laboratory, but the panel thought that "the valid scientific missions to be performed by a manned laboratory of this size could be accomplished using a much smaller instrumented spacecraft." The panel observed that "none of the boosters now planned for development are capable of landing on the moon with sufficient auxiliary equipment to return the crew safely to Earth. To achieve this goal, a new program much larger than Saturn will be needed."

The PSAC report suggested that "certainly among the major reasons for attempting the manned exploration of space are emotional compulsions and national aspirations. These are not subjects that can be discussed on technical grounds... It seems, therefore, to us at the present time that man-in-space cannot be justified on purely scientific grounds." That being said, the panel observed that "it may be argued that much of the motivation and drive for the scientific exploration of space is derived from the dream of man's getting into space himself."

The panel estimated the costs of the Mercury, Apollo circumlunar, and lunar landing programs. It included in the costs of each program the expenses connected with robotic activities undertaken in its support, and used 1975 as the target date for the first lunar landing. The estimates were as follows:

Project Mercury—$350 million
Apollo circumlunar—$8 billion
Manned lunar landing—an additional $26 to $38 billion.

The cost estimates for Project Mercury and the Apollo circumlunar mission came from NASA. The PSAC panel developed the cost estimate for the lunar landing mission on its own; its estimate was dramatically higher than any that NASA was developing at the time.[42]

The panel's report was presented to President Eisenhower at a National Security Council meeting on December 20, 1960. There are various accounts of his reaction, none of them suggesting a positive response. Kistiakowsky reported that Eisenhower "was shocked and even talked about complete termination of man-in-space programs." (Kistiakowsky also says that he had "learned secondhand that the president-elect was also shown our report before the inauguration and had a negative reaction to the moon-landing proposition.") Glennan observed that "The president was prompt in his response: 'He couldn't care less whether a man ever reaches the moon.'" NASA's associate administrator (the agency's number three position), Robert

Seamans, who had joined the space agency in September 1960, said that Eisenhower asked for an explanation of the reasons for undertaking such an ambitious and expensive program, and that one response compared the lunar journey to the voyages of Columbus to America, which were financed by Spain's Queen Isabella. Eisenhower reacted by asserting that he was "not about to hock his jewels" to send men to the Moon.[43]

Reflecting the president's views as voiced at the December 20 meeting, the draft of Eisenhower's last budget message said that there was no scientific or defense need for a man-in-space program beyond Mercury. Glennan went to see science adviser Kistiakowsky on January 3, 1961, to argue that such a statement was "unwise," and succeeded in getting the statement modified. In his final budget message, Eisenhower instead said that "further test and experimentation will be necessary to establish if there are any valid scientific reasons for extending manned space flight beyond the Mercury program."[44] If there was going to be a follow on the Project Mercury, it would be new President John Kennedy who would have to approve it, and it was probable that such approval would have to be based on other than a scientific rationale.

Despite the negative signals from the outgoing administration, NASA's internal planning for Project Apollo and a lunar landing project to follow it had continued. On October 17, 1960, George Low had told his boss, Abe Silverstein, "it has become increasingly apparent that a preliminary program for manned lunar landings should be formulated. This is necessary in order to provide a proper justification for Apollo, and to place Apollo schedules and technical plans on a firmer foundation."[45] Low formed a small working group to develop such a program. The Space Task Group at NASA's Langley Research Center in Virginia, which was in charge of Project Mercury, and the von Braun team in Huntsville also were working on planning a lunar landing mission. The groups presented a status report on their efforts at a January 5 meeting of NASA's Space Exploration Council. There was a "clear consensus" that such planning should continue under Low's direction, but Administrator Glennan reminded his staff that there was no White House support for such an ambitious undertaking.[46] While this was correct for the few days remaining in the Eisenhower administration, NASA's preliminary planning for a lunar landing mission within months became a critical enabler of John Kennedy's decision to go to the Moon.

## Space during the Transition

In the period between the November 8 election and his inauguration, president-elect Kennedy dealt with only one of these pressing space policy issues, asking vice president-elect Lyndon B. Johnson to assume the chairmanship of the National Aeronautics and Space Council. This request was widely viewed at the time as an indication of the secondary priority that Kennedy was assigning to space. Kennedy also was briefed on January 10 on the report of his task force on outer space, which was chaired by MIT engineer Jerome

Wiesner, who had served as a technical and arms control adviser during the presidential campaign and was the top candidate to be Kennedy's White House science adviser. Wiesner's report was very critical of NASA's organization and management and on the emphasis being placed on the Project Mercury human space flight effort; it also had harsh words for the national security space program.

The Weisner task force had prepared its report without any briefings or other input from NASA, and there was no other direct contact between the incoming administration and NASA's managers during the transition period. The position of NASA administrator was one of the few senior positions for which no one had been nominated as Kennedy became president. Robert Seamans, who was the NASA associate administrator during the transition, later commented that "during the interval between Kennedy's election and his inauguration, a sword of Damocles hung over NASA." Seamans added that rumors that the report of the Weisner group would contain ideas such "as a merger of NASA and the military or a transfer of manned space flight to the military, along with hints about the incompetence of NASA leadership," were "quite unnerving."[47]

### *Johnson to Chair Space Council*

During the transition, Ted Sorensen and David Bell, the Harvard economist whom Kennedy had chosen to be his budget director, met several times with BOB deputy director Elmer Staats. Among the many issues discussed, they agreed that the Space Council was not needed and that legislation abolishing it should be reintroduced in the new Congress. As late as December 17, an action to "abolish the National Aeronautics and Space Council" appeared on a "Preliminary Check-List of Organizational Issues" prepared for the transition team by Richard Neustadt.[48]

However, by mid-December, when Kennedy and Johnson met in Palm Beach to discuss the new administration's legislative program, Kennedy had decided to assign the vice president-elect lead responsibility within his administration for space issues. To signify this, Kennedy announced on December 20 that the vice president would replace the president as chair of the National Aeronautics and Space Council.[49]

The precise sequence of events surrounding this announcement remains unclear. On December 17, in preparation for Johnson's meetings with Kennedy, Kenneth Belieu, who was the staff director of the Senate Committee on Aeronautical and Space Sciences that Johnson had chaired, sent a memo to Johnson on "Government Organization for Space Issues." In it, Belieu suggested that Johnson urge Kennedy to "reactivate the Space Council." But he did not suggest that Johnson ask Kennedy to make him the council's chair; rather, he said, "While the law would necessarily need to be changed to include the Vice President as a formal member of the Council, the President could invite the Vice President to attend and preside over the Council meetings in his absence, pending a change in the law." Belieu also told Johnson that "at NASA there has been a continuing lack of leadership and competence" and that "the Air Force can be expected—and apparently already has started—to make a basic power play to grab the entire Space program. This would involve eliminating NASA."[50]

The Space Council assignment was a logical one for Johnson. Beginning soon after the launch of Sputnik 1, he had played a prominent role in shaping the Congressional response to the Soviet space achievements. An October 17, 1957, memorandum from one of his top advisers, George Reedy, pointed out to Johnson that the Soviet achievement "could be one of the great dividing lines in American and world history, the whole history of humanity" and suggested that it offered the Democrats, and specifically Johnson, an opportunity to present themselves as being more in tune with such a development. Following Reedy's advice, Johnson chaired the twenty days of hearings on "satellite and missile programs" held by his Preparedness Subcommittee of the Senate Armed Services Committee between November 1957 and January 1958. As Johnson addressed the Senate's Democratic caucus on January 7, 1958, in an address Reedy characterized as having "compelling power," he claimed that "control of space means control of the world."[51] As the Senate organized itself to deal with space issues, Johnson named himself chairman of the new Committee on Aeronautical and Space Sciences, even as he maintained his position as majority leader.

In a 1969 interview with veteran television journalist Walter Cronkite on the occasion of the launch of the Apollo 11 mission to the Moon, Lyndon Johnson told Cronkite that president-elect Kennedy had asked him whether there was anything "that I thought I could be helpful to the Administration on." Johnson replied that "I would like to work in the field of space," and that Kennedy had replied that "a President had all a person could do and the Law provided that the President would be Chairman of the Space Council." Kennedy told Johnson that if he were "willing to assume that obligation, he would ask the Congress to amend that statue."[52]

Two days after the decision to make Lyndon Johnson chair of the Space Council was announced, Belieu wrote a follow-up memorandum on "Space Problems" that focused on the potential Air Force "power grab." He told Johnson that the NASA-Air Force conflict was "why I have been so firmly convinced that the Space Council needs to be resurrected and reestablished," since "only someone with your force and vigor and understanding can separate the men from the boys. With President's backing and a man you could trust running NASA, and with close liaison and affinity to the Pentagon at the higher civilian levels, the problem can be licked."[53]

Even after the announcement of the space role for Johnson, Richard Neustadt on December 23 sent Kennedy a "memo on space problems for you to use with Lyndon Johnson." Neustadt in a cover letter said that his memorandum was a response to Kennedy for a suggestion, which seemingly must have come before December 20, "on something you could give him to work on and worry about"; he called the memorandum a "quickie," one that "was worked up today in collaboration with the Budget staff, and no doubt could be vastly improved. But the main thing was to get you *something* to use." The Neustadt memorandum noted that "the 'space' program, both civil and military, raises problems of great difficulty" that were "essentially . . . problems of policy direction." The memo noted that "an opportunity now exists to revitalize the National Aeronautics and Space Council under the Chairmanship of the Vice President" in order "to have it operate selectively on the high priority policy issues."[54] Whether the suggestion to make Johnson the Space Council chair came from the Johnson camp or the Kennedy camp is not totally clear. There is, as the preceding paragraphs show, even suggestive evidence that the initiative may have come from Lyndon Johnson himself.

Johnson returned to Palm Beach to meet with Kennedy on December 26, this time accompanied by Oklahoma senator Robert Kerr, who was LBJ's choice as his successor as chairman of the Senate space committee. To make Robert Kerr the new chairman meant bypassing several senators on the committee more senior than Kerr. Of them, only Senator Clinton Anderson of New Mexico was potentially interested in the chairmanship. Anderson was persuaded not to stand in Kerr's way and instead to accept the chairmanship of another committee during a phone call from Kerr as he met with Johnson and Kennedy on December 26; Kerr told Anderson on the telephone that the two men wanted Kerr to have the chairmanship, and that he hoped Anderson would agree. Anderson immediately accepted this request.[55] Until he became committee chair, Kerr had shown little interest in space; his friendship with Lyndon Johnson and his stature as a leading senator were his prime qualifications for the chairmanship. In Kerr, Johnson knew he had a close and powerful ally who would help him push the new administration to propose a larger space program and who would be sympathetic to the political (and pork-barrel) uses of that larger program.

After Kennedy's meeting with Johnson and Kerr, *The New York Times* reported that the three had agreed "on plans to expand the United States' exploration of space . . . reflecting Mr. Kennedy's serious concern over the Soviet lead in this field and his oft-repeated campaign argument that United States prestige had slipped abroad as a result."[56] Whether this was, in fact, what the men had agreed upon was not immediately clear. It was not until four months later, in April 1961, that the Space Council was activated and Lyndon Johnson as its chairman given the task of proposing a way for the United States to enter and win the space race.

## *Wiesner Task Force Critical of NASA and DOD Space Efforts*

One of the twenty-nine task forces advising president-elect Kennedy during the transition focused on "outer space." The task force was led by Jerome Wiesner, a professor of engineering at MIT who had been involved in weapons research during World War II and had had a decade of experience dealing with the national security policy aspects of scientific and engineering issues. Wiesner had been a member of PSAC since its inception in 1957, but had not served on any of the PSAC subcommittees dealing with space issues. He had, however, most likely heard many of the briefings on space issues given to the overall committee. During the presidential campaign, Kennedy had sought his advice primarily on a possible nuclear test ban and other arms control issues. By the time the task force began its work, Wiesner had emerged as Kennedy's most likely choice to be the presidential science adviser.

The other original members of the Wiesner "Ad Hoc Committee on Space" were:

- Edwin Purcell, the Harvard professor who had chaired the first PSAC study on space in 1958 and was chair of the PSAC space flight panel;
- Donald Hornig, professor at Princeton University and the PSAC member who was chair of PSAC's space booster panel and had also led the PSAC "Ad Hoc Panel on Man-in-Space";
- Edwin "Din" Land, President of Polaroid Corporation in Cambridge and an original PSAC member;
- Harry Watters, a top assistant to Land;
- Bruno Rossi, professor of physics at MIT; and
- Trevor Gardner, a former Air Force Assistant Secretary for Research and Development who also was chairing an Air Force committee in support of the service's campaign for a larger role in space.

One indication that the decision to make Lyndon Johnson chair of the Space Council had been taken prior to December 17 was a meeting on that day between Jerome Wiesner, Kenneth Belieu, and Max Leher, who was the Senate space committee's assistant staff director. The latter two, of course, worked for Johnson as committee chair. At that meeting, Wiesner expressed support for reestablishing the Space Council, and invited Belieu and Leher to join his task force.[57] It is unlikely that this invitation would have been offered if it were not for the space policy role that Kennedy envisaged for Johnson.

Most of the members of Wiesner's group were deeply familiar with space issues because of their past involvements. They chose to prepare their report without any briefings or other formal contact with NASA and DOD. The group met together only a few times before issuing its report; Robert Seamans, NASA's associate administrator, notes that "alarming rumors" about what the group might say in its report "kept appearing in journals and newspapers."[58]

On January 10, 1961, president-elect Kennedy met in Lyndon Johnson's Senate office with Johnson, members of the Wiesner panel, Senator Kerr, and Representative Overton Brooks, chairman of the House Committee on Science and Astronautics, to be briefed on the panel's report. Because portions of the report dealt with the military space program and other sensitive policy matters, the entire report was classified "confidential." An unclassified version was made public the next day.[59]

The twenty-four page report was admittedly a "hasty review" aimed at providing a "survey of the program" and identifying "personnel, technical, or administrative problems" requiring prompt attention. It listed five principal motivations for the space program:

1. National prestige
2. National security
3. Scientific observation and experiment
4. Practical nonmilitary applications
5. International cooperation.[60]

The panel recognized that "space exploration and exploits have captured the imagination of the peoples of the world," and that "during the next few years the prestige of the United States will in part be determined by the leadership we demonstrate in space activities."

The task force felt "compelled to criticize our space program and its management" because of "serious problems within NASA, within the military establishment, and at the executive and other policy-making levels of government." The report was critical of the management of both NASA and DOD space efforts. With regard to NASA, it called for "vigorous, imaginative, and technically competent top management," implying that NASA's current top officials had not demonstrated these qualities. The report deplored the tendency of each military service to create an independent space program and called for one service to be responsible for space within the DOD. It was concerned with the lack of coordination among the various agencies involved in space and endorsed the revitalization of the Space Council as "an effective agency for managing the national space program." The use of the word *managing* was particularly noted; some took this as a suggestion that the Space Council would have executive, not just coordinating, responsibilities in space.

Only the scientific portion of NASA's programs was deemed "basically sound." Even so, the report noted that "too few of the country's outstanding scientists and engineers" were working in the space field. Developing boosters with greater weight-lifting capability was "a matter of national urgency," since "the inability of our rockets to lift large payloads into space is key to serious limitations of our space program."

The report recognized that "man will be compelled" to go into space "by the same motives that have compelled him to go to the poles and to climb the highest mountains of the earth." Thus, "manned exploration of space will certainly come to pass and we believe that the United States must play a vigorous role in this venture." The ultimate goal of human space flight, the group recognized, was "eventual manned exploration of the moon and the planets." The panel acknowledged that "some day" humans in space might "accomplish important scientific or technical tasks," but in the short run human space flight "*cannot be justified solely on scientific or technical grounds.*"

The Wiesner task force called Project Mercury "marginal," as had the PSAC report a month earlier, and pointed out that it was "very unlikely" that the United States would be first to send an astronaut to orbit. Echoing the PSAC position, it was critical of the relative priorities given to human and robotic flight: "The acquisition of new knowledge and the enrichment of human life through technological advances are solid, durable and worthwhile goals of space activities... By having placed highest national priority on the MERCURY program, we have strengthened the popular belief that man in space is the most important aim of our nonmilitary space effort."

The task force recommended that President Kennedy not allow "*the present Mercury program to continue unchanged for more than a very few months*" and that he not "*effectively endorse this program and take the blame for its possible failures.*" It suggested that "a thorough and impartial appraisal of the MERCURY program should be urgently made." It recommended that "*we should stop advertising Mercury as our major objective in space.* Indeed, we should make an effort to diminish the significance of this program to its proper proportion before the public, both at home and abroad." Of particular concern was the potential death of an astronaut in a Mercury mission, particularly if he were to be stranded in orbit. Rather than continue to put emphasis on human space flight, suggested the panel, "We should find effective means to make people appreciate the cultural, public service, and military importance of space activities other than space travel." Finally, the panel recommended "a vigorous program to exploit the potentialities of practical space systems" for communications, navigation, and meteorological observation.

After the panel's briefing, the president-elect described the report as "highly informative" and his meeting with the task force as "very fruitful." Once the report became public, it was subject to criticism from space advocates, and at his first post-inaugural press conference Kennedy remarked that "I don't think anyone is suggesting that their [the task force's] views are necessarily in every case the right views."[61] In terms of the space program's substance, in contrast to management issues, the report seemed to endorse the civilian space program that had been pursued by the Eisenhower administration; it certainly was not "the ringing denunciation of Eisenhower's lassitude on space initiatives that Kennedy... might have hoped for." According to one historian, Kennedy "treated the panel's findings "like a skunk at a wedding." There was even some question after the inauguration of whether Wiesner agreed with everything in the report of the task force he had chaired.[62]

The day after he received the report of the panel, Kennedy named Jerome Wiesner as his new science and technology adviser. Wiesner's selection meant that Kennedy would hear both sides of the case for a high profile civilian space program focused on human space flight, one perspective from Lyndon Johnson and the other from Wiesner.

## An Uncertain Future for NASA

In his final weeks in office, NASA administrator Keith Glennan grew increasingly distressed by the lack of any contact from the incoming Kennedy administration. Shortly after the election, Glennan spoke briefly with Jerome Wiesner. He tried to probe Wiesner regarding the schedule for naming the new NASA administrator, but Wiesner only asked whether Glennan was willing to stay on in the job, to which Glennan replied in the negative. By January 3, Glennan noted in his diary that "never in my life have I seemed so frustrated in attempting to bring an important job to a conclusion." He bemoaned that NASA had "been in a state of suspended animation since the election" and that "not one single word or hint of action has been forthcoming from the Kennedy administration."[63]

On January 9, the impatient Glennan called vice president-elect Lyndon Johnson. Glennan told Johnson that he "felt a heavy responsibility in the matter of turning over my job to my successor" and that he "was ready to help in any way desired." He added that he had heard that finding a new NASA administrator was proving difficult, and offered "to help in the process." Glennan noted that this was "his first contact with the new administration, and that he had to initiate it." He reported that Johnson replied to his call by thanking him lavishly for his helpful attitude and then saying "as soon as I have something to tell or discuss with you, I will call you!"[64] Such a call never came.

On January 17, with still no word from the incoming administration, Glennan called the White House to alert the staff there that no senior person at NASA had been asked to stay on, and that Hugh Dryden had expressed his willingness to serve as acting administrator during the change in administrations. An hour later, he was told that Clark Clifford, who was handling personnel appointments for president-elect Kennedy, had indicated to the White House that the new administration did indeed want Dryden to stay on.[65] There still was no direct contact between NASA and the Kennedy team.

Glennan's last day at NASA was January 19, 1961. After sherry with a few of his staff, Glennan left NASA for the final time, intending to begin the drive back to his home in Cleveland that evening. However, Washington was paralyzed by a blizzard, and it was not until 6:30 a.m. on January 20, the day of the Kennedy's inauguration, that he was able to leave. As he drove west and listened to the inauguration ceremonies on his car radio, Keith Glennan reflected on "some 29 months of interesting, exciting, baffling, and, at times, frustrating work in Washington." His somewhat melancholy last entry into the diary of his time in Washington was: "And still—no word from the Kennedy administration!"[66]

Neither Keith Glennan nor anyone else connected with the nation's civilian space program could have anticipated the dramatic changes in the nation's space policy that would emerge over the next few months.

# Chapter 3

# Getting Started

On a frigid and snow-covered January 20, 1961, John F. Kennedy was sworn in as the thirty-fifth President of the United States. In his stirring inaugural address, Kennedy had a particular message "to those nations who would make themselves our adversary." He asked that "both sides begin anew the quest for peace," but that toward potential adversaries, "we dare not tempt them with weakness." So, "let us begin anew, remembering on both sides that civility is not a sign of weakness... Let us never negotiate out of fear, but let us never fear to negotiate. Let both sides explore what problems unite us... Let both sides seek to invoke the wonders of science instead of its terrors. Together let us explore the stars." Kennedy stressed the sacrifices he was asking of Americans in order to lead the global fight for freedom, saying "now the trumpet summons us again—not as a call to bear arms, though arms we need—not as a call to battle, though embattled we are—but a call to bear the burden of a long twilight struggle, year in and year out"; he called his countrymen to action with his much-quoted admonition, "Ask not what your country can do for you—ask what you can do for your country."[1]

Theodore Sorensen, who collaborated with Kennedy in drafting the address, notes that one line in the speech was "a more important statement of his administration's intent" than any other in the speech: "Only when our arms are sufficient beyond doubt can we be certain beyond doubt that they will never be employed." This, says Sorensen, "was the Kennedy approach to war and peace," combining unmistakable strength with a willingness to seek areas of cooperation rather than to focus on areas of conflict.[2] It was an approach that Kennedy was to use with respect to space in all his days in office—preferring to cooperate but being willing to compete if that was the better path to advancing U.S. interests.

The inaugural address, in addition to its soaring rhetoric, reflected Kennedy's world view as he entered office. Thomas Reeves notes that Kennedy brought with him to the White House "the values and many of the ideas his father had instilled in all the Kennedy children." These included "the president's selection of pragmatic advisers, his overall lack of interest in domestic reform, his conservative economic views, his hard-nosed posture in foreign affairs . . . and his intense interest in public relations and his image."[3]

In his September 1960 memorandum on "Organizing the Transition," Richard Neustadt had observed that "one hears talk all over town about 'another Hundred Days' [referring to the beginning of Franklin D. Roosevelt's first term in office], once Kennedy is in the White House." Neustadt felt that "if this means an *impression to be made* on congressmen, bureaucrats, press, public, foreign governments, the analogy is apt." He suggested that "nothing would help the new administration more than such a first impression of energy, direction, action, *and* accomplishment. Creating that impression and sustaining it becomes a prime objective for the months after Inauguration Day." Arthur Schlesinger adds that it was Kennedy's intention in his initial days in office "to create a picture of drive, purpose and hope."[4]

This intention was not realized. Instead, the first one hundred days of the Kennedy administration were marked by slow movement of Kennedy's domestic program through the Congress and immediate challenges from abroad. After an initial victory in the House of Representatives, adding two more liberal members to the southern conservative-dominated House Rules Committee and thus making it more likely that the committee would not block Kennedy's legislative proposals from reaching the House floor, the president found that moving his domestic policy proposals through the Congress was much slower going than he had hoped for.

In the foreign policy and national security fields, even more intractable issues confronted the new president. Kennedy and his close associates were troubled by a January 6 speech by Soviet premier Nikita Khrushchev, in which the Soviet leader projected "bellicose confidence" that international events were trending in the favor of the Communists; Khrushchev had said that "there is no longer any force in the world capable of barring the road to socialism." Kennedy's tough language in his inaugural address was an initial reaction to the Soviet challenge. On February 23, Kennedy sent off a letter to Khrushchev, suggesting an early meeting between the two; it was Kennedy's hope that he could convince the Soviet leader that the United States could not be bullied.[5]

The first crisis of the new administration in foreign policy emerged in mid-March; Kennedy was faced with the decision of whether or not to intervene with U.S. troops in Laos, a small landlocked country in Southeast Asia, where the pro-American government seemed to be on the verge of military defeat by the Communist Pathet Lao forces. The White House interpreted this conflict as one of the "wars of national liberation" that Nikita Khrushchev had said in his January speech would be an important means for spreading Communist values around the world. On March 20 and again on March 21, Kennedy met with his National Security Council to discuss whether immediate intervention was necessary or whether a diplomatic solution was still possible. The joint chiefs of staff, fearing another Korea-like engagement half a world away, urged decisive actions involving 60,000 troops, air support, and possibly the use of tactical nuclear weapons in order to ensure quick success. After these meetings, Kennedy decided not to intervene as yet, but to demonstrate his willingness to do so if the United States and the Soviet Union could not find grounds for compromise on the future of Laos. Kennedy scheduled a press conference for March 23 in order to issue a public warning to the Soviets that the United States would intervene unless an immediate ceasefire could be arranged.[6]

It was in this troubled domestic and international context that the Kennedy administration took its first steps in determining the future of the U.S. civilian space program. Compared to the other issues on his agenda, space remained a relatively low priority item, and Kennedy himself was only occasionally directly involved. However, early attention to a number of issues could not be avoided.

## Finding a NASA Administrator

The most pressing of these issues was finding someone to run NASA. As the new administration took office, no one had been selected as the nominee for the job of NASA administrator, which thus became the most senior unfilled position as the Kennedy presidency began. That no nominee had been named was not for lack of trying. There are several versions of how many people were considered for the position. The number in various accounts ranges from nineteen to twenty-eight.[7]

In their December discussions on space issues, John Kennedy had given Lyndon Johnson the responsibility of identifying the person to be the next NASA administrator. In turn, Johnson asked the staff director of the Senate space committee, Kenneth Belieu, to coordinate the search for the nominee. Belieu had told Johnson on December 22 that "the Administrator of NASA doesn't have to be a technician. He does need to have firm administrative ability, and be able to work with scientists and technicians." Belieu's initial thoughts about people qualified for the NASA position included Karl Bendetsen, an industrialist who had served in the Truman administration; General Maxwell Taylor, retired Army Chief of Staff; and George Feldman, who had been the staff director of the House committee established in 1958 as proposals to create NASA were being considered. Belieu noted that Feldman had been "actively seeking" the NASA job. He also noted that the current Air Force chief of staff, Thomas White, who was soon to retire, "might be interested," and that he had gotten suggestions that Jet Propulsion Laboratory director William Pickering and Marshall Space Flight Center director Wernher von Braun might be good candidates.[8]

Most of these possibilities did not survive a first round of scrutiny. On January 23, Belieu gave a list of possible picks to now–Vice President Johnson. They were Laurence (Pat) Hyland of Hughes Aircraft; Charles (Tex) Thornton of Litton Industries; James Fisk of Bell Laboratories; James Doolittle, World War II hero and former chairman of the National Advisory Committee on Aeronautics; and William Foster of Olin Mathieson

Chemicals. On January 25, Belieu reported to top Johnson assistant Bill Moyers, who was interviewing candidates and then deciding whether or not to send them forward to the vice president, that "we have run through about 25 names to date," and that the 25 did not include "Generals Maxwell Taylor, Jim Gavin, Bruce Medaires [*sic*- the correct spelling is Medaris], Earl Partridge, [and] Thomas White." An unsigned January 26 memorandum, most likely composed also by Belieu, reflected a view that NASA should not be headed by an active military man because "the Communists would scream that this proved our militaristic intentions in space and that NASA was and is a façade"; because "it would have the effect of scaring off allies and neutrals from a program of international cooperation in space"; and because "many of the scientists in NASA might prefer to work elsewhere if NASA took on a military look."[9] According to Lyndon Johnson, at some point Kennedy had suggested appointing retired General James Gavin, who had been a campaign adviser, to head NASA, and Johnson had told Kennedy that "that's the worst thing we could do for the program, would be put a man with stars on his shoulder and a general's uniform, in charge of the space effort of this country, because it would frighten other countries and do a great disservice to our own program."[10]

Belieu reported to Moyers that "at the Vice President's direction" he had called several of the people on his list and would meet with William Baker, head of Bell Laboratories, and Tex Thornton. Belieu also reported that he had interviewed William Pickering, who was "definitely interested," but "we might do better." The head of General Dynamics, Frank Pace, was also involved in the search process. Belieu on January 26 said that Pace "would call me back this afternoon with a check on some of these people and further suggestions." He told Moyers, "it looks as though it will be impossible to find anyone who is completely satisfactory to all factions involved in the space program."[11]

Kennedy, tired of the delay in identifying a candidate for the NASA job, reportedly told Johnson and new science adviser Wiesner soon after his inauguration that he would find someone himself if they did not act soon. On January 25, he told his first press conference that he was "hopeful" that a NASA administrator would be named in the next few days.[12]

One underlying reason for the difficulty in finding a person to take over NASA was the pervasive uncertainty about the future of the agency and of the U.S. civilian space program. John Kennedy had given little indication during the campaign of how he would approach space policy as president. In addition, the Air Force campaign to take over the U.S. lead in space was at a peak, and no individual was interested in presiding over the dissolution of NASA. There were three general perspectives on what kind of person should head NASA. One view favored a person with administrative experience in a science and technology setting; this had been the background of Keith Glennan. Another argued for a top-flight scientist with an academic background. A third argued that political savvy in addition to administrative skill was a more necessary background than either a scientific or engineering background. The first of these positions was supported by Wiesner.[13] The second position was held by many nongovernmental scientists, who wanted NASA priorities determined solely by scientific criteria. The third was the position of Lyndon Johnson and Robert Kerr.[14]

Another significant barrier to getting someone to accept the NASA position was the probability that the NASA administrator would find himself having to work closely with, or even for, Johnson, given LBJ's anticipated new role as Space Council chairman. The new vice president was known for "his tantrums and his wheedling and bullying." Few senior people who had experienced the "Johnson treatment" were eager to undergo it on a continuing basis. According to Wiesner, "no good scientists wanted to take the job on because they didn't want to come under LBJ." Wiesner remembered that "8 or 9 of the best scientists in America were asked to head NASA, and they all said no."[15]

### *James Webb Selected*

In the wake of President Kennedy's pressure, a new name was suggested, apparently to Lyndon Johnson by Senator Kerr and independently to President Kennedy by Wiesner. Wiesner later argued that his suggestion was the one that was decisive, although other accounts suggest that it was Lyndon Johnson who first brought Webb's name to White House attention. The new candidate was James E. Webb, a businessman and lawyer with prior experience in high-level government posts. During the Truman administration, Webb had been head of the Bureau of the Budget (BOB) and then the number two person under Dean Acheson at the Department of State. Webb also had experience in managing large organizations; he had worked in Oklahoma heading one of Robert Kerr's companies from 1953 to 1958. After leaving Kerr's employ, Webb had been active on issues of science and engineering education, in the process becoming well known to many of the leaders of the scientific community, including Wiesner.

According to Wiesner, Kennedy asked him to check whether Johnson agreed that Webb would be a good choice. Johnson did agree, and because he had had such little success with the people he had contacted, asked Wiesner to call Webb. On Friday, January 27, after clearing the contact with the president, Wiesner telephoned Webb, who was at a luncheon in Oklahoma City, and asked him to be in Washington the following Monday to meet with the vice president to discuss the NASA position.

Webb left Oklahoma City on Friday and spent the weekend in Washington discussing the prospects for space under Kennedy with several former associates in the BOB and the Kennedy White House staff and with others whose views he valued. One of them was Webb's longtime friend Lloyd Berkner, who was the current chair of the Space Science Board of the National Academy of Sciences, NASA's primary source of scientific advice; Berkner had himself been approached for the NASA job and had said that he was not interested. By Monday morning, January 30, as he arrived at Johnson's Capitol office, Webb felt that he had a fairly good idea of what was going on with respect to space and had concluded that "I would not take the job if I could honorably and properly not take it."

Before meeting with Lyndon Johnson (whom he did not know well), Webb chatted with acting NASA administrator Hugh Dryden, who was there for the meeting, and Frank Pace, who had been Webb's successor as director of the BOB in the Truman administration. Webb had also known Dryden since the late 1940s. Both Pace and Dryden agreed with Webb that he was not the right man for the job, and Webb asked Pace to convey that view to Johnson. Pace tried to do so, but Johnson was unwilling to listen and in essence threw Pace out of his office. Webb then met with Johnson, who, Webb says, was "very anxious" for Webb to accept the NASA job. Webb made it clear that he would only accept the position on the basis of a direct offer from the president. Arrangements were quickly made for Webb to meet with Kennedy, whom Webb previously had met only once or twice on social occasions.

After lunch with Dryden, Webb met with Kennedy one-on-one in the Oval Office. Kennedy told him that he did not want a technical person for the NASA job, saying that "there are great issues of national and international policy" related to NASA, and that Webb, with his previous government experience, was well qualified to address such issues. Webb felt he could not refuse the president's direct invitation, and so accepted the nomination.

President Kennedy and James Webb on January 30, 1961, as Webb accepted the president's offer to become the second NASA administrator (JFK Library photograph).

He asked Kennedy to keep Hugh Dryden as deputy administrator, and he also asked the president whether he was being hired to implement a predetermined policy. Kennedy assured him that this was not the case and that he was looking to Webb to propose the best direction for NASA. Kennedy then escorted Webb from the Oval Office to the office of press secretary Pierre Salinger, who took Webb to the press room to announce his nomination. Only then could Webb call his wife to tell her what had happened; she had already heard the news on the radio.[16]

Webb's confirmation hearing before the Senate Committee on Aeronautical and Space Sciences, with Robert Kerr in the chair, was held three days later, even before Webb's formal nomination papers reached Capitol Hill; there were no questions after Webb's opening statement. The committee voted unanimously to support the nomination; the Senate followed suit on February 9. Lyndon Johnson swore in Webb as NASA's second administrator on February 14, and Webb set to work with the goal "to end uncertainty, to make unmistakably clear...support for manned space flight, to define necessary additions to the budget for Fiscal Year 1962...and to establish personal and official relationships conducive to effective leadership."[17] It was thus clear from the start of his tenure that James Webb had a different, more ambitious, vision for the future of NASA than his predecessor. Getting this vision accepted would not be an easy task. In preparation for Webb's first meeting with the new director of the BOB, David Bell, on February 16, the BOB staff suggested that "we are pretty much still in the dark as to what position the [Kennedy] administration desires to take in the space field, or whether any general direction has been decided upon."[18]

In addition to getting President Kennedy's agreement to continue Hugh Dryden as deputy administrator, Webb also asked Robert Seamans to stay on as associate administrator. He was happy to learn that Seamans was a Republican, since that would give a bipartisan appearance to the top NASA management team. Webb told Dryden and Seamans that he wanted NASA to be managed jointly by the three of them as a "triad," hammering out the major decisions together. Webb would handle NASA's external political and public relations, Dryden would be the primary link to the U.S. and international science communities, and Seamans would act as NASA's general manager. Seamans describes the arrangement: "Jim was the charismatic leader with long-range vision and a great knack for understanding how policy and politics interacted in Washington. Hugh...possessed a quiet, invaluable sense of practicality...I managed NASA's programs while Jim lined up outside support and Hugh provided sound guidance on our goals."[19]

As he took on the NASA job, Webb was fifty-four years old. He was "stocky and voluble, vigorous, noisily garrulous, and with a broad North Carolina accent." He had a strong physical presence; "though not a tall man, his strong, square head and bullish neck, his sturdy chest, an obstinate jaw and narrowed grey-blue eyes lent him a dominant demeanor." Sorensen notes that "Webb was not...a Kennedy type individual. He was inclined to talk at great length, and the President preferred those who were

The "triad" of men who managed NASA during the Kennedy administration: Deputy Administrator Hugh Dryden (left); Administrator James Webb (center); and Associate Administrator Robert Seamans, Jr. (left) (NASA photograph).

more precise." He adds, however, that "I don't know that the President ever regretted his appointment of Webb." Wiesner remembers that Kennedy "understood that he had somebody with real ability in Webb" and adds that he never heard Kennedy say anything "snide or negative" about him. One account of Project Apollo, however, comments that "only because Kennedy was indifferent to space did Jim Webb end up in the administrator's position." While a number of other men had turned down the job because of NASA's uncertain future, if they had "known that four months later NASA would become a custodian of the nation's honor, most of them would have snapped up the job. If the men in the White House had known, they would not have chosen anyone like Jim Webb." The president's brother Robert Kennedy agreed with this view. He commented in 1964 that if his brother "had realized how much money would be involved and how important it [the space program] was going to be, he never would have made Jim Webb the head of it." Robert Kennedy added that Webb "talked all the time and was rather a blabbermouth . . . The President was very dissatisfied with him."[20]

### Webb Soon Challenged

James Webb faced an almost immediate challenge to his freedom to manage NASA as he saw appropriate, especially in the context of the prevalent NASA–Air Force tensions. The first attempt in July 1960 to launch a Mercury capsule atop an Atlas booster had ended in an explosion. The cause of this failure had been localized to the area where the spacecraft and the capsule were joined. Since the Air Force retained responsibility for booster performance and launching while NASA was responsible for the spacecraft and the overall mission, this meant that both organizations were intimately involved in attempting to correct whatever had caused the failure. A "quick fix" using an improvised steel band was adopted. The NASA top management had agreed to this approach before Webb took office, but the Air Force remained extremely concerned about the possibility of another major accident. There had been a highly visible Atlas failure on December 15, 1960 as NASA attempted to send a robotic spacecraft to the Moon, increasing the level of concern on the part of the Air Force. That worry was linked to the important question of what another failure would communicate about the reliability of the Atlas ICBM, a key element of the U.S. nuclear deterrent force, and thus to the credibility of the U.S. deterrent threat.[21]

Webb was briefed on the situation on February 18 by NASA's Project Mercury managers, who wanted his approval for a launch of the improved Mercury–Atlas combination on February 21. Webb approved the launch, but soon after got a call from the Air Force asking him to reverse that decision. From the White House, Wiesner also expressed his opposition to going ahead. After Webb checked again with knowledgeable people both within and outside NASA, he refused to reverse his decision, although the Air Force "protested vehemently" and made its concerns known to the White House, most likely through one of Kennedy's military aides, Air Force General Godfrey McHugh. The White House decided not to intervene in the dispute, "making the issue a major test for Webb and NASA and their credibility with the president." The February 21 flight was a total success; Webb had passed his first challenge with flying colors.[22]

## NASA–Air Force Tensions Reduced

The Air Force push for a larger role in space had continued after the Kennedy administration came into office; the major aerospace trade magazine *Missiles and Rockets* reported late in March 1961 that "the showdown on who will take charge of the U.S. man-in-space program—and with it the main role in space exploration" would come soon, and that in the choice between NASA and the Air Force, "the best bet on who will win when the cards are dealt: the Air Force."[23] By the time this report appeared, however, Air Force ambitions had been significantly tempered by both the president and the new managers of the Department of Defense.

Webb's conduct with respect to the Mercury–Atlas test flight had demonstrated to the Air Force that he was not easily intimidated. Bolstered by this success, Webb and other top NASA officials embarked on a conscious campaign to establish close working relationships with the new civilian leadership of the Department of Defense. This was an attempt, reported Robert Seamans, to "so handle ourselves that, rather than have things pull further apart, the wounds got healed and things got pulled together."[24] Webb did not know well Kennedy's hard charging and intellectually brilliant secretary of defense, Robert McNamara (and the two apparently did not get along from the start of their relationship), but he knew McNamara's number two person, deputy secretary of defense Roswell Gilpatric, from their time together in the Truman administration. Gilpatric says that one of the reasons he was named McNamara's deputy was "to work with NASA and Webb" and "to help avoid conflicts between DOD and NASA." He also noted that McNamara "was impatient with Webb" and "felt that he talked too much."[25]

### *An Initial NASA–DOD Agreement*

The first result of this "peace-making" effort was a February 23 agreement signed by Webb and Gilpatric, confirming the desirability of a single national launch vehicle program and indicating that neither NASA nor DOD would begin the development of a new space launcher without the written acknowledgment of such a step from the other agency. On February 24, Webb and Dryden met with McNamara, Gilpatric, and director of defense research and engineering Herbert York, an Eisenhower holdover. The group agreed that Webb and Gilpatric would meet "from time to time for lunch and would bring others as needed" as a way to coordinate NASA and DOD space activities at the top level. They agreed on the need for a review in "about four weeks" from the date of the meeting to determine the need for accelerating the existing space program; there was "a general feeling that we should accelerate the booster program." There was discussion of a possible omnibus bill to cover all space activities in both NASA and DOD (an idea which was never implemented). Writing to budget director David Bell a few days later, Webb described the February 24 meeting as "splendid."[26]

Webb's biographer W. Henry Lambright suggests that NASA–DOD agreement was possible because Robert McNamara was already "trying to constrain the expansionist tendencies of the services" and wanted to use NASA "as a check on the air force." In addition, both McNamara and Webb recognized that "if they failed to settle differences at the NASA–DOD level, Lyndon Johnson would have the opportunity to stake out a stronger claim for coordinating them through the National Aeronautics and Space Council." Webb saw this period as part of a process in which NASA and the DOD were "like two strange animals . . . sparring around, smelling each other, seeing what could be done, testing each other out."[27]

Shortly after taking office, McNamara had requested a review examining whether the Wiesner Task Force criticism of a "fractionated military space program" was valid. Based on this review and conversations within DOD, McNamara decided to centralize management of Department of Defense space efforts in the Air Force, and on March 6 issued a directive to that effect.[28] This was not, however, exactly the outcome that the Air Force had hoped for, given the preceding NASA–DOD agreements; from McNamara's perspective, centralizing space activity in one organization made it easier for him to exercise tighter control over that activity.

### *Gardner Report Submitted*

The report of the committee headed by Trevor Gardner that had been intended to map an ambitious future for the Air Force in space was finally submitted on March 20; with its bullish recommendations, the report unintentionally reinforced Robert McNamara's concern about the need to limit Air Force ambitions. The report's conclusions were alarmist; the United States, the report claimed, could not overtake the Soviet Union in space for at least another three to five years; there were "no accelerated and imaginative programs" to close that gap. Thus there was "an impending military space threat," which "endangers our national security and international prestige." The report was critical of the separation between the civilian and military space programs, arguing that there should be one integrated national space program. The report called for a "dramatically invigorated space program" and called upon the DOD to create, and then make available to NASA, a series of "building blocks" of a firm technological foundation for whatever the nation wanted to do in space. It urged the Air Force to "develop the fundamental capability to place and sustain man in orbit," since "the time when man in orbit can be completely, effectively, and efficiently replaced by mechanisms [robotic systems] is beyond today's vision." The report declared that "it is essential that the Air Force play a major support role in manned exploration of the moon and planets," even though "direct contributions to national security cannot be identified." The report devoted significant attention to future human spaceflight efforts, and recommended both that the United States send people to the Moon and develop large space stations in the 1967–1970 period. It suggested that since there was a "military need for a variety of launch vehicles based on the F-1 engine," development responsibility for that engine should be transferred back to the Air Force from NASA, and then the Air Force should "initiate urgent development of a first stage launch vehicle using the F-1 engine."[29] Given the agreements already described and the events of the next two months, the Gardner report had limited impact on both Air Force space activities and national space policy.

### *Kennedy Takes a Position*

The publicity-seeking chairman of the House Committee on Science and Astronautics, Overton Brooks, was aware of and concerned by the Air Force campaign. Unsatisfied by Air Force assurances that it did not hope to take over the lead role in the U.S. space program, Brooks wrote a three-page letter to President Kennedy on March 9, saying that he was "seriously disturbed by the persistence and strength of implications reaching me to the effect that a radical change in our national space policy is contemplated." Brooks told Kennedy that he did not want to see "the military tail wag the space dog."[30] Brooks followed this letter with hearings on the NASA–Air Force rivalry, but he was not able to get top DOD officials to testify, most likely due to an administration decision not to cooperate with the committee. The hearings shed little light on the evolving situation.

Despite the hopes raised by Kennedy's campaign rhetoric, it is unlikely that the new administration could have, or would have, agreed with the Air Force hope for a larger role in space at the expense of NASA. As he left office, Dwight Eisenhower had warned of the increasing power of the "military–industrial complex," and a move in the early months of the administration to increase Air Force activity in a visible area such as space would have been politically very difficult. Secretary of Defense McNamara was trying to get the management of defense activities under centralized control. The top people in DOD were not particularly space conscious, and McNamara and Webb had reached an understanding of the respective roles of their agencies that the White House would have been unlikely to countermand. NASA under Webb's direction seemed to be shaping up in terms both of its organization and program success, and there were no compelling reasons to downgrade the importance of the agency. One of presidential science adviser Wiesner's major priorities was to control the spread of the arms race into new areas; he too was unlikely to have supported an expanded military space effort.

President Kennedy replied to the Brooks letter on March 23, telling the congressman that "it is not now, nor has it ever been my intention to subordinate the activities of the National Aeronautics and Space Administration to those of the Department of Defense."[31] With that response, the controversy over the NASA–Air Force relationship became a secondary policy issue, although it never completely disappeared.

## Project Mercury Reviewed

The Wiesner task force on space had recommended that "a thorough and impartial appraisal of the MERCURY program should be urgently made." Those managing the Mercury effort welcomed this suggestion, but for reasons different than those which had led the Weisner group to call Mercury "marginal." The NASA team was confident that Mercury was a sound program, and feared that without the positive assessment they believed would result from such an independent review, President Kennedy might decide that even the suborbital mission planned in the next few months was too risky, and would not allow NASA to carry it out.

A first step was to inform the White House that there was a rehearsal flight for the suborbital mission, with a chimpanzee named Ham as its passenger, scheduled for January 31. After not hearing from anyone at the White House for several days after the inauguration, acting administrator Dryden was able on January 26 to meet with new science adviser Wiesner to let him know about the upcoming mission; Dryden wanted to make sure that President Kennedy "would not be surprised by reading about it in the morning paper."[32]

Robert Gilruth, the head of the Space Task Group that was in charge of Project Mercury, suggested to George Low at NASA headquarters that NASA push for an early start on the review. Low agreed, and relayed the suggestion to Dryden.[33] Then Dryden in early February met again with Wiesner, who agreed to charter an ad hoc panel of the President's Science Advisory Committee (PSAC), again chaired by Donald Hornig of Princeton University, to conduct the review. Dryden met with Wiesner and Hornig on February 11 to discuss the composition of the panel. They agreed that the basic question the panel would address was: "Was Mercury ready to fly?"[34]

That question was also being debated within NASA. Originally the first suborbital flight with an astronaut in the spacecraft, Mercury-Redstone 3 (MR-3), had been scheduled for late March. But the January 31 flight with Ham aboard had landed 132 miles downrange from its target point and had subjected the chimp to a 14.7 g force on reentry, 3 g more than planned.[35] These deviations from the flight plan were primarily the result of the overacceleration of the Redstone launcher and early firing of the spacecraft escape rocket. Even after these problems, NASA managers at the Space Task Group and some at NASA headquarters were ready to commit an astronaut to the next flight. However, "key members of von Braun's team quickly decided that they wanted another booster test before a man could fly," and von Braun did not overrule them. This decision "likely cost the United States the distinction of putting the first human in space... NASA was more afraid of the consequences of an accident than those of coming in second."[36] On March 3 the first crew-carrying Mercury mission was postponed until late April. The extra Redstone flight was launched on March 24 and went well; from NASA's point of view, there now was no obstacle to launching the first U.S. astronaut on a brief ride through the lower reaches of space.[37]

If the MR-3 mission had gone forward on its original schedule, the astronaut aboard would have been not only the first American, but also the first human, to go into space, albeit not into Earth orbit. If this had happened, it is unlikely that the Soviet launch of Yuri Gagarin into orbit three weeks later would have had such a dramatic impact on U.S. space policy. But of course this was not known in March, and with the review of the PSAC panel not completed and given von Braun's judgment that Mercury was not yet ready to fly with an astronaut aboard, it would have been difficult if not impossible for NASA to get White House permission to go ahead with the mission on its original schedule.

The ten-person Hornig panel spent four days in early March visiting the facilities at the McDonnell Corporation factory in St. Louis where the Mercury capsule was built; the launch facilities at Cape Canaveral, Florida; and the Space Task Group at NASA's Langley Research Center in Hampton, Virginia. Science adviser Wiesner on March 7 thanked Hugh Dryden for the "thorough and candid presentation of all elements of the program," and suggested that "this complete cooperation is evidence of a continuation of the excellent relationship" between NASA and the White House. This was a somewhat ironic suggestion, given the critical tone of the Wiesner task force two months earlier and the lack of preinaugural contact between the incoming administration and NASA. Dryden noted that as the result of the panel's review, "certain members...who, previously, had no contact whatever with the program, changed their minds completely after they visited factories and the laboratories and saw what was going on and talked with the people carrying on the work."[38]

Although the panel had finished most of its work by mid-March, its report was not formally submitted to the White House until April 12. The reason for the delay was continuing reservations by a biomedical subgroup of the panel. The panel report concluded that the planned suborbital flight, which by then had slipped to early May, would be "a high risk undertaking but not higher than we are accustomed to taking in other ventures." The report reviewed the accomplishments and failures of the Mercury program and assessed the risks involved and the probability of success. It noted that "the Mercury program has apparently been carried through with great care" and that "almost everything possible to assure the pilot's survival seems to have been done." The panel rated all aspects of the Mercury system as being more than 85 percent reliable except the booster and telemetry, which were rated 70 to 85 percent reliable. Even these items were "not per se a cause for alarm" for astronaut safety, just for mission success. The probability of the astronaut surviving the suborbital mission "appears to be around 90 to 95 percent although NASA estimates are somewhat higher." The panel noted that "it was too early" to estimate similar probabilities for an orbital flight.[39]

The only serious reservations about the readiness of Mercury to launch an astronaut were expressed by the medical experts on the Hornig panel. They were worried about the fact that the astronaut's blood pressure would not be monitored during flight and that high pulse rates such as those observed on Ham in the January flight, combined with the possibility of low blood pressure during the most stressful parts of the flight, could mean that the astronaut would be near collapse. NASA met with members of the medical panel on March 17 and then again on April 11 together with science adviser Wiesner, who shared the panel's concerns, but were unable to allay their reservations. The experts suggested various additional tests prior to clearing MR-3 for launch, and particularly a high number of ground and flight tests with chimpanzees. Hugh Dryden thought that such a step was "totally unrealistic" and Robert Gilruth facetiously suggested that if so many tests with chimpanzees were needed, the program ought to move to Africa. The panel's final report worried that "it is not known whether the astronauts are likely to border on respiratory and circulatory collapse and shock, suffer a loss of consciousness or cerebral seizures, or be disabled from inadequate respiratory or heat control," and that the degree of risk associated with the mission "is at present a matter for clinical impression and not for scientific projection." Although no additional chimpanzee tests were added to the program, the three astronauts from whom the MR-3 pilot would be chosen did undergo additional runs at a Navy centrifuge in Johnstown, Pennsylvania, that simulated the stresses of reentry.[40]

On April 12, the same day on which the panel's report was delivered to the White House, the Soviet Union launched Yuri Gagarin into Earth orbit, and Gagarin returned to Earth with no obvious ill effects. This feat made moot many of the concerns of the panel's medical experts.

## Planning for a Lunar Landing Mission

Beginning in October 1960, NASA had begun to investigate in a preliminary fashion the technological and budgetary requirements for a lunar landing program. After an interim report on this planning effort at a January 5, 1961, meeting of NASA's Space Exploration Program Council, a small group led by NASA's program chief for manned space flight George Low was chartered to continue further investigation into those requirements. The basic objective of Low's group at that time was to answer the question: "What is NASA's Manned Lunar Landing Program?"[41]

Low submitted his group's fifty-one page report on February 7. Two methods for accomplishing the lunar landing mission were examined: "direct ascent," i.e., launching on one very large rocket the spacecraft, fuel, and other equipment needed to land on the Moon and return safely to Earth, and "rendezvous," i.e., launching separately on smaller boosters the various elements required and assembling them in Earth orbit before departing for the Moon. Significantly, the group concluded that "no invention or breakthrough is believed to be required to insure the over-all feasibility of safe lunar flight." An initial mission to the Moon would be possible in the 1968–1970 period with an average cost of $700 million per year for ten years, or a total of $7 billion. The report noted that the plan it presented "does not represent a 'crash' program, but rather it represents a vigorous development of technology. The program objectives might be met earlier with higher initial funding, and with some calculated risks."[42]

Low's report and the supporting work done by NASA and its contractors on preliminary design of the three-person Apollo spacecraft and the Saturn C-2 launcher (at this point in time, the Nova launcher was only in the early conceptual design phase) would be important to the confidence of the NASA leadership in the following weeks as they responded to President Kennedy's request for a mission that would give the United States an opportunity to claim space leadership. Between NASA's 1959 choice of a lunar mission as the long-term objective of its human space flight program and Low's February 1961 report, NASA had indeed laid the technological foundation for what John Kennedy would soon call "a great new American enterprise."

## Space Science Board Endorses Human Participation in Space Exploration

The Wiesner task force had told John Kennedy that human space flight could not be justified solely on scientific grounds; this had also been the position of most members of the PSAC, who continued their committee membership during the change in administrations, and of many other nongovernmental scientists. This position was counterbalanced in March 1961 by support for the scientific value of humans in space from the Space Science Board of the National Academy of Sciences. The National Academy is a body designated by congressional charter as scientific adviser to the federal government. It is also an honorary body; election to the National Academy is one of the highest scientific honors available to a U.S. scientist. The board was composed of sixteen scientists, many of them National Academy members, and chaired from its inception by Lloyd Berkner, a geophysicist. Berkner was a supporter of a strong space program and a decades-long friend of James Webb. Members of the board in addition to Berkner included Bruno Rossi, Joshua Lederberg, Harrison Brown, John Simpson, Harold Urey, James van Allen, and Donald Hornig, all well known and highly respected members of the scientific community.

At its February 10–11, 1961, meeting, the board discussed the possibility of a national decision on whether or not to send humans to the Moon. Berkner pointed out that "many related decisions and programs are being held in abeyance awaiting this overall national policy decision." Initially, all board members "were strongly for landing someone on the moon." But as the discussion continued, "doubts began to creep in." Berkner was critical of the emerging negativism in the discussion, saying that he had "learned from experience that this is a good way to get left behind" and that there was a need for "a clear-cut national decision now." The board then had a lengthy discussion about issuing a statement on lunar and planetary exploration that included the possibility "that man could be included in the exploration." The board authorized Berkner to "pull together" such a statement. By doing so, they almost guaranteed that the statement would support the human role in space exploration, given Berkner's well-known views on the subject.[43]

The board's statement was formally transmitted to NASA administrator Webb on March 31, although Berkner had informed Webb of the emerging recommendation at least a month earlier. The statement took the form of a position paper on "Man's Role in the National Space Program." It recommended that "*scientific exploration of the Moon and planets should be clearly stated as the ultimate objective of the U.S. space program for the foreseeable future.*" The Board added that it "strongly emphasized that planning for the scientific exploration of the Moon and planets must at once be developed on the premise that man will be included... From a scientific standpoint, there seems little room for dissent that man's participation... will be essential." The board statement went beyond scientific issues to declare that the board "was not unaware of the great importance of other factors associated with the man-in-space program." Among these was "of course, the sense of national leadership emergent from bold and imaginative U.S. space activity." In addition, "the members of the Board as individuals regard man's exploration of the Moon and planets as potentially the greatest inspirational venture of this century and one in which the whole world can share," since "inherent here are great and fundamental philosophical and spiritual values which find a response in man's questing spirit and his intellectual self-realization."[44] James Webb himself could not have written a more positive affirmation of NASA's plans for human space flight.

This ringing endorsement of the human role in exploring space coming from a prestigious group of scientists was controversial in later years. In 1965, a committee on "The Integrity of Science" chartered by the American Association for the Advancement of Science called such advocacy by scientists "closely associated to professional scientific judgments... inherently dangerous both to the democratic process and to science," since its association with organized professional scientific activity, such as the deliberations of the Space Science Board, gave that advocacy "a wholly unwarranted cloak of scientific objectivity." But in 1961, the statement had a "profound influence" on PSAC members and other scientists, making it hard for them to criticize President Kennedy's call two months later for a national effort to send Americans to the Moon.[45]

# Chapter 4

# First Decisions

As planning within NASA regarding future efforts continued, as NASA's relationship with the Air Force was being stabilized, as a White House task force examined opportunities for U.S.-Soviet space cooperation,[1] and as James Webb took the leadership reins at NASA, there were two major issues on which a short-term presidential decision was needed. One was the specific role with respect to space to be assigned to Vice President Lyndon Johnson. Although president-elect Kennedy had indicated in December 1960 that Johnson would have an important space role in his administration, the character of that role and how it would relate to the existing structure for space governance, in which the National Aeronautics and Space Council had played a peripheral role during the Eisenhower administration, remained to be defined. Second, by selecting James Webb as NASA administrator and asking him to recommend what he thought was the appropriate NASA program, President Kennedy virtually guaranteed that he would have to react to an early plea for an accelerated civilian space effort. Webb was a policy activist, and he was unlikely to accept the relatively slow-paced space program that he had inherited from the Eisenhower administration. Although Kennedy and his policy, technical, and budget advisers hoped to delay decisions with respect to the space program until a fall 1961 review, Webb insisted that the president hear his arguments for budget increases as soon as possible.

## Revitalizing the Space Council

After the December meetings in which a new space role for the vice president was discussed, it took several months to sort out just what Kennedy had in mind in this regard. The Bureau of the Budget (BOB) staff in early January drafted a white paper on options for implementing the president-elect's intent. The paper noted that "a way needs to be found to strengthen Presidential leadership of space activities without requiring an inordinate concentration on such matters to the detriment of the performance of other

Presidential responsibilities." It suggested that "the President does not need to delegate decision-making on space matters"; rather, "he needs an assistant of stature, without agency ties, who can take the lead in seeing that plans and policies are formulated in the broad national interest and that agencies work together efficiently to this end." To achieve this, the paper suggested, "the Vice President can provide the necessary assistance by serving essentially as a Presidential assistant on space matters and presiding over the [Space] Council." The staff paper suggested that there was no immediate need for statutory changes; rather, the president could simply assign responsibility for chairing the Space Council to the vice president, even though he was not by law a member of the Council.

### *Johnson Seeks a Major Policy Role*

Acquiring an expanded role with respect to space was just one element in Lyndon Johnson's early push for influence within the Kennedy administration. His ambition was apparently to serve as the president's alter ego with respect to all areas of national security policy, not just the space program. Many prior vice presidents who had served under Democratic presidents, such as John Nance Garner, Harry Truman, and Alben Barkley, were not informed about or involved in national security matters, and Johnson did not want to repeat that pattern.

To achieve this objective, a few days after the inauguration Bill Moyers of Johnson's staff drafted an executive order for President Kennedy's signature that would have given the vice president authority "at all times" for "continuing surveillance and review with respect to domestic, foreign and military policies relating to the national security" and would have allowed the vice president to chair the National Security Council in the president's absence. To exercise this responsibility, the vice president would be "authorized to obtain pertinent information concerning the policies and operations of the Department of State, the Department of Defense, the Office of Civil and Defense Mobilization, the Central Intelligence Agency, the Bureau of the Budget and other departments and agencies affected [*sic*] with a national security interest." The reasons for such an expanded role, suggested Moyers, included that "the nature of our times requires that the Vice President be adequately informed on vital matters" and that "the possibility of immediate succession to the number one job, however remote and however distasteful to think about from the President's viewpoint," would require a fully informed vice president. Even as he prepared the draft order, Moyers recognized that it would likely be opposed by many of JFK's advisers, and suggested to Johnson that "a better way to achieve your objective, perhaps, is for the President simply to issue a directive to you, instructing you to play a greater role in national security."[2]

While the new president and his White House staff were indeed resistant to the kind of publicly visible executive order that Moyers had drafted, they did accept the suggestion of a nonpublic presidential directive. On January 28,

Kennedy signed a letter to Johnson that had been drafted by Moyers, asking Johnson to review policies relating to the national security so that Kennedy could "have the full benefit of your endeavors and of your judgment" and to "maintain close liaison" with "departments and agencies affected with a national security interest." In this letter, NASA was added to, and the BOB deleted from, the list of agencies subject to vice presidential review that had been in the draft executive order. Copies of the letter were sent to heads of all agencies involved in national security matters.[3]

These attempts at the start of the Kennedy administration to give Lyndon Johnson an expanded policy role were not successful. John Kennedy had needed Johnson to attract enough Southern voters to get elected, but Kennedy, and particularly his top aides, had no intention of making Johnson a major player in national security affairs. This quickly became evident. The weeks following the inauguration "were ones of despair for Johnson," according to one of his biographers; "He felt trapped, useless, ridiculed."[4]

### *Reorganizing the Space Council*

In the early days of the eighty-seventh Congress, two bills related to the Space Council were introduced in the House of Representatives. One was the Eisenhower administration bill abolishing the council, on which there had been no action in the previous year; Sorensen and Bell had agreed during the transition to have this bill reintroduced, and apparently that decision had not been countermanded even though president-elect Kennedy had at least tentatively decided to retain the council. The other bill was introduced by congressman Emilio Daddario (D-CT). It would have not only have authorized the president to make the vice president a member and the chairman of the Space Council, but also would have delegated to the vice president executive functions that were assigned by law to the president.[5] No legislative action was taken on either of these proposed bills, but the White House wanted to make sure that the Daddario bill did not move forward, since it would have given the vice president more power than Kennedy desired; in addition, such a delegation of executive power to the vice president was most likely unconstitutional.

On February 14, Lyndon Johnson wrote to Kennedy, telling him that "in accordance with your request, I have made a study of the National Aeronautics and Space Council." That study had suggested that "the Council as it now stands has the power to make certain decisions between agencies and programs which is a power that under the Constitution only the President can have." Thus, "if it is your desire to remove the President as Chairman of the Council, it will be necessary to change the basic structure of the Council so that it coordinates the information—not the activities—of the various space projects and advises you accordingly."[6] Following up on this letter, Moyers asked Richard Neustadt, who was continuing to serve the new administration as a consultant on organizing the presidency, for his thoughts on the organization and operation of the council. Moyers also worked with staff members of the Senate Committee on Aeronautics and Space Sciences to prepare the various documents needed to make the changes in the 1958 Space Act that were thought to be required.

Neustadt responded on February 28. He said that he was "much concerned" regarding "the Vice President's position as a constitutional officer who cannot share, so should not be pressed to take on, operating responsibility" that was assigned to the president. In a memorandum on "Organizing the Space Council," Neustadt noted that "where space programs are concerned, the President should have available the same sort of top-level, politically-responsible *advice* on *policy* (and follow through) that he can claim in other fields from a Cabinet secretary," but that "the Vice President should not be asked to serve as 'Secretary for Space' *except* in the role of senior adviser. It would be unfair to cast him in the role of department head responsible for operations." Neustadt's late February critique of an operational role for the vice president implies that at least some on LBJ's staff, if not Johnson himself, were continuing to push for such responsibility.

Neustadt recommended that there was no need for the council to have the nine members, including three public members, who were mandated in the 1958 Space Act; in his view, "the Chairman, State, Defense and NASA would suffice." (At least one prominent space scientist was interested in becoming a public member of the Space Council. On February 4, 1961, University of Iowa professor James Van Allen, 1958 discoverer of the Earth-circling radiation belts that bore his name, wrote Vice President Johnson, saying that "I would be honored to serve with you on this body [the Space Council] as a vitally interested member of the general scientific community.") Neustadt proposed a small council staff "with broad experience in government, possessed of balanced judgment, keen analytical ability, and a taste for quiet staff work." He suggested that the council's name be changed to either the "President's Advisory Council on Space," which was the term that president-elect Kennedy had used in December as he announced the new role for his vice president, or "President's Space Council." He felt that any modifications, whether through a reorganization plan or through new legislation, "change the law as little as possible." Neustadt's memorandum, which he characterized as "one man's opinion," was followed on March 1 by another BOB staff memorandum. This document stressed that "the Space Council should exist solely to advise the President... The President should retain executive responsibility, and executive functions should not be delegated to the Vice President."[7]

Vice President Johnson again got personally involved in early March, in particular asking budget director David Bell how best to finance the council's staff. Bell told him that since the Space Council was a statutory agency on its own, it was not legal to transfer funds from the NASA or DOD budget to fund its operations, a possibility that had been explored by Johnson's staff. However, Bell said, President Kennedy had agreed to provide funds from his "Special Projects" budget to fund the council's executive secretary and two more staff positions, and that NASA had agreed to delegate three or four of its employees to act as council staff. Bell said that it was his understanding "once you have your initial staff on board, you expect to have them prepare necessary modifications to the National Aeronautics and Space Act."[8]

With the question settled of how staff salaries would initially be paid, Vice President Johnson could recruit an individual to serve as council executive secretary. On March 20, President Kennedy sent the nomination of Dr. Edward C. Welsh to the Senate; Welsh had been actually chosen for the position several weeks earlier and had already been actively working on reorganizing the council. Welsh was a longtime government employee and was at the time a legislative assistant to Senator Stuart Symington (D-MO). He held a doctorate in economics and had been in charge of Symington's hearings on air power in 1956, had helped staff LBJ's Preparedness Subcommittee hearings after Sputnik, and had been the lead staff person for Symington's hearings on government organization for space in 1959. He had also been the executive director of the task force on reorganizing the Department of Defense set up by Kennedy during the presidential campaign. Welsh had been the primary author of the strident October 1960 Kennedy campaign statement on space, which had argued that "control of space will be decided in the next decade. If the Soviets control space they can control earth." In Welsh, Lyndon Johnson got a strong, if somewhat self-important, advocate for a vigorous U.S. space effort.

As he was under consideration for the Space Council position, Welsh met with James Webb, who told Welsh that he believed in "a vigorous role on the part of the operating agency [NASA] and did not want to have a Council or any other interagency group be controlling the operating day-to-day functions." Welsh told Webb he agreed with this point of view.[9] Welsh's confirmation hearings before the Senate Committee on Aeronautical and Space Sciences were held on March 23. After Welsh's opening statements, there were no questions. The Senate voted to confirm Welsh on the same day and he was sworn into office on March 24.

Welsh's first assignment was to draft the changes in the Space Act that were needed to make the vice president its chair and to make other desired adjustments in the council's membership and organizational location within the executive office of the president. By March 30, Welsh had prepared a memorandum noting that there were three options available to change the provisions of the 1958 Space Act—a reorganization plan, an amendment to the then-pending NASA Authorization Bill, or a separate amendment to the Space Act. Welsh had contacted key members of the Senate and House, and had learned that there was a congressional preference for a simple amendment to the Space Act.[10] Within the executive office, the BOB still favored a reorganization plan, but such a plan would have had to wait sixty days to allow any congressional comments before it could be put into place. The congressional perspective prevailed, and a decision was made to move forward with proposing an amendment to the Space Act.

Before the amendment could be approved by President Kennedy and his top advisers, there were two issues to be dealt with. One was whether to propose a name change for the Space Council, a topic that had been discussed ever since December. Neustadt and Welsh discussed this topic at an April 4 breakfast. Apparently Welsh was concerned that the titles that Neustadt had suggested in his February 28 memorandum, which began with the word "President's" rather than "National," would not indicate the intended broad scope of the council's activities. In a follow-up memorandum to Welsh later that day, Neustadt suggested that the issue of the council's name was not "all-important or worth getting into a tizzy about." He added: "I very much appreciate your sensitivity about the change from 'National' to 'President's'... But isn't it possible you are being oversensitive?" Neustadt noted that Lyndon Johnson also chaired the President's Committee on Equal Employment Opportunity, and "no one conceives its title as an attack on him."[11]

The other issue was the wording of the proposed amendment to the Space Act. Edward Welsh and James Webb had not been able to agree on how best to indicate in the amendment the separation of the functions of the Space Council and the functions of the president. While Welsh wanted to put forward a simple amendment that retained the Space Act language that specified the functions of the council, Webb wanted to add a new section to the Act that specified the duties that would remain the president's responsibility. These differences had been discussed in a March 7 meeting between budget director Bell and special counsel to the president Ted Sorensen, and the decision was made to go forward with the Welsh version of the amendment.[12] No change in the name of the National Aeronautics and Space Council was suggested. The secretaries of defense and state, the administrator of NASA, and the chairman of the Atomic Energy Commission remained council members; the one additional government member and the three public members of the council were eliminated, and the council was made part of the executive office of the president.

Welsh told Vice President Johnson on April 6 that his version of the proposed amendment "had been cleared in the Executive Office of the President with Messrs. Bell, Staats, and Neustadt, and that Budget Director Bell had agreed to discuss the paper with Ted Sorensen and President Kennedy." He also noted that Representative Overton Brooks had agreed to schedule a hearing on the amendment and that there had been preliminary agreement to the amendment obtained from the staffs of Senators Kerr, Bridges, and Dirksen and with Speaker of the House Sam Rayburn, Majority Leader John McCormack, and Congressman Thornberry of the Rules Committee.

President Kennedy transmitted the amendment to the Congress on April 10; Welsh testified as the only witness before the House Committee on Science and Astronautics on April 12. This was the day on which the Soviet Union orbited the first human in space, Yuri Gagarin, and that feat, rather than the changes in the Space Act, was the focus of the committee's questions. In his testimony, Welsh noted that "the Vice President is already by statute a member of the National Security Council," and that "to make him a member of the Space Council seems to be a comparable action with suitable precedence." One issue raised by Chairman Brooks during the House hearing was "doesn't the Vice President have some executive authority [under the amendment]? Isn't he for some purposes a part of the executive branch?" Welsh replied that "in this specific instance this responsibility would be advisory and not in a real sense executive."[13]

The House approved the amendment by voice vote on April 17. Welsh then testified before the Senate Committee on Aeronautical and Space Sciences on April 19, again as the sole witness. Committee approval quickly followed and the Senate approved the amendment on April 20. As he signed the bill amending the Space Act on April 25, President John F. Kennedy stated that "Working with the Vice President, I intend that America's space effort shall provide the leadership, resources, and determination necessary to step up our efforts and prevail on the newest of man's physical frontiers."[14] By this time, the Space Council under Vice President Johnson's leadership was already well embarked on a review that would recommend sending Americans to the Moon.

## President Kennedy's Initial Space Budget Decisions

In the weeks after his inauguration, President Kennedy spent very little time in formal consideration of space issues. On February 13, he sent Soviet premier Nikita Khrushchev a congratulatory note on the launch of a Russian mission to Venus on the preceding day. (After returning data during its interplanetary cruise, the *Venera 1* mission ultimately was a failure in achieving its primary objective of returning data from Venus.) Later that same day, Kennedy met with Overton Brooks, chairman of the House Committee on Science and Astronautics, who told Kennedy that "NASA and the civilian space program badly need a shot in the arm." Brooks said that Congress, or at least his committee, "cannot impress upon the Executive too strongly the need for urgency" in the space program, for "if we do not recognize it, we may falter badly in both our domestic and international relationships."[15]

Science adviser Wiesner at some point after the Russian Venus launch discussed with the president the implications of the launch and "our relative positions in the general fields of space exploration and science." The launch had raised security concerns because specialists at the Central Intelligence Agency had suggested that it could be "a step towards creating a capability for achieving a parking orbit with an ICBM warhead"; the spacecraft had in fact gone into orbit around the Earth before ejecting a large probe on a Venus-bound trajectory. The sample questions and answers used in preparing Kennedy for a February 15 press conference suggested that, if asked about the military significance of the launch, he reply that "there is no indication that the Soviet Union plans to use their ability to orbit large payloads to develop any kind of bombardment systems." Such a system would be "inefficient" and would likely "be objected to by all the nations of the world."[16]

In a follow-up memorandum to his discussion with the president, Wiesner noted that "the most significant factor, as we have said many times, is that the Soviets have developed a rocket as part of their ballistic missile program with considerably more thrust or lifting power than anything we have available." He noted that "one of the things we must realize is that in dramatizing the space race we are playing into the Soviet's strongest suit. They are using this accomplishment at home and around the world to prove the superiority of Soviet science and technology." Wiesner told the president that the United States "was superior in most fields to Soviet science" and that "in almost any other area in which we would elect to compete, food, housing, recreation, medical research, basic technological competence, general consumer good production, etc., they would look very bad." He suggested that "we should attempt to point this out rather than assist them by an official... reaction that supports their propaganda."[17] Subsequent actions by President Kennedy demonstrated that he did not accept this advice.

### *NASA's Budget Reviewed*

Budget director David Bell in February issued a government-wide call for a rapid agency review of the proposed Fiscal Year (FY) 1962 budget that had been submitted by the outgoing Eisenhower administration. James Webb's initial examination of the NASA budget convinced him that the Eisenhower proposal was not adequate to support the kind of space program he believed was appropriate for the Kennedy administration; he judged that there was a need to accelerate the pace of the milestones that had been set out in NASA's then-current ten-year plan. For example, he wanted to move the date of a potential lunar landing mission forward from 1973 to 1970; George Low's report had indicated that it would be possible by that year, and perhaps even sooner. On March 17, NASA submitted a request for an additional $308.2 million, a 30 percent increase over the $1,109.6 million budget in the Eisenhower request. In his budget submission, Webb argued that "the civilian space program clearly can achieve a much more substantial contribution in aeronautical research and space exploration and technology if the pace of the program for 1962 is substantially accelerated."[18]

Budget director Bell did not think that either he or the president was ready to decide on such an acceleration of NASA's efforts. He later recalled that "most of us, when we came into office didn't have any notion of what the space program was all about, what the issues were. A lot of people needed to be educated."[19] In response to the NASA request, BOB indicated that it would approve only a $50 million increase in the NASA budget, with any additional increases deferred until fall 1961, after a comprehensive review of the NASA program had been completed. James Webb refused to accept BOB's decision and, as was his right as the head of a government agency, asked for a meeting with the president to appeal his case for a larger increase. The meeting was set for the late afternoon of March 22.

Before meeting with the president, Webb, Dryden, and Seamans briefed Vice President Johnson on their budget request. Budget director Bell, his deputy Elmer Staats, and the top career BOB staff person for the NASA budget, Willis Shapley, explained to the vice president why they had not approved most of the request. In preparation for this meeting, Edward Welsh had prepared a briefing memorandum for Johnson that noted that the "major policy issue" involved in the budget was "does this country want to make an effort to catch up to the Soviet Union in space capability," and particularly in weight-lifting capability. "To a considerable extent, domination of space will belong to those who can put up large manned and unmanned vehicles," suggested Welsh.[20]

After this initial discussion, President Kennedy, Jerome Wiesner, the president's national security adviser, McGeorge Bundy, and chairman of the Atomic Energy Commission Glenn Seaborg joined the meeting in the White House cabinet room. Bundy had not been named to his post until January, making him one of the last people to join Kennedy's White House inner

President Kennedy with two of his top advisors on space issues, special counsel Theodore Sorensen (left image) and special assistant for national security affairs McGeorge Bundy (right image). (JFK Library photos)

circle. He was a Yale-educated member of the Republican Eastern establishment who had become a Harvard professor of government with a widely acknowledged intellect, and then in 1953, at age 34, the youngest ever dean of the Harvard faculty. Bundy during the Kennedy administration was to play an important role in the national security dimensions of the U.S. space effort, including the race to the Moon.

The BOB had prepared an agenda for the meeting that indicated that "the future direction and level of the civilian space program primarily depends on decisions to be made by this administration concerning the rate [at which] it wishes to undertake the following: (1) Increasing the rate of closure on the USSR's lead in weight lifting capability; (2) Advancing manned exploration of space beyond Project Mercury." Each of these issues, said the BOB paper, "merits assessment in relation to its (1) technological significance, (2) impact on the international prestige of the United States, and (3) effect on present and future budget requirements." With respect to agenda item 2, the question was "should we now launch an aggressive program of manned space exploration to follow Mercury, aimed at the progressive goals of (a) multimanned orbital laboratory and later (b) manned circumlunar flight and (c) manned lunar landing?"[21]

James Webb had stayed up late the day before the meeting preparing a six-page talking paper. He argued that "U.S. procrastination for a number of years had been based in part on a very real skepticism by President Eisenhower personally as to the necessity for the large expenditures required, and the validity of the goals sought through the space effort." Webb noted that "in the preparation of the 1962 budget, President Eisenhower reduced the $1.35 billion requested by the Space Agency to the extent of $240 million and specifically eliminated funds to proceed with manned space flight projects beyond Mercury. This decision emasculated the ten year plan before it was even one year old, and unless reversed guarantees that the Russians will, for the next five to ten years, beat us to every spectacular exploratory flight." Webb told Kennedy that "the first priority of this country's space effort should be to improve as rapidly as possible our capability for boosting large spacecraft into orbit." He added that "the Soviets have demonstrated how effective space exploration can be as a symbol of scientific progress and as an adjunct of foreign policy. Without necessarily following the Soviet lead in this kind of exploitation, we should not fail to recognize its potential." Webb closed his paper by suggesting that leadership in space, "pioneering on a new frontier," would help to create "more viable political, social and economic systems for nations willing to work with us in the years ahead."[22] These were themes that Webb realized would resonate with John Kennedy.

Robert Seamans followed Webb's remarks with a concise summary of NASA's budget request. Kennedy told Seamans: "That was very good; I would like your views in writing tomorrow."[23] With that, the hour-and-a-half meeting concluded. During those 90 minutes, John Kennedy, perhaps for the first time, had had the chance to get a clear picture of the space policy and budget issues requiring his decision.

Webb then briefly went into the Oval Office with President Kennedy and Vice President Johnson. He assured them, as the newcomer to the NASA leadership, that he believed "that we were moving right ahead with the things that needed to be done"; Kennedy in return told Webb that Dryden and Seamans had made a positive impression on him at this, their first meeting. Meanwhile, Seamans went home and prepared the memorandum that Kennedy had requested, which Webb quickly forwarded to the White House.[24]

### *NASA Budget Increased*

The next day, President Kennedy met with Vice President Johnson, Welsh, Bell, and Wiesner. No NASA representatives were present. The BOB had prepared a paper for Bell's use at the meeting, which noted that "the case for budget increases . . . was well presented by Mr. Webb and his associates." Bell told the president that he wanted to indicate "some of the points which suggest that the lower alternatives deserve serious consideration when a general decision is made on the course of the space program, either at this time or in the 1963 budget decisions in line with my suggestion that this matter be deferred to that time insofar as possible." Bell noted that even if increases in the budget for boosters were approved, "we will still be in a 'tail chase' and that there is still a strong probability that the Russians will beat us to future spectacular space achievements if they choose, regardless of what we do." He suggested that "the wisdom of staking so much emphasis and money on prestige that might or might not be gained from space achievements in the late 1960s and 1970s appears questionable" and that "it seems virtually certain that alternative, surer, and less costly ways of increasing our national prestige in the world scene could be developed." Bell said that he "cannot help feeling that the total magnitude of present and projected expenditures in the space area may be way out of line with the real value of the benefits to be expected."[25] In support of Bell's memorandum, BOB staff prepared five different proposals for the future NASA program and their budget implications for the coming years. The alternatives ranged from the program NASA was suggesting to much smaller programs with an emphasis on scientific and application objectives, no manned flight beyond Mercury, and cancellation of the Saturn launcher.[26]

The memorandum prepared by Robert Seamans specified the impacts of the budget increases that NASA had requested, which included $98 million for the Saturn C-2 project, $27.5 million for a prototype engine for a nuclear rocket, $10.3 million for the large F-1 engine for use in a Nova launcher, and $47.7 million for the initial version of the Apollo spacecraft for use as an orbital laboratory. Other items in the NASA budget request were not discussed in Seamans's memo. If the requested increases were approved, said Seamans, the orbital laboratory could begin flights in 1965 rather than 1967; circumlunar flights would be possible in 1967 rather than 1969; an initial manned lunar landing might be possible in 1970 rather than 1973.[27]

The president started the meeting by asking Vice President Johnson for his views on the NASA budget. Johnson responded: "Dr. Welsh here knows more about it than I do—let him speak." Welsh told the President that "the main thing to be done was to stimulate the work on boosters; that we were farther behind on our propulsion side of the space program than anything else." This was a refrain that President Kennedy had been hearing repeatedly, and he asked Wiesner if he concurred; Wiesner responded that he did. Bell did not protest, even though his arguments had been overruled; his response was "Whatever the President wants, we will try to get that done."[28]

On the day of the NASA budget meeting, Secretary of Defense McNamara had been consulted by the vice president's office for his views on the NASA budget and policy issues. McNamara's response was that "he was personally unable to assess" the prestige payoffs from human space flight, and would suggest proceeding at "a normal rate of investigation," which was considerably less than a maximum effort. With respect to increases in the NASA budget, McNamara suggested that he would give higher priority to all items in the Department of Defense budget than to increased funding for NASA.[29]

The final increase in the NASA FY1962 budget approved at the meeting was just under $126 million, almost all of it to accelerate the NASA booster effort. No funds for the Apollo spacecraft and thus for human space flight beyond Project Mercury were approved. This was an increase of a bit over 10 percent compared to President Eisenhower's budget submission, but 20 percent less than NASA had hoped for.

In this initial engagement with space policy and program issues, President Kennedy had heard the full range of arguments with respect to the goals and pace of the U.S. civilian space program. He decided that it was a matter of some urgency to begin the process of closing the weight-lifting gap that its powerful rocket had given the Soviet Union, but was not yet ready to commit himself to the use of those new boosters for a post-Mercury human spaceflight program. The expectation as of the end of March 1961 was that this issue would be the focus of a comprehensive review of NASA's future to be conducted by Lyndon Johnson as the new chairman of the Space Council, with a decision coming as the Fiscal Year 1963 budget was being prepared in the coming fall.

One reason for the hesitance at this point to approve any funds for a Mercury follow-on was likely the uncertainty about Mercury's success. The Hornig panel had not finished its work, and its medical experts were very worried about whether an astronaut could survive the stresses of space flight. Jerome Wiesner shared their concern, and had communicated it to the President. In addition, according to Seamans, although Kennedy was tending toward the approval of future human space flight efforts, he "wanted to know more about it. This was all pretty new as far as he was concerned, except in very general terms."[30] Webb recognized that Kennedy "was concerned about a tremendous range of problems as an incoming president," and that he was being asked to make a choice between his budget director, whose judgment he had come to trust, and that of the NASA leadership, whom he did not know well.[31] Added to these factors were the immediate concerns over Laos, which were occupying most of Kennedy's time. The March 23 outcome was thus "deliberately intended as a partial decision which would leave him [Kennedy] free, within a considerable range, to decide later how much of a commitment to make."[32]

As he attempted to resist NASA demands to meet with the president to appeal the original BOB decisions, David Bell had told Hugh Dryden that Kennedy was too busy for direct involvement in decisions on NASA's future. Dryden replied: "You may not feel he has the time, but whether he likes it or not he is going to have to consider it. Events will force this."[33] Dryden's words proved prescient; within three weeks, Kennedy would be faced with a Soviet space challenge that led him to set dramatic new space goals for the United States.

# Chapter 5

# "There's Nothing More Important"

On Monday April 10, 1961, John F. Kennedy threw out the opening day baseball pitch as the Washington Senators played (and lost to) the Chicago White Sox on a chilly and damp afternoon. Baseball was not the only thing on the president's mind that day. Sometime early in the game, Kennedy's deputy press secretary Andrew Hatcher told him that the United Press International news service was about to report that the Soviet Union had successfully recovered the first human to orbit the Earth. Kennedy asked Hatcher to check on the report; he had known for several weeks from intelligence briefings that such a launch was imminent. The Soviet Union had successfully completed one-orbit missions of a spacecraft carrying a dog as a passenger on March 9 and March 25. It was almost certain that the next step would be a mission with a human on board. Hatcher reported back a few innings later that the news reports "have not materialized" and that "elaborate Russian plans to make this anticipated announcement have been abandoned for today." Also, said Hatcher, the Central Intelligence Agency (CIA) "could not confirm or deny the report" of the Soviet launch.[1]

By the end of the next day, April 11, the CIA did report that the Soviet launch was likely within the next few hours. Press Secretary Pierre Salinger prepared and Kennedy approved a statement for the president to issue once the Soviets had announced a successful mission. The president also had an approach ready to take if the launch were unsuccessful and the cosmonaut died. Famed journalist Edward R. Murrow, whom Kennedy had chosen to head the U.S. Information Agency, in an April 3, 1961, memorandum for McGeorge Bundy had suggested that "in the event of a Soviet manned shot failure we should express, with all the sincerity we can muster, the deep regret and distress of the President and the people of the United States." Simultaneously, suggested Murrow, one of the Mercury astronauts might "publicly express the regret of his group" and his confidence that "the Soviet astronaut was prepared," as were the Mercury astronauts, "to give up his life for the advancement of human knowledge." However, "covertly, the

U.S. might encourage commentators in other countries to deplore the low regard for human life which prompted the Soviets to attempt a manned shot 'prematurely.'"[2] As he retired for the evening on April 11, Kennedy told his aides that he did not want to be woken if the Soviet announcement came while he was sleeping.

The expression of regret was not needed. Within a few seconds of the launch of the first human in space at 1:07 a.m. on April 12, Washington time (11:07 a.m. at the launch site in Soviet Central Asia), U.S. intelligence systems knew that it had taken place. They monitored the in-orbit communications during the single-orbit flight and decoded the television transmissions from the spacecraft that showed the cosmonaut moving about.[3] It took several hours for Moscow to announce the successful mission; the Soviet dispatch said that "the world's first space ship *Vostok* with a man on board has been launched on April 12 in the Soviet Union on a round-the-earth orbit. The first space navigator is Soviet citizen pilot Maj. Yuri Alekseyevich Gagarin." Science adviser Wiesner called Salinger at 5:30 a.m. with this news. The president was informed of the Soviet achievement when he woke up around 8:00 a.m.; he authorized Salinger to release the prepared statement, which said: "The achievement by the USSR of orbiting a man and returning him safely to ground is an outstanding technical accomplishment. We congratulate the Soviet scientists and engineers who made this feat possible. The exploration of the solar system is an ambition that we and all mankind share."[4]

Later that morning, NASA administrator James Webb and Senator Robert Kerr came to the Oval Office for a previously scheduled meeting with the president to discuss a planned national conference on space to be held in Tulsa, Oklahoma. Webb brought with him a model of a Mercury spacecraft. Theodore Sorensen recalls that Kennedy, "who had no real grasp of the enormous technology involved, and remained skeptical about the cost and importance of space missions," quipped about the "Rube Goldberg-like contraption" that "Webb might have bought it in a toy store... that morning."[5]

President Kennedy had a previously scheduled news conference on the late afternoon of April 12. Inevitably, the questioning turned to the Soviet space achievement. The first question was relatively friendly, and Kennedy's response predictable:

*Q*: Could you give us your views, sir, about the Soviet achievement of putting a man in orbit and what it would mean to our space program, as such?

*Kennedy*: Well, it is a most impressive scientific accomplishment, and also I think that we, all of us as members of the [human] race, have the greatest admiration for the Russian who participated in this extraordinary feat. I have already sent congratulations to Mr. Khrushchev, and I send congratulations to the man who was involved.

I indicated that the task force which we set up on space way back last January, January 12th, indicated that because of the Soviet progress in the field of boosters, where they have been ahead of us, that we expected that they would be first in space, in orbiting a man in space. And, of course, that has taken place. We are carrying out our program and we expect to–hope to make progress in this area this year ourselves.

Then the questioning became a bit more pointed:

*Q*: Mr. President, a Member of Congress said today that he was tired of seeing the United States second to Russia in the space field. I suppose he speaks for a lot of others. Now, you have asked Congress for more money to speed up our space program. What is the prospect that we will catch up with Russia and perhaps surpass Russia in this field?

*Kennedy*: Well, the Soviet Union gained an important advantage by securing these large boosters which were able to put up greater weights, and that advantage is going to be with them for some time. However tired anybody may be, and no one is more tired than I am, it is a fact that it is going to take some time and I think we have to recognize it.

They secured large boosters which have led to their being first in sputnik and led to their first putting their man in space. We are, I hope, going to be able to carry out our efforts with due regard to the problem of the life of the man involved this year. But we are behind and I am sure that they are making a concentrated effort to stay ahead.

We have provided additional emphasis on Saturn; we have provided additional emphasis on Rover; we are attempting to improve other systems which will give us a stronger position—all of which are very expensive, and all of which involve billions of dollars.

So that in answer to your question, as I said in my State of the Union Message, the news will be worse before it is better, and it will be some time before we catch up. We are, I hope, going to go in other areas where we can be first and which will bring perhaps more long-range benefits to mankind. But here we are behind.

Earlier in the press conference, Kennedy had mentioned one of the areas "where we can be first" and which might bring "more long-range benefits to mankind"—desalinization of sea water. He told the press conference, "we have made some exceptional scientific advances in the last decade, and some of them—they are not as spectacular as the man-in-space, or as the first sputnik, but they are important." For example, added Kennedy, "I have said that I thought that if we could ever competitively, at a cheap rate, get fresh water from salt water, that it would be in the long-range interests of humanity which would really dwarf any other scientific accomplishments."[6] Over the next few days, as he absorbed the political reaction in the United States and around the world to the Soviet achievement, Kennedy would change his mind; by the evening of April 14, he would say "there's nothing more important" than finding a way to overcome the Soviet lead in space.

## Reactions to the Gagarin Flight

Congressional and media reaction to the Soviet achievement on April 12 and the next several days resembled—indeed, in some ways exceeded—the rather hysterical reactions after the launch of Sputnik 1 in October 1957. Clearly, this second Soviet space achievement was a major political setback for the new administration.

The Soviet Union was quick to capitalize on the propaganda significance of the successful flight. In his first telephone conversation with Gagarin after his return to Earth, Soviet leader Nikita Khrushchev boasted: "Let the capitalist countries catch up with our country!" The Central Committee of the Communist Party claimed that the flight "embodied the genius of the Soviet people and the powerful force of socialism." East German Communist leader Walter Ulbricht said that the flight "demonstrates to the whole world that socialism must triumph over the decaying system of yesterday." Reacting to claims such as these, a *New York Times* correspondent suggested that it appeared likely that "the Soviet leaders can further alter the atmosphere of international relations so as to create more pressure on Western governments to make concessions on the great world issues of the present day."[7]

The rest of the world was almost unanimous in its admiration of the Soviet achievement. In Great Britain, "universal praise for the Soviet achievement from Cabinet ministers, diplomats, scientists, and the general public was accompanied by some anti-American barbs from men in the street." The French press "relegated all other news to a secondary position... Even comments and reactions to President De Gaulle's news conference were put into relative obscurity." In Italy, "news of the successful Russian space flight was heralded... in banner headlines." Romans snapped up the papers, emptying kiosks in a matter of minutes, then stood around discussing the event. The Vatican newspaper called the flight "a universal good" and a Geneva paper termed the voyage "the number one event of the twentieth century."[8]

The U.S. Information Agency summarized world reaction to the Gagarin flight in an April 21 report, which noted that "media coverage of the Soviet man-in-space has been extraordinarily heavy," with its initial volume "comparable to that received by Sputnik 1, if not greater." The "general tenor" of the press reports was "to acclaim the first manned space flight as (1) a great event in human history, ( 2) a tremendous scientific and technical achievement, and ( 3) a triumph for the USSR that would have many repercussions in the Cold War," since it would "increase Soviet military, political, and propaganda leverage."[9]

American reaction to the Gagarin flight was characterized by disappointment and chagrin. No high official had prepared the general public to expect the Soviet flight, and thus for many it came almost as much of a shock as the 1957 Sputnik 1 launch. *The Washington Post* commented editorially: "The fact of the Soviet space feat must be faced for what it is, and it is a psychological victory of the first magnitude for the Soviet Union... The general excitement from Europe to Asia, Africa and the Americas will not be diminished by the recognition that no immediate military, commercial or other actual advantage accrues to the Soviet Union. In these matters, what people believe is as important as the actual facts, and many persons will of course take this event as new evidence of Soviet superiority."[10]

*The New York Times* correspondent Harry Schwartz commented that "the President, of course, had attempted to present himself as an image of a young, active, and vigorous leader of a strong and advancing nation... But none of these and other measures have had the effectiveness or the spectacular quality of Soviet efforts. Moreover, since he took office the President's image has been beset by the difficulties he has had with Congress, by his failure to spell out the promised 'sacrifices' to be required of the American people and by the continued recession."[11]

The hawkish *New York Times* military correspondent Hanson Baldwin was even sharper in his criticism.

> This same philosophy, which cost the nation heavily in prestige and marred the political and psychological image of our country abroad, hobbled our space program even before the Russians put the first sputnik in orbit.... It is high time to discard this policy. In fact, if the United States is to compete in space, we must decide to do so on a top-priority basis immediately, or we face a bleak future of more Soviet triumphs.
>
> Even though the United States is still the strongest military power and leads in many aspects of the space race, the world—impressed by the spectacular Soviet firsts—believes that we lag militarily and technologically.
>
> The dangers of such false images to our military power and diplomacy are obvious. The neutral nations may come to believe the wave of the future is Russian; even our friends and allies could slough away. The deterrent, which after all is only as strong as Premier Khrushchev thinks it is, could be weakened.

Baldwin concluded by pointing out that "only Presidential emphasis and direction will chart an American pathway to the stars."[12]

John F. Kennedy was an avid newspaper reader. He very likely had criticisms such as these in mind as he considered how best to respond to this new Soviet challenge.

## *Congress Calls for a Crash Effort*

The most vocal demands for an immediate response to the Soviet flight came from the House of Representatives, and particularly its Committee on Science and Astronautics. Committee hearings took place in a highly charged atmosphere.[13] On April 12, the committee held a previously scheduled hearing on the proposal to revise the 1958 Space Act to make the vice president the chairman of the Space Council. Edward Welsh was the only witness. But the minds of the representatives were on the Gagarin flight, not the Space Council. Committee chair Overton Brooks stated that "we ought to make a determination that we... are going to be first in the future." Republican James Fulton proclaimed that "we in the United States should publicly say that we are in a competitive space race with Russia and accept the challenge." On April 13, James Webb and Hugh Dryden appeared before the committee to defend the NASA budget increases that the president had approved on March 23. The focus of the hearings was not on those additions to the NASA budget to speed up the booster programs; rather, it was on the funds for human space flight on which the president had deferred approval. Fulton told Webb and Dryden that "I believe we are in a race, and I have said many times, Mr. Webb, 'Tell me how much money you need and this committee will authorize all you need.'" Congressman Vincent Anfuso suggested that he was "ready to call for a full-scale congressional investigation. I want to see our country mobilized for war because we are at war. I want to see our schedules cut in half."[14]

### Seamans on the Hot Seat

The next morning, April 14, Robert Seamans and George Low appeared before the committee, and were subject to even more intense pressure. Seamans in particular put himself in a vulnerable position with respect to administration policy, saying that although there were no plans at that time to ask Congress for funding for Project Apollo, the post-Mercury human space flight effort, it might indeed be possible with an accelerated effort to land on the Moon by 1967. Seamans noted that doing so would require "a very major undertaking. To compress the program by 3 years [the date of the first lunar landing in the recently revised NASA planning was 1970] means that greatly increased funding would be required... I certainly cannot state that this is an impossible objective. It comes down to a matter of national policy." Seamans added that he would be "the first to review it wholeheartedly to see what it would take to do the job. My estimate at the moment is that the goal may very well be achievable." Seamans was, of course, well aware of the February report of George Low's committee that had said that a lunar landing within the decade was technologically feasible; he was also aware of Kennedy's decision in March not to approve additional funding for human space flight beyond Project Mercury, even as he provided additional funding for larger space boosters. Seamans's comments, coming just as the committee and the media were calling for an accelerated space effort, appeared to be adding NASA pressure on the president to the pressure coming from the House committee and the media; this was an uncomfortable position for Seamans to be in. He recalls that "it was unwise for an underling to get out ahead of the President."[15]

Indeed, President Kennedy was not at all happy to read in the next day's newspapers that a NASA official had made public statements that seemed to preempt what would necessarily be a presidential decision. *The Washington Post* headline read "Reaching the Moon First Would Cost Billions" and its story began, "A multi-billion dollar crash space program might put an American on the Moon by 1967—a top Government official said yesterday." *The New York Times* headlined its report on Seamans' testimony "costly drive might bring landing by '67." Administrator Webb got both a message from budget director David Bell and a "strongly worded" letter from Kenneth O'Donnell asking about the testimony; O'Donnell was one of the president's top assistants and his policy "enforcer." For a few days, Seamans' job was in real jeopardy, but Webb was able to calm the White House concern. In a letter to O'Donnell, Webb noted that the committee was in a "runaway mood" and that "the members of the Committee, almost without exception, were in a mood to try to find someone responsible for losing the race to the Russians" and were seeking information "that would focus public attention on the Committee, and the role it had chosen for itself as the goad to force a large increase in the program." He defended Robert Seamans, saying that he had done "an exceptionally fine job" of resisting the committee's inquiries with respect to NASA's relations with the Bureau of the Budget (BOB)and the president.[16]

## "There's Nothing More Important"

While the White House had prepared statements to be issued after the Soviet orbital launch, one if it was successful and one if the cosmonaut had died during the mission, there had been no substantial contingency planning on how to respond to the national and international impacts of a Soviet success. The situation was similar to the one after the October 1957 launch of Sputnik 1, when President Eisenhower and his advisers had found themselves unprepared to deal with the Soviet propaganda bonanza resulting from its space success. To address this situation, President Kennedy on late Friday afternoon, April 14, convened a meeting of his advisers on space both to consider his alternatives for responding to the Gagarin flight and to present the image of a president deeply involved in a pressing issue. Kennedy had agreed to an interview with *Time/Life* journalist Hugh Sidey on that day, and he allowed Sidey to attend his meeting with his advisers. Having a journalist present at a presidential meeting was a very unusual occurrence, but Kennedy viewed Sidey as friendly and wanted to portray himself as actively engaged in considering how best to proceed.

Before meeting with the president, Webb and Dryden gathered with Sorensen, Bell from BOB, accompanied by budget staff member Willis Shapley, science adviser Wiesner, and Space Council executive secretary Welsh to review possible actions. Sorensen pressed them on what the United States might accomplish in space before the Soviet Union. The response was that "the Soviet lead in rocket thrust would continue to beat us to every milestone" being discussed. Then Sorensen asked about a human mission to the Moon. Such an undertaking, Webb and Dryden, especially Dryden, agreed "was so large, so complex and so far off, requiring so many new scientific and engineering developments, components, and studies, that the United States might have time . . . to be the first to achieve this goal." To Sorensen, "that ray of hope, after all the previous pessimistic answers, caught my attention . . . The very notion of a manned flight to the moon, as impossible as that seemed, was one that I knew would engage President Kennedy's keen interest."[17]

In requesting the interview with the president, Sidey had prepared a memorandum for Press Secretary Pierre Salinger. He remarked that "there are a lot of good minds in NASA and other dusty offices of the space agencies that think we are still fiddling, haven't made the necessary decisions... They scoff at the theory of some scientists that the Russians have now gone as far as they can for a few years." Sidey noted that "they claim, with compelling logic, that if we are to get in this race at all we've got to declare a national space goal... If we don't do this then we are going to sit here over the next eight years and watch the Soviets march right on ahead." He added, "knowing the President some, I can't believe he hasn't sensed the urgency." Sidey asked: "Why haven't we declared a crash program on one of the big boosters?... Has the President accepted the theory that we can't move faster?... How much of the feeling of no decision is due to the newness of the administration and preoccupation with other things so far?... Will there be a new and tough look followed by some hard decisions soon?"[18]

In preparation for the April 14 meetings, Jerome Wiesner drafted a memorandum for the president discussing the elements required for a review of the space program. He noted that "a crucial decision is now needed as to what degree we are willing to commit our resources, both of talent and of money, to further manned missions. If we, as a nation, want to accomplish the manned exploration of the moon during the next decade, for whatever reasons, the preparations must begin now." Wiesner suggested that "there is general agreement that the application of satellites to provide service functions, such as communication, meteorology, navigation, and mapping, is the field wherein we stand the best chance of gaining a step on the Soviet Union in space achievements during the next few years."[19]

Wiesner prepared a second memo for Kennedy, this time addressing the points raised by Hugh Sidey. He said that "no one in the Administration believes that the Russians are finished with their space exploits or that there aren't exciting space exploits still to be carried out that they will undoubtedly drive hard to accomplish." He added: "We, of course, have no knowledge at all about future Soviet intentions, but it would be surprising if they did not pursue" some of the exciting future prospects. With respect to developing a large booster, "it has become perfectly clear to the Administration that making these decisions [on what booster to develop] had to be faced, and it is our intention to do so. On the other hand, it would have been erroneous to commit very large sums of money without first establishing clear-cut national goals that go beyond present plans."

Sidey had also suggested that the Kennedy administration had advanced the date for the first astronaut-carrying launch in Project Mercury in the hopes of beating the Soviets to a first human launch. Wiesner told the president that this was not the case, but, reflecting his own skepticism, that "some consideration should be given to the question of whether or not the risks involved in a failure don't outweigh the advantages of carrying out of the shot successfully." He was not sure that the Mercury flight was justified in terms of the information it was intended to gather, since "the successful [Soviet] orbiting of man has removed many of the bio-medical questions it was designed to answer."[20]

After the first meeting without President Kennedy present, Sorensen found Sidey waiting in his office. They talked for a few minutes, and then Sorensen invited Sidey to accompany him to the cabinet room, where the participants from the earlier meeting were waiting. At almost the same time, President Kennedy joined the group after a congressional coffee hour. He invited Sidey to sit at the head of the table and to ask the assembled group his questions about the space program. Then Kennedy took over the meeting and went around the table, asking the key participants for their thoughts. This was typical of the way that Kennedy formed his views on complex issues; he "tried to master the substantive content of important issues by asking probing questions." According to veteran diplomat George Kennan, who interacted with Kennedy on U.S.-Soviet relations, "President Kennedy was the best listener I have ever seen in high position anywhere . . . He asked questions modestly and sensibly and listened very patiently."[21]

The discussion soon focused on a human mission to the Moon. Sidey reports that "the main thing everybody was hung up on was the projected cost that might be at the outside as much as forty billion dollars." Dryden suggested that, even with that level of funding, it was only an even chance that the United States could beat the Soviet Union to the Moon. Webb "launched into kind of a bureaucratic paean" in praise of Kennedy's leadership; Kennedy told Webb he was not interested in hearing it, and Webb then "got down to the fact he thought it could be done." Wiesner, "who was slumped down in his chair so far that his head seemed to be at table level," was noncommittal, saying "now is not the time to make mistakes" and asking for three months to complete a review of the program. Welsh reported that Vice President Johnson had long favored a greater effort in space and suggested that any thought that the United States could not catch up to the Soviet Union should be discarded. Bell was the "most pessimistic," with "muted horror at the thought of launching a project that was so ravenous and so vague in promised results." Hearing the discussion, Kennedy's reaction was "the cost—that's what gets me."

Sidey reports that "Kennedy was very anguished during this meeting" and "kept running his hands through his hair, tapping his front teeth with his finger nails, a familiar nervous gesture . . . There was a feeling of impatience." Sidey also got the feeling during the meeting that "Kennedy was beginning to get the feel of the challenge." Theodore Sorensen later told him that "space had not been very much on Kennedy's mind up until the Soviet manned flight," but from that meeting on, Kennedy "was in a race to the moon." According to Sorensen, on that Friday evening, "Kennedy began to really get the feel of what this whole thing might mean to his Presidency and to the United States." To Sidey, the Moon project "was a classic Kennedy challenge. If it hadn't been started, he might have invented it all, since it combined all those elements of intelligence, courage, and teamwork that so intrigued John Kennedy."

At the end of the meeting, Kennedy told the group: "When we know more, I can decide whether it is worth it or not. If somebody can just tell me how to catch up. Let's find somebody—anybody. I don't care if it's the janitor over there." Then, as he returned with Sorensen to the Oval Office, his last remark was "there's nothing more important."

In the Oval Office, according to Sorensen, the president was "incredulous about sending a manned vehicle to the moon, but excited... He immediately sensed that the possibility of putting a man on the moon could galvanize public support for the exploration of space as one of the great human adventures of the twentieth century." Kennedy directed Sorensen to organize a review to answer questions about feasibility and cost. Leaving the Oval Office, Sorensen found Sidey waiting in the hallway, and told the journalist, Sorensen now says "prematurely," that the U.S. response to the Gagarin flight would be "strong and dramatic" and that "we're going to the moon."[22]

In fact, according to Sorensen, "the decision wasn't made then so much as the stage was set for the full-scale inquiry which would be necessary before a final and precise decision could be made." Kennedy in his view had both "affirmative" and "negative" reasons for wanting an accelerated space program:

> Affirmative in the sense that the United States intended to maintain its position of world leadership, its position of eminence in commerce, in science, in foreign policy, and whatever else might develop from space exploration. The United States had to take the lead in this area just as we had taken the lead in other areas. The negative side was that we did not want to have the Soviets dominating space to a point where, in some future time, it could be a military threat to our security or, in any event, cause the rest of the world to draw away from the United States.[23]

Of Kennedy's thinking at this time, Wiesner says:

> The rest of the world had been led to believe by Soviet space accomplishments, and particularly by the U.S. reaction to them, that the scientifically most competent, the technologically most competent nation now was the Soviet Union, not the United States, because they could do this. We were paying a price, all kinds of ways—internationally, politically—and that was the issue the President was dealing with, not was it time to go to the moon or not, but how to get yourself out of this.[24]

If one objective of inviting Hugh Sidey to participate in the April 14 meeting was to give him the image of a president seriously concerned with the situation in space, the White House was successful. Sidey's story appeared in the next issue of the widely read *Life* magazine and said that Kennedy was "gravely concerned" and realized that "it was more urgent than ever to define U.S. space aims."[25]

## "The Perfect Failure"

There was another urgent matter on President Kennedy's mind as he considered options for the future in space. Kennedy had inherited Eisenhower administration planning for an invasion of Cuba, since 1960 under the control of Fidel Castro, by CIA-trained Cuban exiles. Kennedy had reluctantly given the go-ahead to the plan a few days before April 14, and bombing of Cuban airfields by what were characterized as airplanes flown by the exiles but actually flown by U.S. pilots began on April 15. The exiles went ashore at the Bay of Pigs on the morning of April 17. Within thirty-six hours, it was clear the plans had been ill-conceived, and that the invasion was a "perfect failure."[26]

The fiasco in Cuba greatly distressed Kennedy; Sorensen describes him on the early morning of April 19, after the failure was evident, as "anguished and fatigued" and "in the most emotional, self-critical state I had ever seen him. He cursed not his fate or advisors but himself." The days surrounding the failed invasion were "the worst week of his public life." In the following days, "Kennedy's anguish and dejection were evident to people around him." Not only John Kennedy but also his brother Robert was affected. One account of a top-level meeting in the aftermath of the failure reports that Robert Kennedy, just after the president stepped out of the room, "turned on everybody," saying, "All you bright fellows. You got the president into this. We've got to do something to show the Russians we are not paper tigers."[27]

How much Kennedy's emotional state and competitive character determined or merely reinforced his resolve to proceed rapidly in space cannot be definitively known, but most evidence suggests that they were influential but not decisive factors. The failure was never explicitly linked to the review of the space program that took place in the days following the Bay of Pigs; Edward Welsh maintains that the fiasco was "not a factor at all" in that review. But Wiesner says of the Bay of Pigs, "I don't think anyone can measure it, but I am sure it had an impact. I think the President felt some pressure to get something else in the foreground." He adds that, although the failed invasion was never explicitly linked to space, "I discussed it with the President and saw his reactions. I'm sure it wasn't his primary motivation. I think the Bay of Pigs put him in a mood to run harder than he might have." JFK's national security adviser McGeorge Bundy suggests that "it is quite possible that, if the Bay of Pigs had been a resounding success, the President might have dawdled a little longer on the space decision." Sorensen adds that Kennedy's attitude toward the acceleration of the space program was influenced by "the fact that the Soviets had gained tremendous world-wide prestige from the Gagarin flight at the same time we had suffered a loss of prestige from the Bay of Pigs. It pointed up the fact that prestige was a real and not simply a public relations factor in world affairs."[28]

One certain impact of the Bay of Pigs failure was to heighten White House concern regarding a possible failure of the first human launch in Project Mercury, which at that point was scheduled for May 2. A mission failure, especially if it resulted in the death of the Mercury astronaut on live television, was a possibility that the president and his advisers viewed with great concern.

## Vice President Lyndon B. Johnson Leads Space Program Review

It had been assumed since the March meetings on the NASA budget that Vice President Lyndon B. Johnson as the new chairman of the National Aeronautics and Space Council would, once he formally assumed that position, lead a comprehensive review of the U.S. space program that would provide the basis for decisions in fall 1961 on the future of that program, and especially the future of the human space flight effort. But the events of April 1961 drastically shortened the time allotted to that review. Johnson did not attend the April 14 White House meetings on the space program. He was returning from his first overseas trip as vice president, to the West African nation of Senegal, which was celebrating the first anniversary of its independence from France, with stops in Geneva and Paris on the way home. But on April 19, after President Kennedy had in the early hours of the day walked disconsolately with Ted Sorensen and then alone on the south lawn of the White House in the aftermath of the Bay of Pigs failure, he met later in the day with Lyndon Johnson and James Webb to discuss the organization of the accelerated review he had decided was needed. Johnson suggested that he would "have hearings, lay a background and create a platform for a recommendation to Congress." He asked Kennedy to give him a memorandum "that would provide a charter for those hearings" and would be an "outline of what concerned him."[29]

Ted Sorensen drafted a one-page memorandum, and President Kennedy signed it and sent it to Johnson the next day, April 20. The memorandum stands as a historic document; it spelled out the requirements that led directly to the decision to go to the Moon as the centerpiece of an accelerated U.S. space effort. It said:

> In accordance with our conversation I would like for you as Chairman of the Space Council to be in charge of making an overall survey of where we stand in space.
>
> 1. Do we have a chance of beating the Soviets by putting a laboratory in space, or by a trip around the moon, or by a rocket to land on the moon, or by a rocket to go to the moon and back with a man. Is there any other space program which promises dramatic results in which we could win?
> 2. How much additional would it cost?
> 3. Are we working 24 hours a day on existing programs? If not, why not? If not, will you make recommendations to me as to how work can be speeded up.
> 4. In building large boosters should we put our emphasis on nuclear, chemical, or liquid fuel, or a combination of these?
> 5. Are we making maximum effort? Are we achieving necessary results?
>
> I have asked Jim Webb, Dr. Wiesner, Secretary McNamara and other responsible officials to cooperate with you fully. I would appreciate a report on this at the earliest possible moment.

THE WHITE HOUSE
WASHINGTON

April 20, 1961

MEMORANDUM FOR

VICE PRESIDENT

In accordance with our conversation I would like for you as Chairman of the Space Council to be in charge of making an overall survey of where we stand in space.

1. Do we have a chance of beating the Soviets by putting a laboratory in space, or by a trip around the moon, or by a rocket to land on the moon, or by a rocket to go to the moon and back with a man. Is there any other space program which promises dramatic results in which we could win?

2. How much additional would it cost?

3. Are we working 24 hours a day on existing programs. If not, why not? If not, will you make recommendations to me as to how work can be speeded up.

4. In building large boosters should we put out emphasis on nuclear, chemical or liquid fuel, or a combination of these three?

5. Are we making maximum effort? Are we achieving necessary results?

I have asked Jim Webb, Dr. Weisner, Secretary McNamara and other responsible officials to cooperate with you fully. I would appreciate a report on this at the earliest possible moment.

The April 20, 1961, memorandum that led to the decision to go to the Moon (LBJ Library image).

The Kennedy memorandum both contained a very clear statement of a requirement—"a space program which promises dramatic results in which we could win"—and a sense of urgency. Kennedy wanted Johnson's report "at the earliest possible moment." Budget director Bell suggests that "the President would not have made such a request unless he expected a positive answer and a strong program, and therefore he was pretty sure before he made that request that that was what he intended to do." At an April 21 press conference, Kennedy was asked: "Mr. President, you don't seem to be pushing the space program nearly as energetically now as you suggested during the campaign that you thought it should be pushed. In view of the feeling of many people in this country that we must do everything we can to catch up with the Russians as soon as possible, do you anticipate applying any sort of crash program?" Kennedy replied:

> We are attempting to make a determination as to which program offers the best hope before we embark on it, because you may commit a relatively small sum of money now for a result in 1967, '68, or '69, which will cost you billions of dollars, and therefore the Congress passed yesterday the bill providing for a Space Council which will be chaired by the Vice President. We are attempting to make a determination as to which of these various proposals offers the best hope. When that determination is made we will then make a recommendation to the Congress.
>
> In addition, we have to consider whether there is any program now, regardless of its cost, which offers us hope of being pioneers in a project. It is possible to spend billions of dollars in this project in space to the detriment of other programs and still not be successful. We are behind, as I said before, in large boosters.
>
> We have to make a determination whether there is any effort we could make in time or money which could put us first in any new area. Now, I don't want to start spending the kind of money that I am talking about without making a determination based on careful scientific judgment as to whether a real success can be achieved, or whether because we are so far behind now in this particular race we are going to be second in this decade.
>
> So I would say to you that it's a matter of great concern, but I think before we break through and begin a program which would not reach a completion, as you know, until the end of this decade—for example, trips to the moon, maybe 10 years off, maybe a little less, but are quite far away and involve, as I say, enormous sums—I don't think we ought to rush into it and begin them until we really know where we are going to end up. And that study is now being undertaken under the direction of the Vice President.

Then, for the first time in public, Kennedy said: "If we can get to the moon before the Russians, then we should."[30]

# Chapter 6

# Space Plans Reviewed

According to Jerome Wiesner, as of April 1961, "Kennedy was, and was not, for space. He said to me, 'Why don't you find something else we can do?' We couldn't. Space was the only thing we could do that would show off our military power ... These rockets were a surrogate for military power. He had no real options. We couldn't quit the space race, and we couldn't condemn ourselves to be second. We had to do something, but the decision was painful for him." Wiesner added that he and Kennedy

> talked a lot about do we *have* to do this. He said to me, "Well, it's your fault. If you had a scientific spectacular on this earth that would be more useful—say desalting the ocean—or something that is just as dramatic and convincing as space, then we would do it." We talked about a lot of things where we could make a dramatic demonstration—like nation building—and the answer was that there were so many military overtones as well as other things to the space program that you couldn't make another choice.
>
> If Kennedy could have opted out of a big space program without hurting the country in his judgment, he would have. Maybe a different kind of man could have said to the country, "Look, we are going at our own pace. We are going to let the Russians be first. We don't care." But Kennedy said, "If we could afford to do something else, we would do it. If we can't, we had better get back where we belong." I think he became convinced that space was the symbol of the twentieth century. It was a decision he made cold bloodedly. He thought it was good for the country.[1]

## Options Assessed

Lyndon Johnson and his space assistant, Space Council executive secretary Edward Welsh, quickly set to work after receiving JFK's April 20 memo. Welsh was the only staff member of the Space Council at this point. The organization of the review reflected the "Johnson system" of obtaining information through personal contacts rather than formal organizational channels. Johnson consulted many of individuals whom he thought would significantly contribute to examining the space program. He met with officials from NASA, the Defense Department, the Atomic Energy Commission, and Wiesner's office. At the suggestion of Welsh, a Bureau of the Budget (BOB) representative attended most of the meetings, so that the bureau could remain informed of the alternatives under discussion and assess their financial implications.[2]

As the review was getting underway, President Kennedy on April 22 reported to the National Security Council, meeting in the aftermath of the Bay of Pigs failure, that "he had asked the Vice President...to direct an inquiry into our space effort and make a report to me which I hope will constitute the basis of a Presidential Message on this subject to Congress."[3] It is worth noting that even before he received the vice president's report, Kennedy anticipated a positive recommendation justifying a "Presidential Message" to the Congress; it is not clear whether at this point he had also decided to deliver that message in person. One indication that he was being pushed, if not already leaning, in that direction was an April 19 memorandum from Walt Rostow, who as a MIT professor had been a Kennedy campaign adviser and was in April 1961 on McGeorge Bundy's national security staff. Rostow suggested that "as the first hundred days draw to a close, I believe you should consider a major address taking stock of where we are and where we should go, both at home and abroad." Rostow identified an accelerated space effort as one of the potential topics in the speech.[4]

## *NASA and DOD Present Their Views*

Johnson lost little time in getting started with his review. At 10:30 p.m. on April 20, he called Welsh and asked him to arrange a meeting with NASA administrator Webb and "such other NASA people as NASA requires" for 9:30 a.m. on April 22, a Saturday, to outline "what now needs to be done in the space program, what it would cost, and whether more funds are required at this time (FY1962)." A similar meeting with Secretary of Defense Robert McNamara was set up for later the same day; the two organizations were told not to coordinate their views in advance of meeting separately with the vice president.[5]

Hugh Dryden accompanied Webb to the meeting with the vice president and presented the NASA response to the questions in the president's April 20 memorandum. Dryden said that there was "no chance of beating the Soviets in putting a multi-manned laboratory in space since flights already accomplished by the Russians have demonstrated that they have this capability." He told Johnson "with a determined effort of the United States, there is a chance to beat the Russians in accomplishing a manned circumnavigation of the moon," perhaps by 1966. He added, "there is a chance for the U.S. to be the first to land a man on the moon and return him to earth if a determined national effort is made." Dryden thought it "doubtful" that the

Russians had a meaningful head start on a manned lunar landing program and "because of the distinct superiority of U.S. industrial capacity, engineering, and scientific know-how... the U.S. may be able to overcome the lead the Soviets might have up to now." A first landing might be possible in 1967 "with an accelerated U.S. effort." Other areas in which the United States might be first included "returning a sample of the material from the moon surface to the earth in 1964" and "developing communications satellites," which, "although not as dramatic as manned flight," would have benefits to people throughout the world. NASA at this point in the review estimated the cost of an accelerated effort in all areas over the period through 1970 as $33.7 billion, an increase of $11.4 billion over its then-current ten-year plan.[6] Although the potentials of a lunar landing program had been discussed with President Kennedy in the April 14 cabinet room meeting, Dryden's report was likely the first time that Lyndon Johnson had heard a top-level analysis of what it would take to surpass the Soviet lead in human space flight. Although others, especially Wernher von Braun, are often credited with being first to propose a lunar landing to the White House as the "space program which promises dramatic results in which we could win," it seems that honor should go to Hugh Dryden, who had also raised the lunar landing possibility at the April 14 cabinet room meeting with President Kennedy.

Secretary of Defense Robert McNamara's response to Vice President Johnson drew heavily on material provided by John Rubel, deputy director of Defense Research and Engineering, who was the top space official within the Office of the Secretary of Defense. Rubel was an engineer who had worked for Hughes Aircraft before coming to Washington during the Eisenhower administration and who had strong views on how best to organize the national space effort. Rather than provide responses to the questions in President Kennedy's April 20 memorandum, McNamara articulated a particular philosophy with respect to space. He remarked that "all large scale space programs require the mobilization of resources on a national scale. They require the development and successful application of the most advanced technologies. Dramatic achievements in space, therefore, symbolize the technological power and organizing capacity of a nation." For these reasons, "major achievements in space contribute to national prestige" and "constitute a major element in the international competition between the Soviet system and our own." (These words, most likely written by John Rubel, would reappear in a May 8 memorandum to the vice president recommending the lunar landing goal.) "Because of their national importance and their national scope," McNamara added, "it is essential that our space efforts be well planned. It is essential that they be well managed." Effective management was needed so that "engineering resources be focused and not spread too thin," for "our national posture may be worsened rather than improved if added expenditures result in the still greater dispersal of scientific, engineering and managerial talent." McNamara called for an orderly but accelerated program to close the booster gap. With respect to various Department of Defense space programs, he recommended no budget increases above those that had already been approved by the White House the previous month.[7]

### Others Consulted

As part of his review, Vice President Johnson reached out to a number of individuals whom he thought could provide informed advice. For example, on April 24 he received a thirty-four-page report prepared over the weekend by George Feldman, a former staff member of the House space committee who had sought the job of NASA administrator, and Charles Sheldon of the Congressional Research Service. He also asked Welsh to get the views of Cyrus Vance, in April 1961 the top lawyer at the Department of Defense but formerly a member of Johnson's Senate preparedness subcommittee staff.[8]

On Monday morning, April 24, Johnson held a "hearing" to solicit the views on what course of action he should recommend to President Kennedy. Presenting their views were individuals representing the three military services. To get a "keen sense of public reaction" to the kind of accelerated program that was emerging from his review, the vice president invited three prominent businessmen who were also close personal friends to listen to the presentations. They were George Brown of the Houston, Texas construction firm of Brown and Root (who had been a major Johnson campaign contributor); Frank Stanton, president of the Columbia Broadcasting System; and Donald Cook, vice-president of the American Electric Power Corporation. The three service representatives were Vice Admiral John T. Hayward, deputy chief of naval operations for research and development; Lieutenant General Bernard Schriever, commander of the Air Force systems command and the recognized pioneer of the Air Force space program; and Wernher von Braun, who had been transferred in June 1960 from the Army Ballistic Missile Agency to work for NASA but from Johnson's perspective still could represent the Army's views. Johnson had sent each of the three a copy of President Kennedy's April 20 memorandum directly, not through their chain of command. (In fact, when he learned of the meeting after it had taken place, Secretary of Defense McNamara reportedly told the vice president that if he wanted military representation at a future meeting, it would be McNamara who would decide whom to send from the Department of Defense.) Also attending the session were Webb and Dryden from NASA, Rubel from DOD, science adviser Wiesner, and BOB staff.[9]

In opening the meeting, Vice President Johnson told the group that he could not "overstate the fact that 'our Freedom is at stake.' Communist domination of Space could lead to control over men's minds as well as their very existence." Johnson spoke of the "propaganda aspects, as well as the technological and the defense aspects." He said "The President wants the best hard-headed advice he can get—and he wants it now" and "this meeting, complemented by anything you want to send in later, is called to get your specific views on what we can do to get this country into the Space lead."[10]

Hayward told the meeting that he supported a large-scale U.S. space program with a lunar landing mission as a central goal. He believed that, from a national point of view, only the lunar landing mission made sense as a way of accelerating the space program. The Navy was concerned, reported Hayward, that the practical applications of space technology that provided assistance to naval operations, such as the use of satellites for navigation, reconnaissance, communications, and weather forecasting, not be neglected in any accelerated program. He stressed the need for an integrated, orderly space program rather than emphasis on one project at the cost of neglecting others.[11]

Schriever also urged that a program aimed at a lunar landing be adopted, primarily because "it would put a focus on our space program. If we had this sort of an objective, there were so many other things that would be required that you couldn't avoid having a major space program. I felt that we needed a major national space program for prestige purposes, for those things we could see as having national security implications and because of the need for advancing technology." Although the Air Force in 1958 had proposed a very similar idea—a lunar landing effort as a central focus of a national space program—Schriever in 1961 did not suggest Air Force management of such an effort, noting, "that never came up. At that point, there was no argument about who was going to run the program."[12]

In an April 30 follow-up letter to the vice president, Schriever said that he held "a strong conviction that achievements in space in the critical decade ahead will become a principal measure of this nation's position in world leadership—a world in which it is becoming increasingly obvious that there will be no second." This letter addressed directly the questions in President Kennedy's April 20 memorandum. It indicated several areas in which "we have a high probability of scoring a dramatic 'first,'" including a lunar landing by 1967, capturing an object in space and returning it to Earth by 1963,[13] the first flight of a nuclear propulsion upper stage by 1965, and establishing a communication satellite network over the Atlantic Ocean by 1963 and worldwide by 1964. Schriever suggested that

> our currently projected space effort is dangerously deficient. It has been characterized by an attitude of defeatism and a seeming resignation to second place for the United States in the space competition with the Soviet Union. Placing a man in orbit has been called a "stunt."... This negative philosophy places at serious and unacceptable risk both our national prestige and our military security. It fails to recognize the military potential of space and the fact that achievements in space have been the single most important influence in the world prestige equation.
>
> A greatly expanded and accelerated space program can—and should—be undertaken. There is clear evidence that we have the resources to more than double the magnitude of our present space effort. All that is lacking is the decision to do so—a decision comparable to that made by the late President Franklin D. Roosevelt shortly after the attack on Pearl Harbor when he called upon the nation to increase its annual airplane production from a few thousand

to the seemingly impossible figure of 50,000 aircraft per year. The timid souls were routed. The response to the call of our President in that critical hour is a highlight of our nation's history.[14]

Wernher von Braun summarized the views he had expressed in the meeting, which he characterized as "strictly my own," in an April 29 memorandum to Vice President Johnson. He told Johnson that the United States had a "sporting chance of sending a 3-man crew *around the moon* ahead of the Soviets (1965/1966)" and "an excellent chance of beating the Soviets to the *first landing of a crew on the moon* (including return capability, of course)." This was because "a performance jump by a factor 10 over their present rockets is necessary to accomplish this feat. While today we do not have such a rocket, it is unlikely that the Soviets have it." Given this likelihood, said von Braun, "we would not have to enter such a race towards this obvious next goal in space exploration against hopeless odds favoring the Soviets. With an all-out crash program I think we could accomplish this objective in 1967/1968." This estimate reflected von Braun's confidence that with adequate resources his rocket team could develop the large launch vehicle needed more rapidly than could its Soviet competitors. Echoing the call for a more centralized approach to the management of large-scale technological efforts that had been articulated by Robert McNamara on April 22, von Braun noted that "in the space race we are competing with a determined opponent whose peacetime economy is on a wartime footing. Most of our procedures are designed for orderly, peacetime conditions. I do not believe that we can win this race unless we take at least some measures which thus far have been considered acceptable only in times of a national emergency."[15]

Edward Welsh later on April 24 summarized the main points that had come out of the day's meeting. Among them were the following:

- "We have a lot of built-in handicaps which the Russians don't have, i.e., contracting procedures, variety of government agencies and private companies in the act, freedom of the press, etc."
- "A lunar landing and return is not just a 'stunt.' Rather, it should be pushed as a basically important achievement of great technical and scientific importance."
- "The distinction between 'peaceful uses' and defense uses for space is a handicap."
- "We have to have a basic philosophy—make our objectives clear. This means not a 'catch up' philosophy but a leadership philosophy."
- "Idealism is fine, but we have to be realistic in dealing with the rest of the world, as they will align themselves with the leader."
- "The Russians are not going to wait for us, so we should shoot for targets ahead of where they are now."
- "More money, more definite policies, and more effort are needed."[16]

Neither Secretary of State Dean Rusk nor chairman of the Atomic Energy Commission Glenn Seaborg were at the initial meetings organized by the vice president, even though both were members of the Space Council. Johnson apparently did consult with Rusk by telephone to learn if the secretary of state foresaw any negative foreign policy impacts from an accelerated space program and if Rusk agreed that a program aimed at capturing leadership in space for the United States was politically desirable. Rusk did agree to this proposition. The top State Department staff persons on space issues, Philip Farley and Robert Packard, interacted with Welsh during the consultations and attended at least one meeting; also, Richard Gardner, deputy assistant secretary of state for international organization affairs, wrote the vice president on April 24, saying that "I believe the United States could redeem much of its lost prestige in the space race by scoring 'a first' in the field of communication satellites," since such a success "would have very dramatic and obvious practical benefits to millions of people around the world."[17]

There were no dissenters among those consulted to the notion that the country should undertake a vigorous space program funded at a significantly higher level than had been the case under the Eisenhower administration. There was little question, Johnson was told repeatedly, that such a program would have considerable political, strategic, technological, and economic payoffs for the United States. For example, Donald Cook in a letter to Johnson a few weeks later argued that "actions in this field must, I believe, be based on the fundamental premise that achievements in space are equated by other nations of the world with technical proficiency and industrial strength. This proficiency and strength is, in turn, equated with World power. And the conclusion reached by other countries on the question of our position in the world in terms of power is and will be of fundamental importance in their determination as to which group, the West or the East, they will cast their lot."

This view of the world situation—as a bipolar struggle between the United States and the Soviet Union for global leadership at a time that Communist parties were strong in many Western European countries and as newly independent nations were deciding which form of social and political organization to adopt—was at the root of U.S. foreign policy in the early months of the Kennedy administration. Cook's views were commonly shared among the U.S. political and intellectual leadership, including President Kennedy himself. Cook's letter continued: "On this premise, the goal that we must seek is the achievement of leadership in space—leadership which is both clear-cut and acknowledged. Our objective must be, therefore, not merely to overtake, but substantially to outdistance Russia. Any program with a lesser basic objective would be a second-rate program, worthy only of a second class power. And, most important, a lesser program would raise serious questions among other countries as to whether, as a nation, we had the will and the discipline necessary for leadership in the struggle to preserve a free society."[18]

As the Space Council review proceeded, Vice President Johnson kept President Kennedy informed of its progress. At the April 25 ceremony at which President Kennedy signed the bill making the vice president the chair of a reorganized Space Council, Kennedy noted that "enactment of this measure is symbolic of our Government's intention to translate leadership and determination into action... Working with the Vice President, I intend that America's space effort shall provide the leadership, resources and determination to step up our efforts and prevail on the newest of man's physical frontiers."[19]

On April 28, the vice president sent the president a six-page memorandum summarizing his space review to date. Johnson told Kennedy:

- "The U.S. has greater resources than the USSR for attaining space leadership but has failed to make the necessary hard decisions and to marshal those resources to achieve such leadership."
- "This country should be realistic and recognize that other nations, regardless of their appreciation of our idealistic values, will tend to align themselves with the country they believe will be the world leader—the winner in the long run. Dramatic achievements in space are being increasingly identified as a major indicator of world leadership."
- "If we do not make the strong effort now, the time will soon be reached when the margin of control over space and over men's minds through space accomplishments will have swung so far on the Russian side that we will not be able to catch up, let alone assume leadership."
- "Manned exploration of the moon, for example, is not only an achievement of great propaganda value, but it is essential as an objective whether or not we are first in its accomplishments—and we may be able to be first."
- "There are a number of programs which the United States could pursue immediately and which promise significant world-wide advantage over the Soviets. Among these are communication satellites, meteorological and weather satellites, and navigation and mapping satellites."
- "More resources and more effort need to be put into our space program as soon as possible."

Edward Welsh, who drafted the memorandum, has suggested that "the decision to go to the moon was made immediately upon the receipt of the April 28th memorandum." This seems not to have been the case. While President Kennedy at that point in time had given strong indications that he was inclined toward such a choice, two more weeks of review would take place before the decision became final.[20]

### *Some with Reservations*

Although science adviser Wiesner attended some of the meetings Johnson called, at no time during this review was the President's Science Advisory Committee (PSAC) as a body consulted about the wisdom of what was being recommended. Wiesner, reflecting the conclusions of the PSAC report that had been presented to President Eisenhower in December 1960, viewed the decision to accelerate the space program with a lunar landing mission as a central undertaking as "a political, not a technical issue. It was not an issue of scientific versus non-scientific issues; it was a use of technological means for political ends. It was on these considerations that I did not involve PSAC." Wiesner did tell the president that PSAC "would never accept this kind of expenditure on scientific grounds." Kennedy accepted this and in turn promised Wiesner that he would never justify the lunar mission in terms of its scientific payoffs.[21]

Somewhat surprisingly, there was another key individual who was somewhat skeptical of the push for a major acceleration of the space program, with landing on the Moon before the Soviet Union as its central feature—NASA administrator James Webb. Webb described himself as "a relatively cautious person. I think when you decide you're going to do something and put the prestige of the United States government behind it, you'd better be doggone well be able to do it."[22] Webb was reluctant to commit himself to a lunar landing effort until he was convinced that it was technologically sound, that NASA had the capability to execute it, and that it "did not go beyond what I thought Kennedy was willing to approve." Webb wrote Wiesner on May 2, noting that the budget figures that had accompanied NASA's April 22 presentation to the vice president had been put together "in a great hurry" and did not represent the results "of a careful study of the technological bottlenecks or difficulties." Webb asked Wiesner to join him in insuring that the program to be recommended to the president "has real value and validity and from which solid additions to knowledge can be made, even if every case of the specific so-called 'spectacular' flights or events are done after they have been accomplished by the Russians."[23] By acting to emphasize his concern with the underlying validity of the accelerated program, Webb hoped both to maintain his good working relationship with Wiesner and, through Wiesner, the scientific community, and to influence the program recommendations so that if necessary he could later defend the program against charges that it was aimed only at prestige and was fundamentally distorted and unsound. Webb, in essence, "wanted to contain and shape the decision to reflect favorably on NASA, the nation, and himself."[24]

### *Congress Consulted*

The next step in Vice President Johnson's review was to consult key congressional leaders to ensure that they would indeed be willing to support the kind of accelerated program that the president was likely to recommend. The original plan was to meet with the chairman, Overton Brooks, and ranking minority member, James Fulton, of the House Committee on Science and Astronautics on May 1, and the chairman, Robert Kerr, and ranking minority member, Styles Bridges, of the Senate Committee on Aeronautical and Space Sciences on May 2. However, the May 1 meeting had to be postponed so that Johnson could participate in a National Security Council discussion of policy toward Viet Nam. Brooks was unhappy at this turn of events and with Johnson's suggestion that Brooks submit a memorandum on his views in lieu of a face-to-face meeting.[25] The meeting with the two senators and their staff actually took place on May 3. Webb and Dryden from NASA, John Rubel from DOD, Edward Welsh, representatives from the Atomic Energy Commission, the Department of State, and the BOB, and George Brown, Frank Stanton, and Donald Cook were also present.

The two senators whom Johnson invited to the meeting, Kerr and Bridges, were present not only because of their committee positions but also because they were two of the small number of veteran legislators who, together with Johnson, had controlled the Senate during the Eisenhower years. Johnson believed that their support would suffice to carry the rest of the Senate with them. Johnson opened the meeting by saying that "I believe it is the position of every patriotic and knowledgeable American that past policies and performances in space have not been enough to give this country leadership." He noted that "the President has made it clear that he is determined that the United States move into 'its proper place in the space race.' That can only mean leadership. There is no other proper place for our country." Johnson told those at the meeting that "we are here to discuss not *whether*, but *how*—not *when*, but *now*."[26] This strident view of American exceptionalism was a pervasive aspect of LBJ's space review.

Robert Kerr told the others that "we need some cold-blooded decisions, but the Senate can be counted on in the end to face up to whatever is required." Styles Bridges agreed, saying, "it certainly is necessary to attain the highest possible scientific use and to maintain the glory of the United States and its prestige, but basic to the whole matter is the security of the United States." James Webb continued his cautious approach, telling the group that "there is a great deal that must be done before the vice president will be in a position to make the recommendations and the president be ready to go to the Congress and ask for the large sums that will be necessary, so we've got to be very careful now." Johnson reacted negatively to Webb's caution, saying "do you feel that you will not be prepared to give me answers for a month? . . . I am not trying to rush you. But you must not wait a month or Congress will have gone home." He added: ""We'll wait for a month if necessary for people [clearly meaning Webb] to get the guts enough to make solid recommendations." Frank Stanton from CBS added an elitist perspective: "We don't have to be concerned about national support if wise men have decided upon the action necessary in the national interest."[27]

Johnson completed his canvas of Congressional support by telephoning Overton Brooks and James Fulton to inquire whether the House of Representatives would also support an accelerated space effort. Given what these two men had been saying in their recent committee hearings about the need for a faster-paced program, their responses were not surprising. Fulton told Johnson, after checking with some other House Republicans, that he thought Republican support for an accelerated program would be almost unanimous.[28] After the vice president's call, Brooks submitted a ten-page memorandum of recommendations for the space program. Brooks said that he and his committee believed that "the United States must do *whatever is necessary to gain unequivocal leadership in Space Exploration*." He recommended an immediate acceleration of programs for communications, television, weather, and navigation satellites. He said that his committee was "committed to a forceful and stepped-up long range endeavor" and that "we cannot concede the Moon to the Soviets, since it is conceivable that the nation that controls the Moon may well control the Earth."[29]

According to Webb, by the end of the May 3 meeting Lyndon Johnson "was close to demanding" from NASA a specific recommendation on a lunar landing program, not additional study of its requirements. Based on all information available to him, Webb felt that the lunar landing was "the first project we could assure the president that we could do and do ahead of the Russians, or at least had a reasonable chance to do." Johnson pushed Webb, saying that "are you willing to undertake this? Are you ready to undertake it?" Webb replied that he was ready, "but there's got to be political support over a long period of time, like ten years, and you and the President have to recognize that we can't do this kind of thing without that continuing support."[30] The next day Webb wrote to the vice president, telling him that "my main effort yesterday was to be certain that you and the Senators were under no illusions whatever as to the magnitude of the problems involved in carrying out this decision and the absolute necessity, in my opinion, for a decision to back Secretary McNamara and myself to the limit." Six years later, Webb, as he complained to the president about cuts in the NASA budget, was still reminding Lyndon Johnson that he had been "quite reluctant to undertake the responsibility of building a transportation system to the moon" and that Johnson "had almost to drive me to make the recommendation which you sent on to President Kennedy."[31]

As Lyndon Johnson was gauging Congressional support for an accelerated space effort, Kennedy was also independently consulting key members of Congress with respect to what type of enhanced space program would be politically acceptable. In particular, according to Webb, there was "little doubt in my mind" that Kennedy consulted Houston-area Congressman Albert Thomas, who chaired the House subcommittee controlling NASA's appropriations. Thomas's relationship with fellow Texan Lyndon Johnson was "not close," but Kennedy "paid a very great deal of attention to what Thomas told him could be done and what he, Thomas, was prepared to do." Kennedy also needed support from Thomas in other areas of his legislative agenda, and it is likely that Thomas alerted Kennedy to his desire to have an accelerated space program benefit his Houston Congressional district. Webb later recalled that Lyndon Johnson "had more weight in bringing President Kennedy to his decision than the staff around the White House was or is yet willing to recognize... In the end he [Kennedy] was, I believe, as strongly influenced by Johnson and Thomas as by any other two people. Once he felt he had to move ahead, he could proceed vigorously because he knew that these men could maintain a base of support that would give him a chance to succeed."[32]

While James Webb might have wanted more time to have his staff carry out a fuller study of the requirements for sending Americans to the Moon, that was not to be. On May 4, a Thursday, Lyndon Johnson was asked by President Kennedy to embark the following Monday on a tour of Southeast Asia to provide a first-hand assessment of the situation there with respect to Communist-supported insurgencies. The next day, Johnson called Webb and Secretary of Defense McNamara and asked them to "prepare both a program for the President to send to Congress and a message for the President to use in the transmission of the message." The vice president wanted to submit these papers to the president on the following Monday, before he left on his trip.[33]

The same day, Friday, May 5, the first U.S. astronaut, Alan Shepard, was launched on a 15-minute suborbital flight to the lower edge of outer space. Success in that flight was a critical factor in any decision that might follow.

## Mercury-Redstone 3: A Necessary Success

From the start of Project Mercury in 1958, the project's plan called for several brief suborbital flights with an astronaut aboard before committing a human to an orbital mission. The first such flight could have come in March 1961 if it had not been for the combination of some relatively minor problems on the January 31 Mercury-Redstone-2 flight carrying the chimpanzee Ham and the biomedical concerns raised by PSAC about an astronaut's ability to withstand the stresses of spaceflight. An additional test flight without an astronaut (or chimpanzee) aboard was inserted in the Mercury schedule, and the first crewed flight, Mercury/Redstone-3 (MR-3), was slipped until the end of April or early May; to some, that flight seemed rather anticlimactic when Yuri Gagarin orbited the Earth on April 12.

Beginning with the Wiesner task group report in January and extending almost to the day of the flight, there were White House fears that the risks of the MR-3 mission outweighed its benefits. These fears were only amplified by the failure at the Bay of Pigs in mid-April; the possibility that a U.S. astronaut might perish in the full light of the television and other media coverage of the mission so soon after the United States had looked so weak in its unwillingness to support the invasion force it had trained was very troubling to President Kennedy and his top advisers. Sorensen remembers that while "gloating Russians, undecided Third World neutrals, and concerned allies" awaited the outcome of the flight, "untold numbers of the American press insisted for weeks that all their reporters must receive passes to be present." At the same time, there was insistence that "their editorial writers and columnists must be free to deplore the media circus atmosphere resulting from so many reporters being present." The White House concern was "that such a big buildup would worsen our national humiliation [the Bay of Pigs] if the flight were a failure."[34]

Jerome Wiesner had raised similar concerns as far back as March 9. In a memorandum to McGeorge Bundy, Wiesner expressed his worry about the pressure for live TV and press coverage of the MR-3 launch, fearing that there was a danger of the event tuning into a "Hollywood production" that "could jeopardize the success of the entire mission." He suggested that the government should meet such pressures "with firmness." Wiesner on the other hand was concerned about how best to exploit the public relations value of a successful mission to serve administration interests, since "in the imagination of many" the mission would "be viewed in the same category as Columbus' discovery of the new world."[35] Wiesner's hope for a historic first was definitively dashed by the Soviet orbital flight in April.

At the March 22 meeting on the NASA budget, President Kennedy had asked about the risks associated with the first Mercury flight. Hugh Dryden assured him that "no unwarranted risks would be involved" and that "the decision to 'go' was being made by project managers best qualified to assess the operational hazards." The PSAC panel on Mercury agreed with this assessment, saying the flight would be "a high risk undertaking but not higher than we are accustomed to taking in other ventures."[36]

Still, doubts about the wisdom of going ahead with the mission, at least so soon after the Soviet orbital flight and the Bay of Pigs fiasco, persisted. The person who had led the PSAC review of Mercury, Donald Hornig, on April 18 sent a memorandum to Sorensen raising two questions: (1) "Is MR-3 still justified, in view of the risks, after the Russian flight?" and (2) "If so, should the present schedule be maintained or should it be carried out at a later time?" Hornig noted that after the Gagarin flight "the fact that one human can withstand these conditions [of spaceflight] is now established." Hornig's conclusion was that "it seems likely that we should proceed on schedule, particularly since the world already knows that schedule," but that "our estimate of the risk is still that it cannot presently be demonstrated that the likelihood of disaster is less than one in ten or one in twenty."[37]

On April 26, Wiesner told the Space Council's Welsh that his office had been receiving messages from "some of the scientists whom he knows raising a question about the advisability of our going forward with the Mercury man-in-space shots." Their concern, said Wiesner, was that "if these shots were successful, they would still look relatively small compared with what the Russians have done, and, if the shots failed, the damage to our prestige would be serious." In reporting this message to Vice President Johnson, Welsh said that he believed "we should go ahead... Having announced that we were going to make the efforts, I believe that we would suffer seriously if we did not go ahead."[38] Concerns about the wisdom of proceeding were not limited to the White House. Senators John Williams (R-DE) and William Fulbright (D-AK) suggested "that the flight should be postponed and then conducted in secret lest it become a well-publicized failure."[39]

The MR-3 flight was scheduled to lift off on the morning of May 2. In the preceding days, Ted Sorensen and the president's brother Robert Kennedy discussed whether it was worse to postpone the flight after the press buildup had reached such a peak or to go ahead with the flight and run the risks of failure.[40] President Kennedy made the final decision on whether to go ahead with the flight in an Oval Office meeting on April 29. Present at the meeting were Wiesner, Sorensen, Bundy, and Welsh, among others. One of those present raised the point of "maybe we should postpone the Shepard flight, maybe we shouldn't take this risk, something might go bad, there might be a casualty, and we've had a number of things go rather poorly here and maybe we shouldn't do this right now." The majority of the group favored postponing the flight, but Welsh argued that it was no riskier than flying from Washington to Los Angeles in bad weather, and asked the president, "why postpone a success?" According to Welsh, "that ended the discussion."[41]

On May 1, the day before the flight was scheduled, James Webb and White House press secretary Pierre Salinger met with Kennedy for a final review of the press arrangements for covering the launch. Webb assured the president that all precautions had been taken and the flight should go forward as scheduled. Kennedy asked his secretary to place a call to NASA's public information officer in Florida, Paul Haney, to discuss plans for television coverage and to discuss the reliability of the Mercury capsule's escape system. Salinger talked to Haney from the Oval Office and, after Haney reviewed the history of the launch escape system, told Kennedy that he felt that JFK's questions had been adequately answered.[42]

Because of poor weather, the MR-3 flight was postponed on May 2 and again on May 4. Finally, on May 5, astronaut Alan Shepard was launched on what he described as a "pleasant ride." A wave of national relief and pride over an American success swept the country, from the White House down to the person in the street. At the White House, Kennedy's secretary Evelyn Lincoln interrupted a National Security Council meeting to tell the president that Shepard was about to be launched. Kennedy, joined by Lyndon Johnson, Robert Kennedy, Robert McNamara, Dean Rusk, Ted Sorensen, McGeorge Bundy, Arthur Schlesinger, and others crowded around a small black and white television set in Lincoln's office to watch the takeoff. As Jacqueline Kennedy walked by, the president said: "Come in and watch this." Sorensen suggests that the group watching the flight in Lincoln's office "heaved a sigh of relief, and cheered" as Shepard and his spacecraft were pulled from the Atlantic Ocean.[43] After Shepard was safely aboard the recovery ship, Kennedy called him, saying, "I want to congratulate you very much. We watched you on television, of course, and we are awfully pleased and proud of what you did."[44]

The decision to carry out the Shepard flight in full view of the world seems to have paid off. A May 1961 report of the U.S. Information Agency comparing international reactions to the Gagarin and Shepard flights noted that in terms of public reaction "the U.S. reaped a significant psychological advantage over the Soviet Union." This was due in large part to the "openness" surrounding the Shepard flight, plus the flight's "technological refinements and the poise and humility of the U.S. astronaut." People around the world contrasted this "critically to Soviet secrecy and blatant propaganda exploitation, as well as the obviously politically controlled behavior of Gagarin." The report went on to say that "without question the greatest impact on expressed opinion in the Free World was made by the openness of the U.S. both as to the flight itself and to the release of scientific information." In contrast, "Soviet secrecy was deplored and even continued to

President Kennedy in his secretary's office watching the May 5, 1961, launch of the first U.S. astronaut, Alan Shepard. Also visible are Vice President Lyndon Johnson, national security adviser McGeorge Bundy (over Johnson's right shoulder), presidential assistant Arthur Schlesinger, Jr. (in bowtie), chief of Naval operations Admiral Arleigh Burke, and Jacqueline Kennedy (JFK Library photograph).

arouse some skepticism." Some of the commentary "showed a significant tendency to identify itself with the U.S. success."[45]

If the Shepard flight had been a catastrophic failure, it is very unlikely that President Kennedy would have, or politically could have, soon afterward set as a national goal the flight of Americans to the Moon. However, the unqualified success of the flight in both technical and political terms likely swept away any of Kennedy's lingering reservations with respect to the benefits of an accelerated space effort with human space flights as its centerpiece. In a formal statement issued after the flight, Kennedy said: "All America rejoices in this successful flight of Astronaut Shepard. This is an historic milestone in our own exploration into space. But America still needs to work with the utmost speed and vigor in the further development of our space program. Today's flight should provide incentive to everyone in our nation concerned with this program to redouble their efforts in this vital field. Important scientific material has been obtained during this flight and this will be made available to the world's scientific community." At a press conference later in the day, Kennedy announced that he planned to undertake "a substantially larger effort in space."[46]

# Chapter 7

# "A Great New American Enterprise"

With the success of Alan Shepard's flight on May 5, the momentum toward a dramatic acceleration of the American space effort, with its focal point being a "space program which promises dramatic results in which we could win," was becoming unstoppable. Vice President Johnson's consultations over the prior two weeks had revealed almost unanimous support for such an initiative. But before President Kennedy could announce his intent to enter and win the space race, the generalized support had to be turned into specific program recommendations with an accompanying political rationale and budget estimates. Kennedy on April 20 had asked for the conclusions of the vice president's review "at the earliest possible moment." With the vice president leaving Washington on May 8 for almost two weeks, that meant a busy weekend for those charged with putting the results of LBJ's consultations into a form for Johnson to submit with his recommendation for presidential approval.

## Recommendations Prepared

On Saturday morning, May 6, the group charged with preparing the recommendations met at the Pentagon office of Secretary of Defense Robert McNamara. Those present at some point in the day included McNamara, deputy secretary of defense Roswell Gilpatric, Harold Brown, the new director of defense research and engineering, and his deputy John Rubel, from the Department of Defense; James Webb, Hugh Dryden, Robert Seamans, and Abraham Hyatt, director of program planning and evaluation, from NASA; Chairman Glenn Seaborg from the Atomic Energy Commission; and Willis Shapley from the Military Division of Bureau of the Budget (BOB). Neither Jerome Wiesner nor Edward Welsh participated in the weekend sessions, and neither was briefed on their results before LBJ's recommendations were submitted to President Kennedy on May 8.[1]

Seamans reports that "there was unanimous agreement that at this that time we needed to do more in space; that we had a reasonable scientific program; that the military program in space was satisfactory; that we should probably do more in the commercial area, or the civilian area, and specifically, Secretary [McNamara] was very insistent that NASA add to its budget $50 million for communications satellites." John Rubel added that large space projects "reflect the capacity and the will of the nation to harness its technological, economic, and managerial resources for a common goal"; for these reasons, "a successful space program validates your claim to other capacities."

During Vice President Johnson's review over the preceding two weeks, NASA and DOD had never seen each other's responses to President Kennedy's April 20 memorandum, so most of the day was spent with NASA and DOD presenting those responses to the other agency. NASA went first and outlined its proposed initiatives, including setting a manned lunar landing goal with a planning date of 1967 for the first landing attempt. There had been enough analysis, dating back to the work of George Low's task force, to satisfy the NASA leaders that a lunar landing was a technically achievable objective, given a strong commitment of resources and people. That such a program would be recommended to the president was certain enough that NASA had established an internal task force four days earlier to provide a more detailed sense of what would be involved in such an undertaking, but that group's findings were not due until early June. Thus the specifics of the lunar landing plans discussed on May 6 were in a very preliminary state.

In choosing the lunar landing program as the central feature of its recommended program, the group had no firm intelligence regarding whether or not the Soviet Union was already embarked on a similar program. An April 25, 1961, National Intelligence Estimate on "Soviet Technical Capabilities in Guided Missiles and Space Vehicles" had estimated that the Soviet Union could orbit payloads weighing 50 to 100 tons in the "latter part of the decade"; this capability would allow Russia to launch a human mission to circumnavigate the Moon by 1966, to go into orbit around the Moon by 1967, and, "contingent upon success with manned earth satellites and the development of large booster vehicles," to carry out "lunar landings and return to earth by about 1970." The estimate noted: "The Soviet leaders clearly believe that achievements in space enable them to persuade the world that in the realms of science, technology, and military strength, the USSR stands in the very front rank of world powers. In seizing an early lead and following it with a series of dramatic successes, they have sought to bolster their claims of superiority of the Soviet system... Since 1955, the announced goal of the Soviet space program has been manned interplanetary travel." There was, however, no hard evidence that the Soviet Union had initiated a lunar landing program.[2]

Although there is no direct evidence that those preparing the space recommendations had read this intelligence estimate prior to the May 6–7 weekend, they were likely aware of its conclusions. The final recommendations that went to Vice President Johnson the following Monday, May 8, were accompanied by a separate, highly classified, annex on Soviet space capabilities.[3]

The only question about the wisdom of selecting the lunar landing goal as the key step toward space leadership came from Robert McNamara, who wondered, given the lack of specific knowledge about the Soviet space program and its record of achievements to date, whether such an effort was ambitious enough. Might the Soviets already be so far along on a lunar landing program that they could do it in a year or two, he wondered. Might it be better to focus on sending people to Mars, instead, he asked the group. Seamans told McNamara that going directly to a human mission to Mars was technically not feasible, but, he says, "McNamara kept pressing the issue of whether we shouldn't embark on a planetary program."

NASA also listed other areas of space activities that were candidates for acceleration; Shapley suggests that between the items that NASA suggested and those coming from Seaborg regarding nuclear propulsion for space and from DOD with respect to solid-fueled boosters, the result was "a whole smorgasbord." Part of that "smorgasbord" was setting as a goal of creating as soon as possible an operational meteorological satellite system. Such an initiative had been under study within the government and by a President's Science Advisory Committee (PSAC) panel in the previous weeks. The panel was headed by Princeton professor John Tukey; it had concluded that "the usefulness of a meteorological system operating on a continuing basis is unquestionable." Such a system, Wiesner a few days later noted, would "assure to the U.S. the advantages of U.S. leadership in this important peaceful use of outer space and provide opportunities for a significant international cooperative program."[4] Another initiative approved by the group was the creation of an operational communications satellite system, which was listed in a May 5 background paper, apparently prepared by Willis Shapley, as a "first" with "continuing prestige impact." Such a system could be available in about twenty months, suggested the paper, "*provided* the program is directly pushed by the Government, takes full benefit of private commercial developments, and gives priority to time rather than economic factors."[5]

An important reason for Robert McNamara's desire to have a large civilian space program was that it would help him resist Air Force demands for an increase in its space activities. McNamara, aware of the bullish recommendations contained in the Gardner Committee report, believed that the Air Force was "out of control," and that, if NASA was carrying out a large program with heavy involvement of the aerospace industry, it would be more difficult for the Air Force to get political support from the same firms and from the Congress for its space ambitions.[6]

The group then reviewed the Department of Defense space effort. The only new area suggested for the DOD was the development of large solid-fueled rocket motors. The military wanted this capability, Seamans suggested, "for quick reaction, for strategic purposes." The DOD thought that "the solid motor was going to be the wave of the future, easier to store, easier to operate, no need to top off the oxygen as with an Atlas missile." The

Department of Defense wanted to develop a 260-inch diameter solid motor, and NASA had limited interest in such an undertaking. There was agreement that both liquid-fueled boosters, NASA's preference, and solid propulsion boosters would be pursued in parallel, with the choice of which would be used for the lunar landing mission deferred to a future time. The group also agreed that work on a nuclear upper stage for a large rocket should be pursued, but that such a stage was unlikely to be available in time for the initial lunar landing attempts. The major role for nuclear propulsion would be later missions to the Moon and then future human missions to the planets.

On late Saturday afternoon, the meeting then turned to how to proceed in preparing a report containing the recommendations that had been agreed to. The Department of Defense had prepared a draft report, written primarily by John Rubel. NASA had no similar draft, just a listing of the projects it was proposing for acceleration. Robert McNamara suggested using the DOD draft as the starting point and then folding in the NASA recommendations; James Webb agreed to this approach. The meeting then adjourned, with Seamans, Rubel, and Shapley designated as a drafting group to prepare by Monday morning a report for McNamara's and Webb's signatures.

Webb, in fact, had agreed to use the DOD draft without first reading it. When Seamans later on Saturday did read the report, which made the same general points as had the April 22 McNamara presentation to the vice president, but in more detail, he was "very troubled." What bothered Seamans was the approach the draft report took to competing with the Soviet Union in space. The draft claimed that "the government had allowed industry to proliferate to too great an extent," with the result being a diffuse and ineffective effort. It suggested "winnowing" the existing aerospace firms so that only "two or three or four stalwarts" would be allowed to compete for DOD and NASA contracts. The argument was that "free enterprise had gone too far and the government had to take a stronger role." In addition, "there was a lot of philosophical stuff...about the excellent being the enemy of the good." Seamans, a Republican, was "appalled" by the suggestion that the government should "control the industrial complex in a manner that certainly would have represented a very major change in policy." He called Webb on Sunday morning, saying, "we've got a terrible problem with this [DOD] report. I think it would be much better to start all over again." Webb told Seamans that this was not possible, since Webb had already agreed with McNamara to use the DOD report as the starting point; he said "it's up to you to work with John Rubel and revise it until you consider it to be satisfactory." Webb was on Sunday immersed in making arrangements for what would turn out to be a triumphant visit to Washington the next day by astronaut Alan Shepard and his Mercury astronaut colleagues.

Seamans, Shapley, and Rubel and some of their staff worked into Saturday evening and all day Sunday on the report, adding the NASA material to the DOD draft, with Shapley providing budget figures. Seamans told Webb on Sunday afternoon that he still was not satisfied with the report's language, and Webb agreed that he would come to the Pentagon after having dinner with Alan Shepard's family. Webb appeared sometime after 9:30 in the evening; between then and 1:00 or 2:00 on Monday morning, in what Seamans describes as "one of the great experiences of my life," Webb went through the report with Rubel page by page, negotiating changes in language or deletions "clearly to the benefit of the report." Finally, Webb, Seamans, and Rubel agreed that the report was ready for signature and forwarding to the vice president. Then Webb waited until the secretaries who had been typing the revisions finished their work; he insisted on driving the one secretary who did not have a car to her home during a pouring rainstorm, even waiting at the woman's house for a few minutes until the rain let up. Seamans describes Webb's actions as "a remarkable display of a Southerner... being gallant."

Seamans was back at the Pentagon by 7:30 a.m. to make sure the report was in shape for Webb and McNamara's approval. The two men signed a cover letter, and the report was delivered to Vice President Johnson that morning. The cover letter said: "Attached to this letter is a report entitled 'Recommendations for Our National Space Program: Changes, Policies, Goals,' dated May 8, 1961. This document represents our joint thinking. We recommend that, if you concur with its contents and recommendations, it be transmitted to the President for his information and as a basis for early adoption [and] implementation of the revised and expanded objectives which it contains."[7]

## Alan Shepard Visits Washington

On the morning of May 8, Alan Shepard and the six other Mercury astronauts were flown from Grand Bahama Island, where Shepard had been brought after his recovery from his suborbital mission, to Andrews Air Force Base outside Washington and then by helicopter to the White House lawn. They were met in the Rose Garden by a gathering that included President Kennedy and his wife Jacqueline, members of Congress, NASA leaders, and others. Awarding the NASA Distinguished Service Medal to Shepard, the president said: "how proud we are of him, what satisfaction we take in his accomplishment, what a service he has rendered to his country." He noted that "this flight was made out in the open with all the possibilities of failure, which would have been damaging to our country's prestige. Because great risks were taken in that regard, it seems to me that we have some right to claim that this open society of ours which risked much, gained much."[8]

After the award ceremony, the seven astronauts and others in the gathering joined President Kennedy in the Oval Office; the group totaled 20 to 25 people, including Vice President Johnson, the chairs of the Senate and House space committees, and several managers from NASA. The astronauts sat on couches on either side of the president, who "gushed with questions." He and Shepard discussed how Shepard's flight had demonstrated the ability of a human not only to survive, but also to carry out various functions in space; Kennedy seemed well aware of the reservations of his science advisers on this point. Alan Shepard recalls that "everybody certainly was running over with confidence at that time because the flight had gone so well and we had proved our point...that a man can operate effectively in space." Robert Gilruth, the director of NASA's Space Task Group that was managing Project Mercury, was present. He remembers Kennedy saying, "Look, I want to be first." Gilruth replied: "Well, you've got to pick a job that's so difficult, that it's new, that they'll [the Soviets] have to start from scratch. They can't just take their old rocket and put another gimmick on it and do something we can't do." Gilruth added, "it's got to be something that requires a great big rocket, like going to the moon. Going to the moon will take a new rocket...and if you want to do that, I think our country could probably win because we'd both have to start from scratch." Kennedy's reply was "I want to go to the Moon." Gilruth, himself only five years older than Kennedy, added that while Kennedy seemed to accept Gilruth's view, "he was a young man; he didn't have all the wisdom he would have had. If he'd been older, he probably would never have done it."

After leaving the White House, Shepard was taken by President Kennedy to a meeting of the National Association of Broadcasters; this was not on the planned schedule for the day, but Shepard's surprise visit provoked a tumultuous welcome. After his stopover at the broadcasters' meeting, Shepard and the other astronauts, accompanied by the vice president, paraded up Pennsylvania Avenue to the Capitol as thousands, assembling with little advance notice, cheered. Shepard suggests that "these two things—the successful demonstration of man's capability and the public support of a program which immediately became to them a very thrilling, exciting program—affected him [President Kennedy] in his decision-making process." After a "throng-packed, pulsing" meeting with members of Congress, the group went to the State Department for a luncheon hosted by Vice President Johnson; then Shepard held a press conference.[9]

As he left the luncheon to go first to the White House and then to the airport to catch the plane that would take him to Southeast Asia for two weeks, Lyndon Johnson carried a large manila envelope. In it was the Webb-McNamara report recommending sending Americans to the Moon.

### "Part of the Battle along the Fluid Front of the Cold War"

The thirty-page report, classified "Secret," was titled "Recommendations for Our National Space Program: Changes, Policies, Goals."[10] It called for an additional $686 million for the space program above the increases that President Kennedy had already approved in March; all but $137 million of that amount was for NASA. In particular, "to achieve the goal of landing a man on the moon and returning him to earth in the latter part of the current decade requires immediate initiation of an accelerated program of spacecraft development"; the report called for adding $210.5 million dollars for developing the Apollo spacecraft. At the time in mid-1960 when it was first identified publicly as the project to follow Project Mercury , the objective of

Project Apollo had been to support a three-person crew either in Earth orbit or on a circumlunar flight; now the Project Apollo was to carry Americans to a lunar landing. An additional $112.5 million was requested to allow NASA to accelerate development of the large F-1 liquid-fueled rocket engine and related facilities; $62 million was requested for DOD to develop a large solid propellant rocket motor in parallel to F-1 development. Another $15 million was allocated to DOD for a back up to the Centaur upper rocket stage that NASA was developing. Other increases included an additional $50 million to NASA for communication satellites; $75 million for meteorological satellites, $22 million of that amount for NASA and $53 million for the Weather Bureau; and $30 million for nuclear rocket development, $23 million for NASA and $7 million for the Atomic Energy Commission.

The specifics of what programs would receive additional funding was NASA's primary input into the report; the second section was based on John Rubel's draft material. That section argued that "projects in space may be undertaken for any one of four principal reasons." These included "gaining scientific knowledge," "commercial or chiefly civilian value," "potential military value," and "national prestige." The report noted that the United States was not trailing the Soviet Union in the first three categories, but that "the Soviets lead in space spectaculars which bestow great prestige." The central argument of the report was:

> All large scale space projects require the mobilization of resources on a national scale. They require the development and successful application of the most advanced technologies. They call for skillful management, centralized control and unflagging pursuit of long-range goals. Dramatic achievements in space, therefore, symbolize the technological power and organizing capacity of a nation.
>
> It is for reasons such as these that major achievements in space contribute to national prestige. Major successes, such as orbiting a man as the Soviets have just done, lend national prestige even though the scientific, commercial or military value of the undertaking may by ordinary standards be marginal or economically unjustified.
>
> *The nation needs to make a positive decision to pursue space projects aimed at enhancing national prestige.* Our attainments are a major element in the international competition between the Soviet system and our own. The non-military, non-commercial, non-scientific but "civilian" projects such as lunar and planetary exploration are, in this sense, part of the battle along the fluid front of the Cold War.

In order to undertake such projects, suggested the report, "what was needed were management mechanisms capable of centralized direction and control." It was "particularly vital" that the United States avoid the "error of spreading ourselves too thin." The report analyzed the results of the rapid build-up of defense capabilities in the 1950s, suggesting that "we have over-encouraged the development of entrepreneurs and the proliferation of new enterprises." While the report did not suggest that the United States should "apply Soviet type restrictions and controls," it said that "our American system can be and must be better utilized in the future." It added that "we must stress performance, not embellishment. We must insist from the top down, that, as the Russians say, 'the better is the enemy of the good.'"

The final section of the report spelled out the specific new space goals that were being recommended. They included the following:

- "Manned Lunar Exploration": Webb and McNamara recommended "that our National Space Plan include the objective of manned lunar exploration before the end of this decade ... The orbiting of machines is not the same as the orbiting or landing of a man. It is man, not merely machines, that captures the imagination of the world." The report noted that there was no information about Soviet plans for a similar program, but suggested "even if the Soviets get there first, as they may, and as some think they will, it is better for us to get there second than not at all."
- "Worldwide Operational Satellite Communication Capability": Webb and McNamara noted that while "advances in technology will make it possible to set up an operational satellite-based telecommunications capability within a few years," it was "too early to be sure what kind of capability we should create." Even so, they were "confident that an operational satellite capability can have far reaching applications and implications for the U.S."
- "Worldwide Operational Satellite Weather Prediction System": Such a system, Webb and McNamara suggested, "would be of great value to people in every country, to public and private interests in the U.S., and to our military forces."
- "Scientific Investigation": Webb and McNamara suggested that it was "essential that the national space sciences program be broad and comprehensive both in content and in participation by the scientific community of the world."
- "Large Scale Boosters for Potential Military Use": Webb and McNamara noted that while "the military potential and implications" of space technology were "largely unknown ... without the capacity to place large payloads reliably into orbit, our nation will not be able to exploit whatever military potential unfolds in space."

The Webb-McNamara report was necessarily vague with respect to whether the Soviet Union was already embarked on a lunar landing program. It noted that while the United States was "uncertain of Soviet intentions, plans, or status," the Soviet Union had announced a lunar landing as a "major objective of their program" and that the Soviet Union "may have begun to plan for such an effort years ago" and "may have undertaken important first steps which we have not begun." The memorandum suggested that Soviet successes in space were the result of "long-range planning" and that the slow pace and disappointments in the U.S. space effort "are symptoms of the lack of adequate national planning and guidance for the long pull." It concluded that "even if the Soviets get there [to the Moon] first... it is better for us to get there second than not at all... If we fail to accept this challenge it may be interpreted as a lack of national vigor and capacity to respond." These words were certain to resonate with President Kennedy.

## "A Great New American Enterprise"

In his May 8 cover letter transmitting the Webb-McNamara report to the president, Vice President Johnson noted that the document revealed "an agreement between the two major agencies involved in the space picture and points up clearly what they consider should be done and how much funding is needed at this time. I am much impressed with the thoroughness and the sense of urgency reflected in this document."[11]

### *Kennedy Accepts Recommendations*

The recommendations contained in the Webb-McNamara report did not stay secret for long. In a story dated May 9, *The New York Times*, based on a leak from Senator Robert Kerr, headlined a page one story: "600 Million More Planned to Spur Space Programs." The story reported in some detail the specific recommendations of the report, but did not mention the proposal to set a lunar landing as a national goal.[12]

By the time President Kennedy met on the morning of May 10 with his advisers, including Sorensen, Wiesner, and Bundy; budget officials Bell and Staats; Webb and Dryden from NASA; and Welsh from the Space Council, to review the Webb-McNamara report, his decision to accept the report's recommendations was almost foreordained. McGeorge Bundy, who was somewhat skeptical of the validity of the arguments in support of setting the lunar landing goal, suggests that Kennedy "had pretty much made up his mind to go" and was not particularly interested in hearing arguments to the contrary.[13] It was thus at this meeting that Kennedy finalized his policy decision to go to the Moon. Kennedy did ask the BOB to carry out its normal assessment of the financial and policy implications of his decision before committing to the specific programs and budget recommendations contained in the report.

In parallel with Lyndon Johnson's review of the space program, others within the Kennedy administration had been reviewing issues related to the defense budget, military assistance programs, foreign aid, civil defense, and overseas information programs. Sorensen says that "since space, like these other items, obviously did have some bearing upon our status in the world, it was decided to combine the results of all those studies with the President's recommendations [on space] in the special message to Congress," which was billed as a second State of the Union Address on "Urgent National Needs." Sorensen checked with the Library of Congress regarding whether past presidents had addressed a joint session of Congress at times other than the annual State of the Union speech, and was told that while President Eisenhower had done so only once in his eight years in office, President Harry Truman had made eight such speeches, and President Franklin D. Roosevelt five. There was this ample precedent for a second address to Congress less than four months after Kennedy had spoken on the State of the Union. The Kennedy address was originally scheduled for May 23, but then was postponed for two days, until May 25, 1961.[14]

There is some evidence that during the two weeks after Kennedy approved the Webb-McNamara memorandum his economic advisers evaluated the likely impact of the increased space spending Kennedy would propose. Their conclusion was that these expenditures were neither large enough nor properly designed to inject enough stimulus into the economy to by themselves mitigate the ongoing recession. Walter Heller, chairman of the Council of Economic Advisers, and Secretary of Labor Arthur Goldberg proposed that Kennedy approve a substantial public works program rather than (or in addition to) new space spending. Such a program, they thought, would provide the needed stimulus. Kennedy turned down this suggestion; it is reported that Heller viewed Kennedy's decision to spend money on the space program rather than on public works as one of Heller's worst defeats during the Kennedy administration.[15]

### Bureau of the Budget Review

After helping prepare the Webb-McNamara report over the May 6–7 weekend, Willis Shapley found himself in charge of carrying out a review of that report from the BOB perspective. The draft of the internal BOB review was completed on May 18, and BOB director Bell sent it to Secretary of Defense McNamara, NASA administrator Webb, and Atomic Energy Commission chairman Glenn Seaborg with a request for immediate comments. Welsh of the Space Council also received a copy in the absence of Vice President Johnson, who was still touring in Asia. The final version of the review was dated May 20.

The review was the kind of thorough "due diligence" assessment that was the BOB's responsibility, pointing out the implications of the decisions being proposed and examining potential obstacles to their successful implementation. With respect to the magnitude of resources that would be needed, the review pointed out that what was being proposed was an increase of over $2 billion per year—perhaps even $3 billion per year—over previously planned budgets. It recognized that the budget estimates being used were subject to upward revision, suggesting that "the funds required by the manned lunar landing objective may have been underestimated by as much as $200 million in 1962 and perhaps $1 billion per year in future years." What would be needed was "a commitment to a long term-effort and to provide the resources it requires. Starts and stops, changes in goals, or failure to provide the required level of budgetary support would impair the success of the program." It noted that "the commitment actually extends beyond the achievement of the manned lunar landing... By 1967 we will have geared

James Webb with Willis H. Shapley (on left), the Bureau of the Budget staff person who played a key role in space decisions during the Kennedy administration (NASA photo).

the nation up to an annual space effort of almost $7 billion per year; it is unrealistic to assume that an effort of approximately this level would not continue for many years."

The draft BOB report pointed out the need to consider "the implications of likely and possible outcomes other than complete success" of the lunar landing program; interestingly, this discussion was missing from the final version. The draft noted that "the magnitude of the effort required for the manned lunar landing program is so great and the proposed schedule so tight that it will place a major strain on our capabilities in the space and related fields." It also noted that "increases in the space programs of the magnitude proposed cannot help having the effect of diverting scientific and technical manpower from other areas of national need" and might cause "a major and continuing distortion in the utilization of our scientific and technical resources which will have detrimental effects in other areas of serious national concern." The commitment to space would "also reduce our flexibility as a nation to undertake large scale, all-out efforts in other areas not now foreseen which may suddenly appear to be of comparable national importance."

The review recognized that using 1967 as the internal planning date for the first lunar landing would "necessitate a rapid build-up," but recommended that this planning date be maintained. With respect to making the target date publicly known, the BOB recommendation was to "make a major effort to avoid any public commitment to [a] specific target date."[16]

Whether President Kennedy or his closest advisers read the BOB analysis cannot be known with certainty. If they did read it, they would have had the benefit of a comprehensive and thoughtful analysis of the implications of the decisions that Kennedy was about to announce.

### *How Much Would Landing on the Moon Cost?*

The BOB review did not attempt to assign a cost to the overall Apollo project through the planned first landing. In preparing NASA administrator Webb for possible questions at the press briefing planned to follow President Kennedy's May 25 speech, NASA's public affairs chief Bill Lloyd suggested that the answer to the questions "What is your best estimate? How many billions of dollars would the lunar landing program cost?" should be "our best guess is in the neighborhood of $20 billion."[17] The origins of this $20 billion figure apparently lie with James Webb. Robert Seamans reports that the NASA staff estimate for the additional cost of the lunar landing program above what had been previously planned was in the range of $10 to $12 billion; Hugh Dryden had used an $11.4 billion increment in his April 22 presentation to Vice President Johnson. According to Seamans, "Jim Webb put an 'administrator's discount' on our ability to predict costs precisely." Lambright suggests that Webb's administrative discount applied both to announcing a date for the first landing attempt and for a precise cost of the project. With respect to the landing date, Webb wanted "a margin of flexibility weighted against what the technical experts thought was possible, just in case something went wrong. He did not want the prestige of the nation (much less his own reputation) resting on an overly optimistic deadline." With respect to the projected costs, the $10 to $12 billion estimate "looked much too low to Webb. Because no one could anticipate all the contingencies, he enlarged the figure NASA sent Kennedy to $20 billion for the first lunar journey." There are stories, apparently apocryphal, that Webb doubled the Apollo cost estimate during a ten-minute car ride from NASA headquarters to Capitol Hill; Seamans's account suggest that there was substantially more thought given to the cost estimate than such stories would suggest.[18]

### *James Webb Has His Own Agenda*

As Lambright notes in his biography of James Webb, "while the decision to go to the moon was unfolding, a separate decision process—mainly in Webb's own mind—was unfolding. This was the personal agenda Webb had brought to NASA—"a mission to use science and technology . . . to strengthen the United States educationally and economically." Webb's objective was to maximize the benefits of an accelerated space program for Earth in terms of research, education, and regional economic development. Walter McDougall in his award-winning book... *the Heavens and the Earth* described the totality of Webb's vision as "Space Age America," a term that indeed Webb sometimes used.[19]

While the final review of the accelerated program was underway in the White House, Webb was consulting with his colleagues outside the government and those whose support he thought might be important to the public acceptance of the new effort. For example, on May 15, he wrote to Vannevar Bush at the Massachusetts Institute of Technology; Webb had known Bush since his time in Harry Truman's BOB. Bush had been President Roosevelt's top technical advisor during World War II and his report *Science, the Endless Frontier* had laid the foundation for postwar government support of science.[20] Webb told Bush that he regretted that the two "find ourselves on somewhat different sides of the complex question of manned space flight"; they earlier in May had had a confrontation at a Washington social function over the value of humans in space. Webb noted that "no one could have ridden down Pennsylvania Avenue with Commander Shepard without feeling the deep desire of those lining the Avenue for something to be proud of and a hero. At the moment I believe this feeling is somewhat expanded to include a desire for a real effort in the space field." Webb assured Bush that "in the programs that are now underway and which will shortly be put forward, I expect to do all that I can to build up the university research, teaching, and graduate and post-graduate quality and quantity of education... If we do not find ways to make the major program carry a burden in each of these fields, we simply are not going to meet the challenge of our times."[21]

The most sweeping version of Webb's vision can be found in a May 23 memorandum he prepared for the vice president as Johnson returned from his Asian trip. Webb told him that Houston Congressman Albert Thomas

> has made it very clear that he and [Houston construction magnate and Johnson campaign contributor] George Brown were extremely interested in having Rice University make a real contribution to the effort, particularly in view of the fact that some research funds were now being spent at Rice, that the resources of Rice had increased substantially, and that some 3,000 acres of land had been set aside by Rice for an important research installation. On investigation, I find that we are going to have to establish some place where we can do the technology related to the Apollo program, and this should be on the water where the vehicle can ultimately be barged to the launching site. Therefore we have looked carefully at the situation at Rice, and at the possible locations near the Houston Ship Canal or other accessible waterways in that general area. George Brown has been extremely helpful in doing this.

In essence, Webb was preempting the decision on where to relocate the Space Task Group as it took on the lunar landing assignment, even though the process through which that decision would be formalized extended for four more months. But Webb did not stop with Houston; he now broadened his horizon to the whole region. His vision of using centers of excellence in areas like Oklahoma, Missouri, and Texas to spur regional development had been developed during his years working for Robert Kerr in Oklahoma. In January 1956, Webb in a letter to his former boss, President Harry Truman, had laid out his concept of using an Oklahoma-based "Frontiers of Science Foundation" to stimulate science, technology, and industry in that state and beyond.[22] Now he told Johnson:

> No commitments have been made, but I believe it is going to be of great importance to develop the intellectual and other resources of the Southwest in connection with the new programs the Government is undertaking. Texas offers an unusual opportunity at this time due to the fact that [long-time Webb friend and chairman of the National Academy Space Science Board] Dr. Lloyd Berkner...is establishing a Graduate Research Center in Dallas with the backing of Eric Johnston, Cecil Green, and others in that area (estimated at about one hundred million dollars) and in view of the fact that Senator Kerr and those interested with him in the Arkansas, White, and Red River System have now pushed it to the point that it is opening up the whole area related to Arkansas, Oklahoma, and in many ways helping to provide a development potential for Mississippi. If it were possible to get a combination where out-in-front theoretical research done by Berkner and his group around Dallas in such a way to strengthen all the universities in the area, and if at the same time a strong engineering and technical center could be established near the water near Houston and perhaps in conjunction with Rice University, these two strong centers would provide a great impetus to the intellectual and industrial base of this whole region.

Webb was still not done. He related his vision for Southwest regional development to the nation as a whole. Developing a strong technical competence in the Southwest

> would permit us to think of the country as having a complex in California running from San Francisco down through the new University of California installation in San Diego, another center around Chicago with the University of Chicago as a pivot, a strong Northeastern arrangement with Harvard, M.I.T., and like institutions participating. Some work in the southeast perhaps revolving around the research triangle in North Carolina (in which Charlie Jonas as the ranking minority on Thomas's Appropriations Subcommittee would have an interest), and with the Southwestern complex rounding out the situation.[23]

This "grand mix of noble vision and pork-barrel politics"[24] went well beyond anything that President Kennedy had in mind as he approved an accelerated space program, primarily as a foreign policy response to Soviet space successes and their political impacts. But the space program buildup over the next few years that resulted from Kennedy's decision allowed Webb the room needed to put his agenda into practice, and Webb, a New Deal Democrat, had little hesitation in using his position at NASA to implement his vision of an improved America.

### "Before This Decade is Out"

Preparation of Kennedy's message to Congress began in mid-May; the section on accelerating the space program was first drafted by the BOB's Willis Shapley; NASA, DOD, AEC, and the Space Council provided their comments. The overall theme of the speech was the need for U.S. citizens to make sacrifices to meet the challenges facing the country and to insure the U.S. position as the leading power in the world by addressing "urgent national needs," the title given to the address. After Kennedy received messages from Moscow suggesting that he was likely to encounter a belligerent Nikita Khrushchev in their meeting in Vienna, the speech was "quickly redrafted" and "the language toughened to signal his [Kennedy's] resolve to Khrushchev."[25]

On May 25, in a nationally televised address,[26] President John F. Kennedy told the assembled lawmakers that "these are extraordinary times. And we face an extraordinary challenge. Our strength as well as our convictions have imposed on this nation the role of leader in freedom's cause." He noted that "there is no simple policy that meets this challenge." But "there is much we can do—and must do. The proposals I bring before you are numerous and varied. They arise from the host of special opportunities and dangers which have become increasingly clear in recent months." Then the president turned to his specific proposals, which included measures to continue economic recovery from the recession the new administration had inherited; measures to help developing nations make economic and social progress; cooperation in terms of military alliances and military assistance to U.S. allies; an enhanced overseas information program; an additional build-up of U.S. military power beyond what Kennedy had requested just two months earlier; a strengthened civil defense program; and an increased emphasis on disarmament negotiations. All of these initiatives, Kennedy said, would involve substantial costs. Echoing his inaugural address call to "ask not what your country can do for you, ask what you can do for your country," Kennedy argued that "our greatest asset in this struggle is the American people—their willingness to pay the price for these programs—to understand and accept a long struggle." He warned that "this Nation is engaged in a long and exacting test of the future of freedom."

After listing all of the other areas he was recommending for new action, the president turned last to space. As he did, he deviated from his prepared text to emphasize the sacrifices involved and the commitment he was requesting; Sorensen says that this departure from the text was "the only time I can recall his doing so in a formal address."[27] Kennedy's words as they deviated from the prepared text are indicated in bold italics below. Kennedy also skipped a few portions of the prepared text or deleted passages by hand. These deletions from the prepared text are in brackets:

> Finally, if we are to win the battle for men's minds, the dramatic achievements in space which occurred in recent weeks should have made clear to us all ***as did Sputnik in 1957***, the impact of this adventure ***on the minds of men everywhere who are attempting to make a determination of which road they***

***should take.*** Since early in my term, our efforts in space have been under review. With the advice of the Vice President, **who is *Chairman of the National Space Council,*** we have examined where we are strong and where we are not, where we may succeed and where we may not. Now it is time to take longer strides—time for a great new American enterprise—time for this nation to take a clearly leading role in space achievement, ***which in many ways may hold the key to our future on earth.***

I believe we possess all the resources and all the talents necessary. But the facts of the matter are that we have never made the national decisions or marshaled the national resources required for such leadership. We have never specified long-range goals on an urgent time schedule, or managed our resources and our time so as to insure their fulfillment.

Recognizing the head start obtained by the Soviets with their large rocket engines, which gives them many months of lead-time, and recognizing the likelihood that they will exploit this lead for some time to come in still more impressive successes, we nevertheless are required to make new efforts ***on our own.*** For while we cannot guarantee that we shall one day be first, we can guarantee that any failure to make this effort will ***make*** [find] us last.

We take an additional risk by making it in full view of the world—but as shown by the feat of astronaut Shepard, this very risk enhances our stature when we are successful. But this is not merely a race. Space is open to us now; and our eagerness to share its meaning is not governed by the efforts of others. We go into space because whatever mankind must undertake, <u>free</u> men must fully share.

I therefore ask the Congress, above and beyond the increases I have earlier requested for space activities, to provide the funds which are needed to meet the following national goals:

I believe that this nation should commit itself to achieving the goal, before this decade is out, of landing a man on the moon and returning him safely to earth. No single space project in this period will be more [exciting or] impressive ***to mankind*** or more important for the long-range exploration of space; and none will be so difficult or expensive to accomplish. [Including the necessary supporting research, this objective will require an additional $531 million this year and still higher sums in the future.] We propose to accelerate the development of the appropriate lunar spacecraft. We propose to develop alternate liquid and solid fueled boosters much larger than any now being developed, until certain which is superior. We propose additional funds for other engine development and for unmanned explorations—explorations which are particularly important for one purpose which this nation will never overlook: the survival of the man who first makes this daring flight. But in a very real sense, it will not be one man going to the moon—it will be an entire nation. For all of us must work to put him there.

* * *

Let it be clear—***and this is a judgment which the Members of Congress must finally make. Let it be clear*** that I am asking the Congress and the country to accept a firm commitment to a new course of action—a course which will last for many years and carry very heavy costs, ***531 million dollars in fiscal 1962***—an estimated $7–9 billion additional over the next five years. If we ***are***

[were] to go only halfway, or reduce our sights in the face of difficulty, in my judgment it would be better not to go at all.

*Now this is a choice which this country must make, and I am confident that under the leadership of the Space Committees of the Congress, and the Appropriating Committees, that you will consider the matter carefully.*

*It is a most important decision that we make as a nation. But all of you have lived through the last four years and have seen the significance of space and the adventures in space, and no one can predict with certainty what the ultimate meaning will be of the mastery of space.*

*I believe we should go to the moon. But I think every citizen of this country as well as the Members of Congress should consider the matter carefully in making their judgment, to which we have given attention over many weeks and months, because it is a heavy burden, and there is no sense in agreeing or desiring that the United States take an affirmative position in outer space, unless we are prepared to do the work and bear the burdens to make it successful. If we are not, we should decide today and this year.*

[Let me stress that more money alone will not do the job.] This decision demands a major national commitment of scientific and technical manpower, material and facilities, and the possibility of their diversion from other important activities where they are already thinly spread. It means a degree of dedication, organization and discipline which have not always characterized our research and development efforts. It means we cannot afford work stoppages, inflated costs of material or talent, wasteful interagency rivalries, or a high turnover of key personnel.

New objectives and new money cannot solve these problems. They could, in fact, aggravate them further—unless every scientist, every engineer, every serviceman, every technician, contractor, and civil servant involved gives his pledge that this nation will move forward, with the full speed of freedom, in the exciting adventure of space.[28]

Sitting near the rostrum as Kennedy delivered his speech, Sorensen "thought the President looked strained in his effort to win them over." Kennedy apparently had sensed that some in the Congressional audience were "skeptical, if not hostile, and that his request was being received with stunned doubt and disbelief"; this likely led to his decision to deviate from his prepared text to try to convince the congressmen of the need for what he was proposing, and to skip a few passages toward the end of his address. Returning to the White House, Kennedy remarked to Sorensen that "the routine applause with which the Congress greeted" his proposal for a national commitment to go to the Moon had struck him as "less than enthusiastic."[29] Indeed, during the speech the Senate and House Republican leaders "took notes, inspected their fingernails, brushed their hair back and joined in the almost complete Republican silence."[30]

*The New York Times* the next morning had a banner headline saying "Kennedy Asks $1.8 Billion This Year to Accelerate Space Exploration, Add Foreign Aid, Bolster Defense" and also "Moon Trip Urged." The newspaper reported that "Members of Congress embraced with some warmth today the objectives outlined in President Kennedy's speech, but shied at providing all the funds to meet them," and that "some fears were expressed by Democratic liberals, however, that the huge spending in the effort to reach the moon...might divert funds from programs such as aid to the aged." The *Times* editorialized that "there is an element of 'race' involved cannot be denied, but that is only secondary to the main purpose" and that "it is in the spirit of free men, and the cherished traditions of our people, to accept the challenge and meet it with all our resources, material, intellectual and spiritual." The editorial thought that "the majority of our people will agree" with the lunar landing goal.[31]

Kennedy need not have worried about congressional, and particularly Senate, support of the accelerated space effort he had proposed. Lyndon Johnson's earlier consultations with congressional leaders had helped lay the foundation for strong bipartisan support of the initiative. The Senate on June 28 took up a House authorization bill passed the day before the May 25 speech, and amended it to include the full $1.784 billion for NASA that White House had requested for Fiscal Year 1962; the bill passed without even the formality of a roll-call vote. The House of Representatives agreed to this increase in a conference committee, and the authorization bill passed the House on July 20 by a 354 to 59 vote. The appropriations bill containing NASA funds had a similarly easy ride through Congress; it passed both houses on August 7 and contained a $1,671,750,000 NASA appropriation for Fiscal 1962, only $113 million less than Kennedy had requested.[32] This amount represented an 89 percent increase over the previous year's NASA budget, the last one enacted during the Eisenhower administration.

Kennedy's speech called for sending Americans to the Moon "before the decade is out." There is some uncertainty on how the decision was made to use this phrase, rather than the 1967 date for the first landing attempt that was being used in NASA's internal planning. The May 8 Webb-McNamara report had suggested an "end of the decade" target for the first lunar landing. The BOB review of the Webb-McNamara memorandum had suggested that there should be "a major effort to avoid any public commitment to a specific target date." Robert Seamans reports that the first draft of Kennedy's speech "called for a lunar landing by 1967" and that NASA was "aghast" at specifying a particular year. He says that James Webb "called Ted Sorensen and convinced him, and later the President, that the stated goal should be *by the end of the decade*." Sorensen, by contrast, says that Kennedy's "self-imposed deadline, 'this decade,' was chosen and inserted by JFK himself to exert pressure on NASA. The phrase deliberately left some flexibility—it could mean within the decade of the sixties, or within the next ten years."[33] A "within the next ten years" interpretation was never acknowledged; "before this decade is out" was universally seen as setting a target of the end of 1969 (or, for some, before the end of 1970) for the initial lunar landing. Indeed, as the Apollo program gained momentum, John Kennedy pushed for a landing as soon as possible, in 1967 or even late 1966.

## The Race Begins

In the six months between December 1960 and May 1961, the status of the U.S. civilian space program was elevated from a scientifically oriented effort with an uncertain future for human space flight to a key instrument of national strategy. This shift was the end result of a process in which many factors were involved. The change in administrations was clearly vital. In addition to putting a new president and his advisers into the White House with a clearly different set of values and objectives than their predecessors, the new administration meant new leadership for NASA. NASA planners convinced James Webb, who probably needed little convincing, that human space flight was key to the agency's future, and Webb became an effective advocate of NASA's interests. The support of the Space Science Board helped allay some of the scientific criticism that the human space flight program had little scientific value. The success of Alan Shepard's flight demonstrated both human capability to survive and function in space and the great public enthusiasm for human space flight. The ability of NASA to withstand an Air Force and industry challenge to its role as the nation's primary space agency strengthened NASA's claim that it could undertake new, ambitious missions. Lyndon Johnson's personal conviction about the strategic importance of space, coupled with his assignment as head of the Space Council, placed a forceful advocate of a larger space effort at the side of the president. The consistent call from the Congress, particularly from the House Committee on Science and Astronautics, loosened one constraint on the president's freedom to choose a bold course of action. The flight of Yuri Gagarin and the world's reaction to it provided a strong impetus to make space decisions quickly; the Bay of Pigs added to that pressure.

Walter McDougall suggests that "how this change occurred in so short a time is not a mystery," but rather an "overdetermined event."

> New men arrived and brought with them those ideas of the "seed time" of the 1950s. Among those ideas were the notions that the Third World was the main theater of the Cold War and that in that contest prestige was as important as power. Their ideas validated a far greater role for government in planning and executing social change. The new men also cared more for imagery and felt increasing pressure to display their control over affairs in the wake of early setbacks in foreign policy. Finally, each of the major figures in space policy—Kennedy, Johnson, Webb, Dryden, McNamara, Welsh, Kerr, and others—saw ways in which an accelerated space program could help them solve problems in their own shop or serve their own interests...They were technocratic, applying command technology to political problems.
>
> * * *
>
> We will probably never know precisely what was in Kennedy's mind when he decided that Americans should go to the moon. What may have tipped the balance for him and for many was the spinal chill attending the thought of

> leaving the moon to the Soviets. Perhaps Apollo could not be justified, but, by God, we could not *not* do it.[34]

All of these factors converged on the White House and particularly on John F. Kennedy. In the weeks between the Gagarin flight and his May 25 speech, Kennedy had "fired off" to his advisers "a constant stream of written questions...on costs, risks, manpower, alternatives, and administrative responsibility. He had heard from hundreds of individuals in the process of making his decision—scientists, engineers, experts of all kind—and became convinced that the United States must not remain second in this race." From "a tentative premise" in the aftermath of the Gagarin flight there emerged in Kennedy's thinking a "firm conclusion" about the importance of space achievement, but "only after it had been carefully studied, the estimated costs calculated, the risks weighed, and the responsibilities allocated."[35] Robert Kennedy commented that his brother thought that winning the space race was "very important. As he used to say, it compared to the explorers in our country, Lewis and Clark... He thought we needed to do it for our position throughout the world, that our efforts should be for excellence and that we should do whatever was necessary."[36] Willis Shapley, the longtime staff person from the BOB who was directly involved in the decision process, suggests that "after having been through quite a few major decisions, there was never a major decision like this made with the same degree of eyes-open, knowing-what-you're getting-in-for" character.[37] President Kennedy, at first uncertain but finally convinced that the United States should accept the Soviet challenge in space, decided that "whatever mankind must undertake, free men must fully share."

# Chapter 8

# First Steps on the Way to the Moon

In anticipation of President Kennedy's decision to approve a lunar landing project as a top priority national undertaking, NASA on May 2, 1961, had begun in earnest to examine just what would be required to carry out the president's mandate. That examination revealed the immense dimensions of the task. New facilities would be needed, new approaches to space flight would be required, and new hardware would have to be developed. In his book *Digital Apollo,* David Mindell observes: "For the first couple of years the Apollo project was largely undefined, the money flowed freely, and the nerve-racking deadlines seemed far in the future."[1] This was certainly not the perception of those directly involved with the mobilization of human and financial resources required to carry out the lunar landing project. The second half of 1961 and most of 1962 were marked by a rapid series of decisions. To many of those in the White House and NASA concerned with attempting to meet the late 1967 target date that NASA had set for the first attempt at a lunar landing, there was a sense of urgency in getting a fast start on the needed buildup of people, facilities, and hardware; to them, "nerve-racking deadlines" were a daily reality.

Between the final Eisenhower budget request that went to the Congress in January 1961 and the Fiscal Year 1964 request that President Kennedy sent to Congress in January 1963, the NASA budget grew from $1.1 billion to $5.7 billion, an increase of over 400 percent in just two years. NASA in 1961 and 1962 chose the locations for the facilities that would be required for the lunar landing mission, selected and contracted for the launch vehicles and spacecraft to carry out the mission, and decided on the technical approach to landing on the Moon, this last decision resulting in an intense conflict with the White House science advisers. NASA reorganized itself for the task of managing Project Apollo while carrying out Project Mercury and getting started on an intermediate human space flight effort, Project Gemini. The NASA workforce increased from the 10,200 civil servants as John F. Kennedy came into the White House to 23,700 at the end of 1962; the total would ultimately increase to 34,500 by the end of 1965. The associated contractor workforce grew at an even more rapid rate. At the end of 1960, it totaled 36,500; two years later, it was 115,000, and by the end of 1965, it was 376,700.[2] This was truly an unprecedented warlike, albeit peaceful, mobilization of national resources.

President Kennedy and his White House associates viewed this rapid buildup with mixed emotions. On one hand, Kennedy made it very clear that getting to the Moon before the Soviet Union was one of his top policy priorities, and by the end of 1962 appeared willing to provide even more resources to NASA if doing so would increase the chances of achieving that objective. While Kennedy focused his interest on the lunar landing program, NASA argued that across-the-board preeminence—a clearly leading position in all areas of space activity—was the fundamental goal, and the lunar landing only the most visible element of achieving it.

On the other hand, the budget projections that had accompanied Kennedy's May 1961 decision to go to the Moon and to accelerate the other elements of the NASA program were admittedly highly uncertain, and the Bureau of the Budget (BOB) had warned Kennedy that the decisions he was making would lead to a very expensive space effort in the coming years. In his desire to get the country on the path to space leadership, Kennedy seemingly did not pay very much attention to that warning. Theodore Sorensen comments that although JFK was a "fiscal tightwad," at the time of the lunar landing decision " I'm sure he had no idea what the whole effort was going to cost."[3] Kennedy soon became concerned about the space program's rapidly growing costs, and this concern intensified during 1962 as the full scope of the lunar landing effort became clearer. Kennedy pressed his staff to make sure that the related costs were fully justified. Even so, at the end of 1962 Kennedy's determination to win the race to the Moon remained firm; his desire to be first in space justified in his mind the high costs of achieving that position.

## White House Oversight of NASA

One impact of his decision to send Americans to the Moon "before this decade is out" was to make the progress of the space program of direct personal interest to President Kennedy. As he planned for his presidency, Kennedy could not have anticipated making such a decision within four months of taking office, and thus he had been willing to delegate to Vice President Lyndon B. Johnson the administration's lead role in space policy, and to symbolize that role by making the vice president the chairman of a revitalized National Aeronautics and Space Council. When asked whether President Kennedy would have assigned such an apparently central role with respect to space to the vice president if he had realized how significant space efforts would become in the months immediately after his inauguration, Theodore Sorensen in 2009 responded, with a note of skepticism, "that's a very interesting and thoughtful question, to which I don't know the answer . . . I think, being very frank about it, that at the time he gave Johnson the chairmanship of the Space Council it was not a position that was among the highest priorities he faced, or among those that most interested him."[4]

### *The Diminishing Role of LBJ and the Space Council*

On April 20, 1961, President Kennedy asked Johnson, as chairman of the Space Council, to carry out a survey of the status of the space program. Johnson did indeed take charge of the review, but on a highly personal basis. There were no formal meetings of the National Aeronautics and Space Council before Johnson forwarded the May 8 Webb-McNamara recommendations to Kennedy with his endorsement, and in the April–May 1961 time period, there was only one Space Council staff person, executive secretary Edward Welsh.

Kennedy made the final decision to approve the recommendations of the Webb-McNamara report and to announce them before a joint session of Congress while the vice president, at Kennedy's request, was traveling in Asia. The absence of Johnson during this critical period was symbolic of his declining role after May 1961 as the first among equals with respect to advising the president on NASA's programs. It is very likely that Johnson's urging Kennedy to approve a major acceleration of the civilian space program was an important influence on JFK's decision to go to the Moon. However, once Kennedy had made the basic decision, he relied primarily on his White House policy, technical, national security, and budgetary staff and on the NASA leadership to provide him information on the implementation of that decision and to relay his concerns and decisions to the NASA workforce. Kennedy soon realized that what he was likely to get from Johnson and the Space Council staff was unquestioning advocacy for a strong space effort, not the kind of dispassionate analysis that he most welcomed. Science adviser Jerome Wiesner and his staff and David Bell and his BOB staff thought it was their role, whatever their personal views, to raise questions about the choices NASA was making with respect to various aspects of the space effort. The predictability of the positions that Lyndon Johnson and the Space Council staff would take on most space issues limited their influence on Kennedy's policy choices.

Thus—with one important exception, developing during 1961 the administration's approach to bring communications satellites into early practical use—the Space Council as a body was not central to any of the civilian space decisions of the Kennedy administration. James Webb characterized the council's role as different from "the popular image that the president had turned everything over to the vice president" with respect to space; that was "simply . . . not true." Rather, Kennedy "wanted to control the agenda of the council" and did not want "to abandon the normal budgetary process by having the Space Council make the space budget." On the other hand, Kennedy was "very happy" to have Johnson "to take the lead in talking about things and making speeches and participating actively in carrying out things that the president decided he wanted." Johnson may have wanted "different instructions" from Kennedy about his responsibilities for the space program, but in practice Webb "never saw him go beyond what Kennedy had indicated he wanted done."[5]

In his role as chairman of the Space Council, Vice President Lyndon Johnson "appeared to take his duties seriously, even if his responsibilities were only advisory and minimal." Johnson had originally hoped that the budget for the Space Council staff would be $1 million; "the bigger the budget, the bigger empire Johnson could build." But Congress, sensing "that Johnson's role in the space effort was not a significant one," cut the budget request in half; this left Johnson "bitter and hurt." Even so, by the end of 1962, the council staff had grown to twelve professional staff members, several consultants, and a large support staff; the total staff complement was twenty-eight. This made the council a sizeable element of the executive office of the president, comparable in size to the Office of Science and Technology, which had been created in 1962 to give an organizational foundation for Kennedy's science adviser Jerome Wiesner and which dealt with all other areas of science and technology. The Space Council budget during the Kennedy administration grew to just over $500,000 per year.[6]

The main role of the council during the Kennedy administration "was to keep a dialogue going between the people responsible for the military, peaceful and diplomatic uses of space...If there was a dispute between any of the Council members and between their agencies, Johnson was to umpire that dispute." Johnson and the council staff in practice seldom intervened in disputes between operating agencies, especially NASA and the Department of Defense (DOD). In addition to this mediating role, Johnson saw the Space Council as a vehicle for explaining to the American public the importance of the U.S. effort in space; he suggested to Welsh that the council should enlist "the cooperation of the various agencies to produce a series of comprehensive reports to inform the public upon the actualities of the space program." When Welsh responded, suggesting a series of "Vice President's Reports on Space," Johnson's reaction was "I sure like that. Get on it." There were fourteen formal meetings of the Space Council during the Kennedy administration, but most were devoted to discussion of already-decided activities, rather than to formulating recommendations for presidential decision or for settling disagreements. During 1962 and early 1963, the council staff devoted a great deal of time and effort to drafting a statement of national space policy, but this initiative was abandoned when both NASA and the Department of Defense opposed issuing such a statement.

In the fall of 1961 Johnson toured a number of space installations, "trying to seek out problems and to boost morale." Johnson had been advised by his press secretary George Reedy that such a tour "would attract tremendous attention" and "provide a natural basis for a series of reports which would be extremely helpful in informing the public and clarifying the program." Johnson frequently gave speeches prepared by the Space Council staff on space issues and allowed staff-prepared articles to appear under his name in various publications. At White House breakfasts before presidential

Chairman of the National Aeronautics and Space Council Lyndon Johnson (seated) listens to a space program briefing. Standing are (left) Space Council member and chairman of the Atomic Energy Commission Glenn Seaborg and (right) Space Council executive secretary Edward C. Welsh (NASA photo).

press conferences, Johnson came prepared with responses to potential press questions. The vice president was eager to associate himself with the public attention given to the Mercury astronauts; these occasions "were important to him because they were almost the only times he received national attention and he wanted to keep his name before the public."[7] As Johnson biographer Merle Miller suggests, "the Space Council chairmanship was...only briefly satisfying for Lyndon. Once the present program had been evaluated, new directives issued and the revised program set in motion, there was little for him to do."[8]

Kennedy's approach to governance placed heavy responsibility on the line officials in charge of the executive agencies of government; in the case of space, that meant that Kennedy delegated most decision-making authority for the civilian space program to NASA administrator James Webb, and backed Webb up when Webb's choices were challenged by JFK's White House advisers. As a veteran of Washington bureaucratic politics and as the head of an agency newly charged with an effort that was a top presidential priority, administrator Webb was insistent on having direct access to the president. Webb was scrupulous in keeping Johnson informed, but he made it clear that he worked for the president, not the vice president. This also left little room for the vice president and the Space Council staff to play a central role in most of the decisions on how best to move forward in sending Americans to the Moon.

### *The Role of White House Staff*

The burden of White House oversight of NASA and its plans for implementing the lunar landing program and the other activities that were part of the accelerated space effort thus fell on various members of the White House staff and those career bureaucrats supporting them.[9] Although most of those individuals have been mentioned previously, it may be useful to depict the structure of White House decision-making for space before discussing the specific actions taken during the June 1961 to December 1962 period.

The recommendation that President Kennedy approved in accelerating the U.S. space program suggested that the prestige associated with space achievements was "part of the battle along the fluid front of the Cold War." Kennedy defined the lunar landing program primarily as a national security effort, and that meant that his special assistant for national security affairs, McGeorge Bundy, played an increasingly important role in space policy discussions between 1961 and 1963. Bundy's deputy, Harvard economist Carl Kaysen, and National Security Council career staff member Charles E. Johnson played key roles in supporting Bundy on space issues; Kaysen also had a direct personal relationship with the president, particularly on arms control issues, and on occasion reported directly to Kennedy rather than through Bundy. On technical issues, Kennedy relied on his special assistant for science and technology, Jerome Wiesner, and various panels of the President's Science Advisory Committee (PSAC). From the start of 1962, Wiesner's principal staff assistant on space matters was Nicholas Golovin, a physicist who had left NASA at the end of 1961 on less than harmonious terms. Another of Wiesner's staffers, Eugene Skolnikoff, dealt with the international aspects of the space effort. In August 1961 Wiesner was designated the White House official (instead of Welsh, the Space Council executive secretary) "to review and consult with relevant agencies of the Federal government on organizational planning for the expanded space activities of the Federal government."[10] As planning for the accelerated space program moved forward, the president became increasingly concerned with its exponentially increasing costs. He leaned heavily on his director of BOB, David Bell, for careful assessments of the budgetary implications of the fast-paced space program. Bell's deputy, Elmer Staats, and especially career BOB senior staffer Willis Shapley were deeply involved in space matters. Shapley was central to framing policy and budget issues as he drafted various policy papers for presidential review and decision.

Kennedy's top adviser on most domestic policy matters, in addition to his duties as Kennedy's speechwriter, was special counsel Theodore C. Sorensen. Kennedy in April 1961 had asked Sorensen to organize the review of the space program that was carried out by Vice President Johnson. Sorensen remained involved in space policy decisions as the president's alter ego on most policy matters, but he seldom got directly involved with NASA oversight as the Moon program evolved. On politically sensitive matters, such as the allocation of NASA contracts and the location of NASA's facilities, Kennedy's special assistant Kenneth O'Donnell became involved. Although he was the president's closest confidant on most policy and political matters, Kennedy's brother Robert seemingly had only limited involvement on space issues, although it is impossible to know how frequently space matters were discussed between the two brothers. It thus fell to Bundy, Wiesner, and Bell and their staff to be the primary points of contact between the White House and NASA as the U.S. space effort took its first steps toward a landing on the Moon.

## Locating the Facilities

It was clear to NASA managers that as part of the decision to start a rapid space buildup, NASA would have to quickly create several large new facilities. The decision on what kind of facilities to build and, politically more important, where to locate them, was thus a high priority issue in the months immediately following President Kennedy's May 1961 speech. Although a formal announcement of facility decisions could not be made until August 1961, after the Congress had actually appropriated the increased FY1962 budget that the president had proposed, planning for the facility buildup began in earnest even before the speech. While most decisions with respect to launch vehicle production, testing, and launch were made without significant White House involvement, such was not the case with respect to locating the new NASA "field center" which would train the men who would go to the Moon and oversee the development and operation of the spacecraft that would carry them there.[11]

Well before President Kennedy's approval of the lunar landing mission, it had been clear to the NASA leadership that, if there was to be a follow-on

President Kennedy meeting with his senior advisers for a late 1961 budget review at his father's house in Hyannis Port, Massachusetts. From left to right: Secretary of Defense Robert McNamara, the president, budget director David Bell, deputy secretary of defense Roswell Gilpatric, science adviser Jerome Wiesner, special counsel Theodore Sorensen, and national security adviser McGeorge Bundy (JFK Library photograph).

effort to Project Mercury, the Space Task Group, which was managing Mercury, needed to be moved from its location at the Langley Research Center in Hampton, Virginia. This move was necessary because human space flight required engineering development, flight operations, and especially project management skills, not the engineering research-oriented approach that was characteristic of Langley.

As early as NASA's creation in 1958, there was a specific view on where to locate the next new NASA installation; that view came from a politically powerful source. Soon after NASA opened its doors for business on October 1, 1958, Administrator Glennan heard from Congressman Albert Thomas, whose district included Houston, Texas. Thomas was chairman of the House of Representatives appropriations subcommittee that controlled NASA's funding. Glennan learned that Thomas "was anxious that his district...should benefit from the space program." Thomas suggested that Houston's Rice University was willing to give NASA 1,000 acres of land as a location for a new NASA "laboratory." Glennan told him that NASA was "not about to build any new laboratory facilities beyond the one already authorized and on which construction had begun." (This would become the Goddard Space Flight Center in Greenbelt, Maryland.) Thomas responded "somewhat peevishly" that the decision to locate the new facility in Maryland "had gone through without his sanction since he had been absent" from Washington. Thomas persisted in his advocacy. He made several more calls to Glennan in late 1958 and finally told the NASA administrator: "Now look here, Dr., let's cut the bull. Your budget calls for $14 million for Beltsville [actually Greenbelt], and I am telling you that you won't get a god-damned cent of it unless the laboratory is moved to Houston." Glennan was able to fend off this threat, but when it became evident in 1961 that there would be a new NASA installation for human space flight and that many locations in the United States would compete for hosting it, Glennan, who by then was back at Case Institute of Technology in Cleveland, Ohio, advised Ohio state officials not to waste their time in that competition because "Houston would be the site chosen."[12]

In late April 1961, as it was becoming clear that President Kennedy was likely to approve a major acceleration of the NASA human space flight program, NASA administrator Webb recognized that a separate, new NASA center would indeed be needed to manage the effort, and instructed his staff to begin the site selection process. He put $60 million in the NASA budget estimates being prepared for the White House as a down payment on constructing the new center. The site selection team considered locations in Florida and California, but was also well aware of Representative Thomas's long-standing interest in having a new NASA installation in Houston. Thus NASA representatives visited Houston on May 16, even before the president's announcement of the lunar landing goal. They were met there by George Brown of the Houston-based Brown & Root construction company, a Lyndon Johnson ally who had been consulted by the vice president during Johnson's recent space review, and by a representative of the Houston Chamber of Commerce.[13] But this turned out to be a false start in the site selection process; it was not restarted in earnest until August 7, when the Congress passed the appropriations bill that included the $60 million in funding for the new center.

As it became widely known in May that the president was going to propose a major acceleration of the NASA program, Representative Thomas made it clear to NASA administrator James Webb that the 1958 Houston offer, and the implied threat of problems for the NASA budget if it were not accepted, still stood. On May 23, Webb reported to Vice President Lyndon Johnson that Thomas had "made it very clear" that he was "extremely interested" in Houston being selected as the location for the new manned spacecraft center that clearly was going to be needed to manage the lunar landing program.[14] In 1961, being the location for this center was a much more attractive proposition than it had been earlier, since the demands of Apollo would clearly require a major facility with many jobs created and thus a significant demand for housing and services in the areas adjacent to the new NASA installation. Rice University was still prepared to donate to NASA a sizeable plot of land some thirty miles south of downtown Houston as the location for the new center; Houston construction, real estate, and other business interests recognized that the facility would generate a wide variety of economic opportunities for the area. Humble Oil, the Houston corporation that had donated the land to Rice now to be transferred to NASA, still owned most of the surrounding property, and realized that its value would increase substantially if a major new government facility were located on the university's land. George Brown's construction company, Brown & Root, hoped to get the contract to build the new NASA installation. Brown was chairman of the Rice Board of Trustees, had been a major contributor to Lyndon Johnson's senatorial campaigns, and was also closely allied with Albert Thomas. Securing NASA's agreement to locate its new center in the Houston area thus became a political issue of the first order.[15]

The opportunities presented by the decision to develop a new NASA center of course did not go unnoticed in other parts of the country, and both the White House and NASA were bombarded after the president's May 25 speech by communications from members of Congress and state and local officials suggesting that the area they represented would be an ideal location for the new installation. Pressing the case for California was Representative George Miller, who was the acting chairman of the House Committee on Science and Astronautics after Overton Brooks fell ill. Missouri directed its advocacy through powerful Senator Stuart Symington. Making the case for Texas in addition to Thomas were Representative Olin "Tiger" Teague, chairman of the House Subcommittee on Manned Space Flight, and Representative Joe Kilgore; in addition, Speaker of the House Sam Rayburn and Vice President Lyndon Johnson advocated the Texas cause. Johnson and Albert Thomas were not political allies within the fractious Texas Democratic Party, but they were united on this issue.

Of particular political concern to President Kennedy and his top political assistant Kenneth O'Donnell was continuing pressure from the governor of Massachusetts, John Volpe, to locate the new center at Hingham Air Force Base near Boston. Volpe wrote Kennedy on July 19, before the site selection process had formally begun, saying that "as one Bay Stater to another," he wanted to call the advantages of the Massachusetts location to the president's attention. By this time, NASA had made public its preliminary criteria for deciding on the location for the new center, and Volpe outlined the ways in which the Hingham location met those criteria, conveniently omitting the requirement set by NASA for "a mild climate permitting year-round, ice-free, water transportation; and permitting out-of-door work for most of the year." Volpe closed his letter by saying to Kennedy, "may I urge your help in bringing this project to Massachusetts."[16] In the succeeding two months, O'Donnell and NASA's Webb had a series of interactions reflecting Volpe's hope that Boston would be chosen as the location for the new center.

The final criteria for site selection, including both eight "essential criteria" and four "desirable criteria," were approved by top NASA managers in mid-August. Before that approval, conscious of the Massachusetts interest, Administrator Webb had reviewed and specifically reiterated the "mild climate" requirement as being essential. In a September 14 memorandum to the president discussing the site selection process, Webb provided five justifications for the climate requirement, concluding that "selection of a site in an area meeting the stated climate criterion will minimize both the cost and the time required for this project" and noting the many ways in which the Boston area failed to meet the requirement.

Upon an initial assessment by the NASA site selection team, nine potential sites, notably not including Houston, met all or most criteria, and arrangements were made by the team to visit those areas. While visiting the original nine locations, an additional fourteen sites were brought to the attention of the team; the Rice University site favored by Representative Thomas was one of those additions to the list. In all, the site selection team visited twenty-three potential sites; they were located in Florida (2), Louisiana (3), Texas (9), Missouri (4), and California (5). At each site, the routine was similar: an afternoon arrival and greeting by state and local dignitaries, a meeting to explain the selection criteria, a breakfast meeting with local representatives, and a visit to the proposed site and a nearby college or university. The site selection team "felt that locations north of the freezing line were unlikely to meet the requirements" and thus did not originally plan to visit any such site.

Delegations representing sites in Virginia and Rhode Island not being considered by the selection team pleaded their case in presentations directly to NASA's James Webb and Hugh Dryden. Also, on September 1, a Massachusetts delegation headed by Governor Volpe and Senator Benjamin Smith, John Kennedy's former college roommate who had been appointed in December 1960 to fill JFK's Senate seat, met with the two NASA leaders to argue for consideration of the Hingham site and to ask that the site selection team at least visit Massachusetts. A large meeting of Boston business interests sponsored by the leading local newspaper, *The Boston Globe*, also called upon the president to select the Massachusetts site. On September 8, Governor Volpe called James Webb, again asking whether the team would visit Massachusetts. The phone conversation was described by Webb's biographer as "acrimonious."[17] Volpe told Webb that "great political pressure was building up" for such a visit. Webb responded that "it was most difficult to promise this without doing so in many other cases," but told Volpe that he could make public his intervention with Webb in order to relieve some of the political pressure on the governor. Webb told President Kennedy in a somewhat self-congratulatory way that he believed that "it was an eminently fair proposal for me to have put to him." Then, on September 13, without notifying Volpe or any other Massachusetts official, the site selection team did visit the Hingham site "for an inspection of the terrain and existing buildings." The only other site visited on the basis of political intervention was in St. Louis, to satisfy Senator Symington's request.

John F. Kennedy followed this process closely, and Webb kept him informed and then on September 11 briefed him on the likely outcome. According to Webb's public statements at the time of the site selection, Kennedy told the NASA administrator then that even though there had been pressures on him to intervene in the process, he expected Webb "to make this decision in the light of the national interest." Webb noted that Kennedy had "intervened in no way to try to favor his own state of Massachusetts, or to rule it out of the game." Rather, the president wanted NASA to have full responsibility for the site selection decision. Webb later revised this account of the selection process, saying that Kennedy had at some point called Albert Thomas to seek his support for several bills before the House of Representatives. Thomas had been vague about his willingness to support the bills until Kennedy told him: "Now, you know Jim Webb is thinking about putting this center down in Houston." From that point on, Thomas supported the three bills and "felt that he had a commitment from Kennedy" about the location of the new center.[18]

The site selection team reported its findings and recommendations in the second week of September. The team's first choice, flying in the face of the political pressures from Texas interests, was MacDill Air Force Base near Tampa, Florida, which was scheduled for closure as a Strategic Air Command airfield. It is interesting to speculate whether Webb and Dryden would have accepted this recommendation, given their broader perspective and the need to consider political factors. But at the last minute the Air Force changed its mind about closing MacDill, and the team's second preference, the Houston site associated with Rice University, became the top-ranked choice of the site selection group.[19]

Webb and Dryden met on the evening of September 13 and again on the morning of September 14 to review the site selection team report and hear the results of its last minute visit to Massachusetts. That visit did not change their assessment of the team's ranking, and Webb and Dryden decided that "this laboratory should be located at Houston, Texas, in close association with Rice University and the other educational institutions there and in that region." The new installation was designated the Manned Spacecraft Center. (It was renamed the Johnson Space Center after Lyndon Johnson's death in 1973, even though it was Albert Thomas, not Johnson, who had the greater influence on the decision to locate the center in Houston.)

Webb informed the president of this decision in a September 14 memorandum, noting that "a press release has been prepared announcing this decision, and we are holding it for issue after the White House notification of those who your staff feels should have advance information."[20] Kenneth

O'Donnell remembered President Kennedy as saying, after reading Webb's two September 14 memos, "It's a good decision. Let's go through with it." The public announcement of the selection of the Houston location came on September 19. With that announcement and the choice of sites in Florida, Louisiana, and Mississippi for launch-related facilities, the arc of new NASA installations along the Gulf of Mexico coast in the southeastern region of the United States that James Webb had advocated in his May 23 memorandum to Vice President Johnson had come into being.

## A White House Status Check

While the locations for new Apollo facilities were settled expeditiously, NASA was slower in selecting what approach to use in getting to the Moon and thus what spacecraft and launch vehicle would be needed for the lunar landing. On November 20, 1961, almost six months after President Kennedy's May 25 speech, Jerome Wiesner provided to Theodore Sorensen an "outline of major problems" with respect to NASA's progress in implementing the lunar landing program.[21]

Wiesner noted that "six months have elapsed since the decision was announced to put man on the moon, yet none of these crucial hardware programs have progressed beyond the study phase. Lead times on these development and construction programs are of critical importance." He also noted that "it is hoped that there will be no further field stations beyond these already announced. However, there are major problems related to the activation of these centers."

NASA was aware of these White House concerns. Webb told Dryden and Seamans that he had "scouted around" and had discovered that President Kennedy "has some concern as to whether we are proceeding rapidly enough and with enough procurement and program commitment activity to accomplish the goals he has set for the nation." NASA had issued a contract on August 10 to the MIT Instrumentation Laboratory for the Apollo guidance system, and on November 20 was one week away from contracting for the Apollo command and service module, one element of the spacecraft needed for the lunar landing. However, delays in selecting the design of launch vehicle for the lunar mission had meant that its procurement had had to wait. While NASA had chosen the locations for its Apollo-related facilities, Webb also reported that "there is some evidence that the President has had some doubts raised as to whether our decisions with respect to the Cape Canaveral, Michoud, and Houston installations were based on the needs of the program or had political overtones."[22]

### *Delays in Picking a Launch Vehicle*

Wiesner in his memorandum to Sorensen noted that "the major decisions have not been announced as to what extent rendezvous will be employed, what Advanced Saturn vehicle will be built (probably C-4), and what will be the characteristics of the so-called Nova that could put man on the Moon by direct ascent. The relative emphasis of rendezvous versus direct ascent is a key to the entire program."

There were two reasons for the delay in selecting the launch vehicle for the lunar mission. One was that the "national space plan" called for in the May 8 Webb-McNamara memorandum had anticipated a collaborative NASA-DOD effort to define a family of launch vehicles that could meet both agencies' requirements and advance the development of both liquid fuel and solid fuel propulsion systems. The focus of this planning effort was a "NASA-DOD Large Launch Vehicle Planning Group." The group was directed by Nicholas Golovin, then with NASA; its deputy director was Lawrence Kavanaugh of DOD. The group started work in July 1961, and by the fall had become bogged down in very detailed studies and deadlocked over the relative roles of liquid-fueled and solid-fueled boosters in the lunar landing program. Rather than come up with an integrated plan, the group had suggested a new Air Force-developed launch vehicle, called Titan III, with lift capabilities closely resembling the Saturn 1 vehicle that NASA was developing. The group's final recommendations attempted to satisfy both NASA and DOD, and ended up pleasing neither agency.[23]

The second reason for the delay in selecting a launch vehicle for the lunar mission was NASA's difficulties during the May–November period in deciding its preferred approach to sending men to the Moon.[24] Indeed, this uncertainty would continue well into 1962 and become a focus of NASA-White House controversy.

Beginning on May 2, even before a final decision on whether to approve a lunar landing effort had been made, there were a series of NASA studies examining alternatives for accomplishing the lunar mission. The first of these studies took as its starting point a "direct ascent" approach, in which the spacecraft for the lunar mission would be launched by a giant booster with eight F-1 engines in its first stage. The spacecraft would fly directly to the Moon and land intact on the lunar surface. A portion of the spacecraft would then take off from the Moon after the astronauts had completed their exploration, and return directly to Earth. This approach meant designing a seventy-five-ton spacecraft, almost forty times the weight of the Mercury capsule, that would "back down" to a lunar landing, using rocket firings to slow the craft to landing speed; during the landing, the astronauts would be on their backs at the other end of the craft, more than eighty feet above the surface and with no or very limited direct visibility of the landing site. The direct ascent approach also required that the fuel for the return journey and the heat shield needed for reentry into the Earth's atmosphere, both of which were heavy, would have to be carried to and then launched from the lunar surface. All of this would require a large and heavy spacecraft, and thus a very powerful booster NASA called Nova to send it to the Moon in a single launch. The more NASA, and especially von Braun and his team at Huntsville, thought about the technological leap required to develop the gigantic Nova vehicle, the more it looked for alternatives to making such a jump.

NASA during the summer of 1961 began to look harder at an approach called Earth-orbit rendezvous (EOR). This approach would allow the lunar spacecraft and the rocket stage needed to send it toward the Moon to be divided into two or more pieces, each piece launched separately. One or more rendezvous in Earth orbit would then be needed to assemble the pieces into a single spacecraft. An alternative EOR approach was to send the complete spacecraft and its Earth departure stage fueled with light liquid hydrogen into Earth orbit with one launch. Then a second launch would carry into orbit the comparatively heavy liquid oxygen used as the oxidizer for burning the hydrogen fuel; the oxygen would then be transferred to the lunar-bound rocket stage. Using an EOR approach meant that a launch vehicle significantly smaller than the Nova could be developed for the lunar mission. However, it did not solve the problem of how to land a single large spacecraft on the lunar surface.

Several versions of a smaller launch vehicle were proposed during the 1961 studies. The Saturn C-2 that had been part of NASA's plans in the spring was soon abandoned, and an "Advanced Saturn" with several powerful F-1 engines in its first stage became the focus of attention; the issue was how many of the large engines to use. The two-engine version became known as the Saturn C-3 and the four-engine version the C-4. This was the vehicle Wiesner mentioned in his November 20 memorandum to Sorensen.[25]

When he wrote that memorandum, Wiesner was apparently not aware of the latest NASA study of the launch vehicle issue. On November 6, Milton Rosen of NASA headquarters, reflecting the deadlock in the Golovin-Kavanau group, had organized a separate two-week study to recommend to the NASA leadership "a large launch vehicle program" which would "meet the requirements of manned space flight" and "have broad and continuing national utility." Rosen reported that "to exploit the possibility of accomplishing the first lunar landing by rendezvous," NASA should develop an "intermediate vehicle" that had five F-1 engines in the first stage, four or five J-2 engines in its second stage, and one J-2 in its third stage. (The J-2 was an engine powered by high-energy liquid hydrogen fuel that would have the capability to be stopped and restarted.) The four-engine Saturn C-4 had a "hole" in the center of its four first-stage F-1 engines; adding a fifth F-1 would thus be relatively straightforward. Rosen argued that NASA should build the most powerful rocket possible short of a Nova, and von Braun agreed that "the hole in the center was crying out for another engine." Adding a fifth engine would increase first stage thrust at liftoff to 7.5 million pounds. Since a direct flight to the Moon was at this point still NASA's officially stated preference for the lunar landing mission, Rosen also recommended that "a NOVA vehicle consisting of an eight F-1 first stage" should be developed on a "top priority basis." He added that "large solid rockets should not be considered as a requirement for manned lunar landing." The recommendation for a five-engine first stage for the launch vehicle, soon called the Advanced Saturn C-5 and ultimately the Saturn V, was quickly accepted by the NASA leadership. Within a few weeks, some form of rendezvous using the Saturn V replaced direct ascent as NASA's preferred approach to getting to the Moon, although design work on the Nova vehicle continued for some months.[26]

Wiesner likely also was not aware of a November 6 meeting between the NASA and Department of Defense space leadership at which there was agreement to "cancel the development of very large (240" class) solid rocket as a backup for NOVA," since "the work of the past six months shows that the reliability and potential of NOVA will be sufficient to make unnecessary the parallel development of the large solids on identical time scales," as had been called for in the May 8 Webb-McNamara memorandum.[27] Overall, the situation with respect to a launch vehicle for Apollo was not in as bad a shape as the Wiesner memorandum suggested; however, Wiesner was correct in his assessment that "the relative emphasis of rendezvous versus direct ascent is a key to the entire program."

### *How Much Would Apollo Really Cost?*

Wiesner noted in his November 20 memorandum that the NASA budget being discussed for Fiscal Year 1963 was 50 percent greater than had been projected just six months earlier. Indeed, the budget projections accompanying the accelerated space effort that President Kennedy had approved in May had projected a NASA budget of $3.029 billion for FY1963. However, when NASA submitted its FY1963 request to the BOB several months later, that total had grown to $4.238 billion, which was a 40 percent increase over the May estimate. By the time NASA and the BOB in December 1961 completed their negotiations over what the president would request for NASA in Fiscal Year 1963, the budget level had been reduced from $4.238 billion to $3.787 billion; this was still more than a 125 percent increase over the final FY1962 level. In a memorandum to President Kennedy summarizing the state of the NASA program, budget director David Bell provided a comprehensive overview of the evolution of the program since May and recommended to the president "that it would be desirable for you and the Vice President to have in the near future a short briefing and discussion of the status and plans for the civilian space programs, especially the manned lunar landing program." Bell noted that NASA administrator Webb had concurred in this recommendation and on the budget figures outlined in his memorandum.[28] The proposed briefing did not take place until Kennedy and Johnson visited NASA's Apollo facilities in September 1962.

In his review of the situation for the president, Bell said that even in May the BOB had thought that NASA's projections for future budgets "appeared to us to be understated," and that BOB had anticipated a $3.5 billion budget request for FY1963, although it had not used that number in any official documents. Bell told the president that "the cost estimates for the manned lunar landing program appear to have been underestimated to an even greater degree than anticipated." This was because "earlier estimates were of necessity made in advance of detailed technical plans and cost studies," but such studies were now "showing clearly that the costs of the principal elements of the program will be greater than anticipated." Bell estimated that the NASA budget would continue to increase, reaching $4.9 billion in FY1964, $5.3 billion in FY 1965, and $5.6 billion in FY1966. This meant that at the end of 1961 the accelerated space effort was projected to cost at least $2.8 billion more than had been estimated just six months earlier. That such cost growth was worrying to the White House was clear; Webb told Dryden and Seamans on November 21 that "there is some evidence that the President is concerned about the cost of our program."[29] Webb was correct; the president was indeed concerned.

## Space Programs Reviewed

The rapidly increasing costs of the U.S. space program, and particularly its civilian component, continued to trouble President Kennedy after he sent his $3.787 billion Fiscal Year 1963 request for NASA to the Congress in early 1962. There was no parallel single national security space budget request; Department of Defense and intelligence space programs were incorporated into the general DOD budget, rather than receiving separate budget treatment. However, increasing DOD expenditures for space were also of concern to the president. To obtain a total overview of the U.S. space program, Kennedy asked the BOB in June 1962 to carry out a comprehensive review of all U.S. space efforts.

### *Initial Budget Concerns*

After NASA administrator Webb met with Kennedy on May 3, 1962 to deliver a copy of NASA's revised long-range plan, he reported that Kennedy "was quite concerned about the high level of expenditures involved in our program, plus the military program, and urged that everything be done that could possibly be done to see that we accomplished the results that would justify these expenditures and that we not expend funds beyond those that could be thoroughly justified." Webb also reported that the president "had expressed some concern" about the geographical distribution of NASA funding; Kennedy noted that he had received complaints from states such as "Michigan, Pennsylvania, and the eastern states" that NASA was focusing its expenditures on California, Florida, Mississippi, and Louisiana. Kennedy was "quite anxious" that NASA "maintain the best geographical distribution of contracts and still get the most efficient job done." To provide the White House with its own channel of information on NASA procurement actions, Kenneth O'Donnell sent Richard Callaghan, a Kennedy loyalist and congressional staffer, to NASA as a special assistant to Webb. According to one account, "Callaghan's job was to arrange for a more equitable distribution of contracts, which would relieve congressional pressure on Kenny O'Donnell, and find out whether [Senator Robert] Kerr and [Vice President] Johnson were pulling strings for their friends at NASA." With respect to this latter mission, Callaghan found no evidence of undue Kerr or Johnson influence on NASA's contract awards.[30]

Webb responded to Kennedy's concern regarding geographical distribution in a June 1 letter. He told Kennedy that during 1961, states west of the Mississippi River received 56 percent of NASA prime contracts; states east of the Mississippi, 44 percent. One reason for this distribution was that "major aerospace and electronic companies have concentrated their growth within a few areas of the country." However, Webb continued, when both prime contracts and first-tier subcontracts by the prime contractors were considered, 53 percent of the work was in the East and 47 percent in the West. Webb also noted that in the second half of 1961 Massachusetts had received 64 percent more in NASA funding than it had received in the first half of the year. In summary, Webb told the president, "the NASA effort is being spread broadly throughout the United States."[31]

### *Special Space Review*

Beginning in late June 1962, the BOB began a review that was intended to lay out in a consistent format the five-year space programs of the Department of Defense, NASA, the Atomic Energy Commission, and, although it could not be acknowledged at the time, the National Reconnaissance Office, the organization developing and operating U.S. reconnaissance satellites, the very existence of which was highly classified. This review was in response to President Kennedy's specific request "for a consolidated presentation of the space programs and estimates of all agencies" and "that 1964 estimates for space programs be given an especially critical review."

According to Willis Shapley of the BOB, who was in charge of the review, one question that prompted the review was a White House "reevaluation of whether the Apollo program should really proceed." It is not clear whether it was the president himself who was raising this question; given his interest just a few months later to push for an earlier date for the first lunar landing attempt, this seems unlikely. More probable was that his budget, technology, and policy advisers, who were in general more skeptical of the value of Apollo than was the president, were making sure that Kennedy recognized the full implications of his space commitment. In addition, there were short-term concerns in mid-1962 about a possible recession, talk of a temporary tax cut, and a desire to avoid an unbalanced federal budget; this meant that Kennedy was paying particular attention to controlling rapid increases in spending in discretionary areas such as space and defense.[32]

By August 15, the BOB had compiled some 250 "data sheets," one for each of the principal space projects of the government. These were put into two loose-leaf binders and, because intelligence satellite programs were included, classified at such a high level that only relatively few people inside the government were cleared to possess the binders. Shapley recalls that he was "not too proud" of the review, "because it was really pretty bureaucratic."

The BOB did prepare a late August "draft staff report" based on the review. The report noted that "the central decision to be confirmed or modified is whether the manned lunar landing program should proceed at an optimum pace as contemplated in present NASA plans, or whether a decision should be made to stretch out the program to avoid as great an increase in expenditures in 1964 and 1965." The report examined the short-term budget impacts of slipping the target date for the first landing attempt until late 1968. It concluded that "under all feasible alternatives, barring a complete reversal of the MLL [manned lunar landing] and other augmented space program decisions of May 1961...substantial increases in expenditures appear unavoidable in 1964 and 1965." The staff report mentioned that this situation had been pointed out as the decision to accelerate the space program was being made a year earlier; the BOB had noted that the decision "was a long term commitment involving increasing expenditures for a period of several years." At the end of the section of its report dealing with NASA, the BOB recommended a course of action that recognized "that neither the total fiscal situation nor the space program alternatives and implications are clear enough now to permit a definite decision on the program and budgetary guidance to be given to NASA." The BOB recommended that "the issue should remain open until the final 1964 budget decision period in November or early December."[33]

As the BOB was finishing its review, James Webb was once again reminded of President Kennedy's concern about the rising costs of the NASA program. On August 15 Kennedy sent a brief memorandum to Webb, asking him about press reports that the cost of the new Manned Spacecraft Center in Houston had increased from $60 million to $123 million. Kennedy asked: "Is this correct? Who are the architects and the builders and under whose control is the Space Center building to be put up?" Webb replied on August 18, saying that the costs Kennedy was quoting were those in a statement by Senator William Proxmire (D-WI), who was questioning "the prudence with which the space program is being administered" and implying that "the increase in cost...resulted from a lack of budget discipline." Webb said that "the figures quoted are substantially correct; the implications are not," and provided a lengthy explanation of the reason for the higher cost figure. This explanation did not satisfy President Kennedy. In a note to budget director Bell, the president said that it seemed to him that the cost of the new center was "excessive," and the cost increase "does raise the question of funding of the entire program. This needs the most careful continuing scrutiny." Kennedy asked Bell for his "suggestions on the recent appropriations for this space program—what programs are essential and desirable and how we can make them meet the cost estimates more precisely." He added: "This program has so much public support that unless there is some restraint there is a possibility of wasting some money."[34] This tension between Kennedy's desire to be first in space and his concern over the very high costs of Apollo was to run throughout 1962 and 1963.

## Soviet Successes—and Failures

In mid-August 1962, the United States was also reminded that the space race with the Soviet Union was still on. A year had passed since the second Soviet orbital flight of cosmonaut Gherman Titov on August 6, 1961. That flight had had its share of troubles; in particular, Titov, unlike Yuri Gagarin, had experienced significant motion sickness during his seventeen-orbit, day-long flight; in contrast, Gagarin had completed his one orbit with no ill effects. On August 11, 1962, the Soviet Union launched its third human space flight and then, on the next day, much to the surprise of Western observers, launched a fourth human mission. In the United Kingdom, astronomer Bernard Lovell, who was a year later to become involved in a controversy over whether the Soviet Union intended to send people to the Moon, called the two launches "the most remarkable development that man has ever seen." The two Soviet spacecraft passed close to one another early in their joint mission; the two cosmonauts communicated using their on-board radios, and, according to at least some reports, saw each other's spacecraft, but they did not have the maneuvering capabilities required for a rendezvous attempt. Until the lack of that capability became evident to U.S. observers, there was concern that the Soviet Union had beaten the United States to another important milestone, the ability to carry out a space rendezvous.[35]

After rejecting the suggestion that he make a formal statement on the space competition with the Soviet Union at the start of his August 22 press conference, President Kennedy chose instead to respond to an inevitable question about the Soviet feat. His response suggested both his continuing commitment to catching up with the Soviet Union and his recognition of how expensive the space effort was becoming.

> *Q*: Mr. President, the Soviet Union's latest exploit, the launching of two men within 24 hours, seems to have caused a good deal of pessimism in the United States. You hear people say that we're now a poor second to Russia. How do you size up the situation, Mr. President, for the present and the future?
>
> *Kennedy*: We are second to the Soviet Union in long-range boosters. I have said from the beginning—we started late, we've been behind. It's a tremendous job to build a booster of the size that the Soviet Union is talking about, and also have it much larger size, which we are presently engaged in the Saturn program. So we are behind and we're going to be behind for a while. But I believe that before the end of this decade is out, the United States will be ahead. But it's costing us a tremendous amount of money...And it's going to take us quite a while to catch up with a very advanced program which the Soviets are directing and there's no indication the Soviets are going to quit.
>
> We're well behind, but we're making a tremendous effort. We increased after I took office, after 4 months, we increased the budget for space by 50 percent over that of my predecessor. The fact of the matter is that this year we submitted a space budget which was greater than the combined eight space budgets of the previous eight years. So this country is making a vast

> effort which is going to be much bigger next year and the years to come and represents a very heavy burden upon us all. But we might as well recognize that we're behind now and we're going to be for a while. But what we've got to do is concentrate our efforts.[36]

While publicly President Kennedy was acknowledging the continued Soviet lead in space, behind the scenes the White House was debating whether to counter the public awareness of Soviet space successes with what the U.S. government knew about Soviet failures. The Soviet Union had attempted on August 25 to send a spacecraft to Venus, but the Central Intelligence Agency (CIA) informed Carl Kaysen, McGeorge Bundy's deputy, that "the evidence points to a failure of the probe to eject from earth orbit." By contrast, the U.S. launch of its Mariner II spacecraft to Venus on August 27 went well; for the first time, the United States was on its way to another planet, and the White House was anxious to contrast the U.S. success with the Soviet failure. On August 31, Kaysen sent a memorandum to White House press secretary Pierre Salinger discussing how best to announce this and prior Soviet failures without revealing the classified means through which the information had been acquired. On September 5, Lieutenant General Marshall Carter, acting director of the CIA, provided to President Kennedy a "fact sheet" on six Soviet failures of probes to Venus or Mars. He noted that "the information from which the fact sheet was developed has been obtained from many intelligence sources, some of them our most sensitive," but that "there is enough collateral information available to warrant unclassified publication of this fact sheet without blowing the cover of our sensitive sources." Carter was worried about such a release, however; he told the president, "I am concerned over the opening up of this entire matter of our knowledge of Soviet activities to the general scrutiny of the public, and particularly the probing press," who might be able to discover "our entire box of tricks."

The White House decided to accept this risk, and on September 5 James Webb sent a letter to the chairmen of the Senate and House Space Committees detailing the Soviet failures; the letter was intended to be leaked to the media, and the press soon picked up the story. *The New York Times* on September 9 reported the release of information on Soviet failures and commented that "this week the Administration finally decided that the information was too good—from the standpoint of embarrassing and deflating the Russians—to keep secret any more," and that the release was a "neat propaganda ploy."[37]

### One More NASA Center?

In his November 20, 1961, status report on NASA's Apollo buildup, Jerome Wiesner had noted that "it is hoped that there will be no further field stations beyond these already announced." This turned out to be a false hope. As the NASA leadership assessed the various capabilities it would need to manage Apollo effectively, it concluded that NASA was sorely lacking in high-quality electronics competence. This presented a problem with respect to NASA's ability to manage its contracts with industry and academia, since NASA calculated that 40 percent of the cost of launch vehicles was related to their electronic components; for spacecraft, the cost was 50 to 70 percent. In addition, experience with early robotic spacecraft suggested that there were significant differences in the reliability requirements for electronic components in space as compared to on Earth. Most NASA employees at the time were more interested in the structural and propulsion aspects of spacecraft and launch vehicle design and development than their electronic aspects.[38]

In January 1962 associate administrator Robert Seamans asked the NASA staff to investigate what could be done to address this situation. Albert Kelly, director of electronics and control in NASA's Office of Advanced Research and Technology, spent the next ten months preparing a detailed report on the issue; the November 1962 report concluded that the best approach to gaining the needed competence was to create a new NASA laboratory, or "field center," dedicated to managing NASA's electronics research. The NASA leadership had in fact several months earlier reached the same conclusion; the issue then became where to locate the new center. Webb, Dryden, and Seamans gave greatest weight to two criteria in making this decision: (1) a location near one or more universities involved in advanced electronics research, and (2) a location where the industrial community was also working on electronics and was research-oriented.

Another consideration, according to James Webb, was President Kennedy's questioning "why some of the best brains in the East were not working more actively in our program." Webb told Kennedy that "a new Electronics Research Center in the eastern part of the country" would not only satisfy a specific NASA need, but would also "kill several birds with the same stone by making this Center a focal point of contact between some of our ablest people and some of the ablest ones working in advanced fields in universities." Kennedy told Webb that "while he felt that this was certainly an important objective, he was going to leave the decision to me but would like to be kept informed." By October 1962, Webb told Kennedy that he, Dryden, and Seamans had decided to locate the Center in Boston, "making it clear that the geographic proximity to Harvard, MIT and the brilliant researchers and scholars in the electronics and associated fields in the city was one of the major bases for our judgment." In fact, said Webb, NASA wanted "to put it [the new Center] within walking distance of both Harvard and MIT."[39]

There were two political problems with that decision. President Kennedy was of course from Massachusetts, and thus such a decision could appear as if it had been influenced by his desire to bring some of the benefits of the space buildup to his home state. Even more problematic was the fact that the President's youngest brother, Edward "Ted" Kennedy, was in 1962 running to fill the remaining two years of President Kennedy's Senate term, and his campaign argued that he "could do more for Massachusetts." If NASA had announced, without any prior notice or competition, a decision to locate a major new facility in Massachusetts, the political reaction likely would have involved NASA in a tightly contested election, a situation both President Kennedy and James Webb wanted to avoid.

When Webb on October 16, 1962, told Kennedy of NASA's plans for locating the new center in the Boston area, he also said that it was extremely important from NASA's "image of careful professional work and decisions made on a technical basis that this should not become a matter under discussion in the then ongoing campaign in Massachusetts where his brother was running for the Senate." Kennedy's response was that "he approved the concept of the Electronics Research Center." Kennedy also "stated that he was prepared to accept it in his budget" and "agreed that it should not be introduced into public discussion until the budget was to go to Congress" in early 1963, after the Senate election. To avoid the appearance of Kennedy's political influence on the decision, Webb buried the initial funding for the new center in the NASA budget request submitted to the BOB in September 1962; this was not difficult to do, since the initial request of $5 million was very small compared to the overall $6.2 billion NASA budget request. Even the BOB was not informed of NASA's intentions. As a former director of the BOB, Webb was well versed in ways to manipulate the normal process of BOB review. In Fall 1962, according to Webb, "the only persons who knew we were planning this Center outside of NASA" were President Kennedy and his top political operative, Kenneth O'Donnell."[40]

Once the election was over and Ted Kennedy had won the Senate seat, NASA was ready to let the BOB in on its plans. Before meeting with budget director Bell on December 13 to finalize the NASA FY1964 budget request, Webb asked Kenneth O'Donnell to check with the president to make sure that Kennedy still agreed with NASA's decision to develop the new center. Assured that this was indeed the case, NASA and the BOB inserted into the president's budget message notice of the decision to create a new Electronics Research Center and to locate it "in the Greater Boston area." Kennedy directed that "this matter should be handled with the most complete discretion." There was no leak to the press of this decision until the budget became public with its submission to the Congress in mid-January 1963.

The Congress, and particularly NASA's House of Representatives oversight committee, the Committee on Science and Astronautics, was not pleased to learn that NASA had made this decision without prior consultation with the committee. Over the next several years, the committee and NASA remained at loggerheads over whether NASA could proceed with its plans. There was also opposition from some Senate members who believed that the areas they represented should have been able to compete for the new NASA center. At one point, President Kennedy got personally involved, meeting on June 11, 1963, with Webb and Representative Joseph Karth (D-MN), who thought that the center was not really needed, but if NASA went ahead with its plans to create it, the new center could very well be located in his state. Kennedy was "very gracious," but he was unable to change Karth's mind regarding the issue.[41]

By the time Congress finished work on the NASA FY1964 budget in December 1963, there was tentative agreement to allow NASA to proceed with its plans, subject to Congressional review of several required studies. Even after those studies were completed, there was continued questioning of NASA's plans for the center; "the fight for and against the Center raged on through 1965." It was not until 1966 that the Congressional opposition died down, even though the Electronics Research Center had become operational in 1965. This was too late for the center to have much of an impact on the Apollo program. The Electronics Research Center was to have a short lifetime; NASA announced in December 1969 that as part of its post-Apollo retrenchment, it had decided to close the facility.[42]

## Conclusion

While Jerome Wiesner at the end of 1961 might have been concerned by what appeared to be too slow a pace in NASA's implementation of the lunar landing decision, to those at NASA involved in the effort the rate of activity during 1961 and 1962 seemed extremely rapid. In the weeks following Wiesner's November 20, 1961, memorandum, NASA chose the contractors for the Apollo spacecraft and the first and third stages of the Saturn V vehicle. By the start of 1962, construction had begun at all the new facilities that would be required for Apollo. A major NASA reorganization to prepare the space agency for managing Apollo was announced on November 1, 1961; among the changes made was the creation of a separate Office of Manned Space Flight as one of the major program units at NASA headquarters. Chosen as its head, with the title associate administrator for manned space flight, was a dynamic young executive from RCA named Brainerd Holmes. Webb and Seamans had thought briefly about asking Wernher von Braun to become the human space flight manager, but that possibility disappeared when Hugh Dryden said that he would retire if it became reality. (Dryden was apparently one of those at NASA who resented von Braun's involvement with the Nazi regime in Germany before and during World War II.) On April 11, 1962, President Kennedy assigned to Project Apollo the highest national priority, designated DX; this gave the undertaking first call (together with some defense and a few other space efforts) on whatever human and physical resources were needed for its accomplishment.[43]

It was thus clear by mid-1962 that the mobilization of the resources needed to accomplish a lunar landing was well underway. President Kennedy had warned the Congress and the American public on May 25, 1961, that achieving the lunar goal "would take many years and carry very heavy costs." In the fifteen months following his May 25 speech, the realism of that warning became increasingly evident; by September 1962, President Kennedy concluded that it was time for him to take a first-hand look at the unfolding effort.

# Chapter 9

# "I Am Not That Interested in Space"

One critical decision with respect to the lunar landing project still remained unsettled as President Kennedy prepared in September 1962 to make an inspection tour of the facilities being developed for the accelerated space effort. That decision was the best approach to getting astronauts to the lunar surface. NASA in July 1962 had selected as its preferred approach rendezvous in lunar orbit, a way to the Moon that had emerged in its consideration only at the very end of 1961. Kennedy's science adviser Wiesner and his staff did not agree with this choice, and were actively pressuring NASA to reverse it in favor of an Earth orbit rendezvous approach. The president's September 11–12 visit to NASA and industry installations was intended to give him an overview of the human space flight effort in preparation for the hard budget decisions that all knew were upcoming later in the year; it also exposed him to the ongoing argument about the choice of how best to fly to the Moon. It was also on this trip that John F. Kennedy at Rice University in Houston gave his most memorable address on the reasons why he had chosen to accelerate the U.S. space effort.

As he toured the NASA facilities, Kennedy, as was his style, asked many questions, and learned that some within NASA believed that the first landing on the Moon could come as much as a year sooner than the late 1967 date that was at that point NASA's target. To advance the landing date by that many months would require requesting from Congress an extra short-term supplement to the NASA budget, and there was considerable debate during October and November 1962 about the wisdom of that action. Adding more money to the human space flight budget was strongly advocated by associate administrator for manned space flight Brainerd Holmes, but equally strongly resisted by NASA administrator James Webb. Their disagreement escalated into tensions that culminated in Holmes leaving NASA in mid-1963.

The debate over extra money for NASA led to a November 21 White House meeting in which President Kennedy and Webb disagreed about the priority of the lunar landing program compared to other NASA activities. In the aftermath of this meeting, President Kennedy reluctantly decided not to try to accelerate the Apollo schedule, and to continue on the path of requesting funds for NASA adequate to maintain the late 1967 target date for the first lunar landing attempt. Even pursuing that path required a NASA Fiscal Year 1964 budget request of $5.712 billion, an increase of 55 percent over the NASA FY1963 budget of $3.674 billion but almost a half billion dollars less than what NASA had requested in September 1962. The continuing exponential increase in NASA funding came at a time when Kennedy and his White House advisers were striving to limit overall budget growth, even as the financial demands of Apollo approached 4 percent of the total federal budget.

## A White House-NASA Argument: How Best to Go to the Moon

The first stop on President Kennedy's two-day tour to inspect the facilities being developed for Apollo was the Marshall Space Flight Center in Huntsville, Alabama. There he was briefed by Wernher von Braun on the approach that NASA had finally chosen for carrying out the lunar landing mission, called lunar orbit rendezvous (LOR). This approach had emerged in the first half of 1962 as NASA's preferred approach to getting to the Moon, but the White House Office of Science and Technology and its external advisers on the President's Science Advisory Committee (PSAC) were at the time of JFK's trip embroiled in a dispute with NASA over the wisdom of that choice. As von Braun described the LOR approach, Kennedy interrupted, saying that "I understand that Dr. Wiesner doesn't agree with this," and calling his science adviser to join the discussion. "Some lively dialogue" then ensued among Wiesner, von Braun, Webb, Seamans, and others; this discussion was just out of the earshot of reporters. According to Seamans, the reporters "obviously knew we were discussing something other than golf scores." Wiesner suggests that he and von Braun were having a "friendly" talk in front of the president when Webb, who thought the two were arguing, "moved right in," so that "what had been a friendly discussion became a real argument" as the press "watched, heard, and listened." Kennedy ended the five-minute discussion by saying, "Well, maybe we'll have one more hearing and then we'll close the books on the issue."[1]

Key to the LOR concept was separating the functions of the lunar landing mission between two spacecraft, rather than the single heavy spacecraft contemplated in both the direct ascent and the Earth orbit rendezvous (EOR) approach to the lunar landing. One spacecraft, designated the command and service module (CSM) , would carry the crew to lunar orbit and back to Earth; the other, designated the lunar excursion module (LEM), would carry two crew members from lunar orbit to the Moon's surface, and then, after they had finished their exploration, back to a rendezvous in lunar orbit with the CSM. After the crew and lunar samples were transferred to the CSM, the LEM could then be discarded. Since the LEM would operate only in the atmosphere-free vicinity of the Moon and would not have to carry the

The presidential party as President Kennedy toured the Marshall Space Flight Center in Huntsville, Alabama on September 11, 1962. Identifiable in the image, in addition to President Kennedy are (left to right) center director Wernher von Braun, NASA administrator James Webb, Vice President Lyndon Johnson, Secretary of Defense Robert McNamara, presidential science adviser Jerome Wiesner, and director of defense research and engineering Harold Brown. NASA associate administrator Robert Seamans, Jr. is partially visible behind von Braun. Most of those in this photograph participated in a brief but spirited debate about the wisdom of the lunar-orbit rendezvous approach to the lunar landing mission (NASA photograph).

heavy fuel and even heavier heat shield required for the return to Earth, it could be much lighter than a spacecraft that would both land on the Moon and also have to carry the crew back to Earth. This weight reduction resulting from this separation of functions made it possible to launch the whole lunar landing mission with one Saturn V booster, rather than the two that would be required by the Earth orbit rendezvous (EOR) approach.

The LOR concept had been brought to Seamans's attention in an impassioned nine-page November 15, 1961, letter from John Houbolt, an engineer at NASA's Langley Research Center, who had bypassed several layers of the NASA chain of command in sending the letter. NASA in late 1961 was focusing on some form of EOR as its preferred approach to the lunar mission, but Houbolt argued that the LOR approach was the better way to getting to the Moon before the end of the decade, was safer and less expensive, and required only one launch.[2]

After extended analysis of the concept, NASA's top leaders by early July 1962 had agreed that LOR was indeed the best choice for achieving the lunar mission by the end of the decade and were preparing to announce their decision at a July 11 press conference. In anticipation of the announcement, NASA on July 3 sent a summary of its comparison of the various options to science adviser Wiesner. By the end of the day, Wiesner called Webb "in a highly emotional state" to say that "L.O.R. is the worst mistake in the world." Webb asked Joseph Shea, the NASA systems engineer who was leading the effort to select the lunar landing approach, to go to the White House immediately; when he met with Wiesner, the science adviser called LOR a "technological travesty."[3]

There were several reasons for Wiesner's reaction.[4] One was the intuitive sense that a mission that depended for its success, and for the crew's survival, on a rendezvous in lunar orbit, 240,000 miles from Earth, would be excessively risky. This was especially the case since there had been no experience with rendezvous, and Project Gemini, the just-initiated effort to gain that experience, was at that time not scheduled to have its first flight until late 1963. (The first flight actually did not come until early 1965.) Given the end of the decade deadline, the choice of mission approach would have to be made before the feasibility of its key element, rendezvous, had been demonstrated. (This of course was also true for Earth orbit rendezvous, but if a problem developed in an Earth-orbiting mission, the astronauts could easily return home.) NASA's engineering analyses showed that LOR was safer than EOR, but Wiesner and his staff did not trust those analyses. The principal staff person supporting Wiesner on space issues was Nicholas Golovin, who had been forced to resign his position at NASA the preceding fall and left the space agency with "bitter gall" in his throat, only to be quickly hired by Wiesner as his top staff person for space issues. Golovin was tenacious in his criticism of NASA's choice of LOR during 1962 and became a particular irritant to the space agency as NASA attempted to move forward.

Wiesner was also hearing from the space vehicle panel of PSAC that it had serious reservations about the LOR choice. The panel had followed NASA's planning for the lunar mission throughout the year. Members of the panel together with Wiesner met with the NASA leadership on July 6 to outline their concerns. As a result of this meeting, NASA was forced to change its message for the July 11 press conference to saying that the LOR choice was tentative and subject to further study and review.

Jerome Wiesner transmitted the PSAC views as an attachment to a July 17 letter to James Webb; in that letter Wiesner also set forth his own views. Like the PSAC panel, he was concerned about "which mission mode is most consistent with the main stream of our national space program, and therefore the one most likely to be useful in overtaking and keeping ahead of Soviet space technology." In particular, Wiesner wanted more attention paid to "the question of which mode is likely to be most suitable for enhancing our military capabilities in space, if doing so should turn out to be desirable." Wiesner told Webb that he had "reported the results of our discussions to the President."[5]

Webb replied to Wiesner's letter on August 20, listing the various studies that NASA was undertaking based on the Space Vehicle Panel's suggestions and assuring Wiesner that NASA had thought carefully about the issues raised in Wiesner's letter. He told Wiesner that "it is our considered opinion that the LOR mode...provides as comprehensive a base of knowledge and experience for application to other possible space programs, either military or civilian, as either the EOR mode or the C-5 [Saturn V] direct mode." Webb thanked Wiesner for his and PSAC's efforts, saying that "this constructive criticism by eminently qualified men is of tremendous value."[6] This final comment may not have been totally sincere. The pressure resulting from President Kennedy's "end of the decade" deadline was being felt within NASA, and continued criticism from the White House science office was a barrier to NASA's moving ahead with its plans.

Wiesner and his assistant Golovin continued to intervene in NASA's decision process during July and August, and on September 5, a few days before the president was to visit various NASA installations, Wiesner once again wrote Webb reiterating his concerns regarding the LOR choice. Wiesner sent a copy of his letter to the president. In this letter, Wiesner called the contribution of the LOR choice to the overall space capabilities of the United States "seriously questionable," noting that any military applications of those capabilities "will be in the near earth environment" and that "whether manned military missions are to be either of a defensive or offensive character...the obvious needs for maneuverability and reasonable stay-time in orbit would require that refueling techniques be developed more or less contemporaneously with those for rendezvous and docking." While James Webb was sympathetic to using Apollo to build up overall U.S. space capabilities, his main focus in 1962 was selecting the approach that gave the best chance of getting to the Moon before the Soviets; Wiesner appeared less interested in the race to the Moon and more focused on developing near-Earth space capabilities with military relevance.[7]

This is where the situation stood as President Kennedy visited Huntsville and was party to the Wiesner-NASA debate. As Kennedy flew to Cape Canaveral, the next stop in his tour, he was asked what the likely outcome of the controversy would be. "Jerry's going to lose, it's obvious," replied the president; "Webb's got all the money, and Jerry's only got me."[8]

Wiesner and Golovin did not give up their fight easily and continued to contest NASA's choice and supporting analyses in September and October. NASA finally assembled all of its analyses into a document that it hoped would be a final comparison of the approaches to carrying out the lunar landing mission and sent it to Wiesner on October 24. In that report, NASA estimated that the probability of success for any one lunar landing attempt was 40 percent for the LOR approach, 36 percent for the direct ascent approach, and 30 percent for EOR; probabilities of crew safety were 85 percent for LOR, 83 percent for direct ascent, and 89 percent for EOR. Wiesner and Golovin continued to question the NASA figures, but their fight was coming to a conclusion.[9]

The NASA report was accompanied by a "peevish" letter from Webb to Wiesner. Webb told Wiesner that "my understanding is that you...will examine this and you will let me know your views as to whether we should ask for an appointment with the President." Webb's view was "we should proceed with the lunar orbit plan, should announce our selection of the contractor for the lunar excursion vehicle, and should play the whole thing in a low key." Webb said that unless he heard to the contrary, he would advise President Kennedy's appointments secretary Kenneth O'Donnell "that neither you nor the Defense Department wishes to interpose a formal objection" to NASA's going ahead with LOR. "In that case," wrote Webb, "I believe Mr. O'Donnell will not feel it wise to schedule the President's time and that the President will confirm this judgment."[10]

As Webb wrote his October 24 letter, President Kennedy was totally involved with dealing with the problem of Soviet missiles in Cuba, and most certainly was not going to take time to referee the NASA dispute with his science adviser. Webb and Wiesner talked by telephone on October 29, the Monday after the weekend during which the Cuban missile crisis was resolved. Wiesner said that his message to the president would not be to overrule any decision NASA might reach, but rather to be sure that a full and honest assessment had been made of all the options; Wiesner still questioned whether this was the case. Webb told Wiesner he "thought it better not to go to a formal hearing or involve the President personally in the decision," but Wiesner thought that "involving the President couldn't be avoided" because someone was sure to ask Kennedy whether the decision was made after the best possible analysis. On November 2, Wiesner and three PSAC members met with Webb and his senior staff to go over once again the White House objections to LOR; Wiesner recommended that NASA select instead a new alternative, a two-person flight using the direct ascent approach that was being strongly advocated by the builder of the Apollo command and service module, North American Aviation. Choosing this approach would mean that North American would manufacture spacecraft that would land on the Moon, not just the ship that would ferry astronauts from the Earth to lunar orbit and back. Again NASA refused to alter its position that LOR was the preferred approach. After discussing his reservations with the president, Wiesner decided not to insist on a formal meeting with Kennedy to make a final decision.[11]

On November 7, the day that NASA confirmed its tentative choice of lunar orbit rendezvous and announced its intent to issue the contract for the lunar lander to Grumman Aerospace, McGeorge Bundy wrote Wiesner, saying that the "President's conclusion on the moon method is that he would like a last letter from Wiesner to Webb saying that Kennedy "thinks the time is coming for a final recommendation and relies on Director Webb to review all the arguments and to produce that recommendation." Bundy also told Wiesner that "we should make Webb feel the responsibility for a definite decision and the importance of weighing all opinions, without trying to make his decision for him." This communication was somewhat after the fact, given the NASA announcement that day.[12]

Webb did write the requested letter to President Kennedy. He told the president that "by adopting LOR, the mission can be accomplished at least one year earlier in comparison to the EOR mode." By this time, Webb was aware of Kennedy's desire to accomplish the initial lunar landing at the earliest possible date in order to give maximum assurance of accomplishing the landing before the Soviet Union did. The cost of LOR approach, suggested Webb, would be "10% to 15% less than for the EOR approach." He told Kennedy that the decision to go with the LOR mode "had to be made at this time in order to maintain our schedules, which aim at a landing attempt in late 1967," and that, with this decision made "we intend to drive forward vigorously on every segment of the manned lunar landing program."[13]

With NASA's November 7 announcement and Webb's letter to Kennedy, NASA now indeed had all the pieces in place to "drive forward vigorously" on Project Apollo. Kennedy's approach with respect to the LOR decision was again typical of his management style as president. He welcomed a wide variety of views being expressed while decisions were being made, but seldom if ever forced on an operating agency of the executive branch a course of action with which its leadership disagreed. If Kennedy felt that a wrong policy was being pursued, he was more likely to remove the agency head than insist on his carrying out a White House–imposed perspective. Now it was up to NASA to deliver on its commitments.

## "Not Because They are Easy, but Because They are Hard"

Kennedy's September 1962 trip to space installations was intended to give him a sense of the character and scope of the accelerated space effort he had approved in May 1961. At Huntsville, he saw one of the very large F-1 engines that would be used to power the Moon rocket, and witnessed a test firing of the engines of the first stage of the Saturn 1 booster that generated 1.5 million pounds of thrust. At the Cape Canaveral Air Force Station and the NASA Launch Operations Center on Merritt Island in Florida, he saw the launch pad from which two Saturn 1 boosters had already been successfully launched, and was briefed on the Gemini and Apollo programs. He told the NASA staff at the Launch Operations Center that "as long as the decision has been made that our great system and others will be judged at least in one degree by how we do in the field of space, we might as well be first."[14] On the flight from Florida to his next stop in Houston, Kennedy spent more than an hour in informal conversation regarding both the civilian and the military space programs with BOB director Bell, director of defense research and engineering Harold Brown, and Seamans of NASA. Kennedy was surprised to discover that he and Seamans had both been members of the Class of 1940 at Harvard. (James Webb was on a separate plane with Vice President Johnson.)

The trip also provided an opportunity to make a major speech on the reasons behind his decision to go to the Moon; that speech is often confused with the May 25, 1961, address to a joint session of Congress during which Kennedy had announced his decision. At 10:00 a.m. on September 12, 1962, on a day that was oppressively hot and humid, President Kennedy addressed a Houston crowd of more than 40,000 people, mostly students, in the Rice University football stadium. Like many other of the president's speeches, various government agencies had suggested to the White House what Kennedy might say; in this case, material had been submitted to JFK's top speechwriter Theodore Sorensen from at least NASA, BOB, the State Department, and by Sorensen's brother Tom at the U.S. Information Agency. Sorensen drew on ideas, phrases, and words from most of these inputs in drafting Kennedy's Rice University speech, but he was a consummate master of spoken rhetoric, and his drafts transcended the sometimes pedestrian content of the agency submissions. Even so, most of Kennedy's speeches as president, including this memorable space address, were group products.

The Rice University speech is perhaps most remembered for the line "we choose to go to the moon in this decade and do the other things, not because they are easy, but because they are hard." The original version of this sentence was actually suggested by NASA; the agency's lengthy input into the speech preparation included the sentences: "We chose to go to the moon in this decade not because it will be easy, but because it will be hard. It will bring out the best in us." These words formed the basis of a much more eloquent declaration. As Kennedy spoke, he substituted the word "choose" for the word "chose," and said, "we choose to go to the moon in this decade and do the other things, not because they are easy, but because they are hard, because that goal will serve to organize and measure the best of our energies and skills, because that challenge is one that we are willing to accept, one we are unwilling to postpone, and one which we intend to win." Among the other memorable sentences in Kennedy's Rice University address are the following:

> "The exploration of space will go ahead, whether we join it or not, and it is one of the great adventures of all time, and no nation which expects to be the leader of other nations can expect to stay behind in the race for space."
>
> "This generation does not intend to founder in the backwash of the coming age of space. We mean to be part of it—we mean to lead it." (This last sentence was extemporized rather than in the prepared text of the speech.)
>
> "We set sail on this new sea because there is new knowledge to be gained, and new rights to be won, and they must be won and used for the progress of all people."
>
> "But why, some say, the moon? Why choose this as our goal? And they may well ask why climb the highest mountain. Why, 35 years ago, fly the Atlantic? Why does Rice play Texas?" (This last sentence was written into the speech's reading text in Kennedy's hand.)

> "Many years ago the great British explorer George Mallory, who was to die on Mount Everest, was asked why did he want to climb it. He said, 'Because it is there.' Well, space is there, and we're going to climb it, and the moon and planets are there, and new hopes for knowledge and peace are there. And, therefore, as we set sail we ask God's blessing on the most hazardous and dangerous and greatest adventure on which man has ever embarked."[15]

President Kennedy completed his tour of space installations later in the day, first visiting the temporary quarters of the new Manned Spacecraft Center in Houston, then flying to St. Louis to visit the plant of the McDonnell Aircraft Corporation, where the Mercury and Gemini spacecraft were developed. By the time he returned to the White House, Kennedy had had an apparent change of mind; rather than giving high priority to controlling the costs of sending Americans to the Moon, Kennedy wanted to know "how soon can we get there?"

## To the Moon in 1966?

At some point during his tour of NASA's installations, President Kennedy asked NASA administrator James Webb whether it was possible to get to the Moon earlier than the late 1967 target date that NASA was using in its planning. This question was most likely prompted by Kennedy's conversation with manned space flight head Brainerd Holmes, whom Kennedy met for the first time during the Huntsville stop on the tour. Holmes had come to NASA in October 1961, and a year later was becoming impatient with the pace of the Apollo program and James Webb's resistance to Holmes's suggestion that it might be accelerated.

Seamans, Holmes's immediate supervisor, describes Holmes as having "blinders on... All he was going to do was just move to the best of his ability on manned flight, but the devil take the hindmost on anything else." Holmes also did not operate on the same wavelength as the verbose Webb; Holmes told Seamans, "I don't understand what the hell the boss is talking about a lot of the time on these general [management approach] things, and I could care less." Holmes also "was a very exciting person for the news people... He had a way of expressing himself that made news because it was a little bit controversial." *Time* magazine featured Holmes on the cover of its August 10, 1962, issue; in the issue's six-page story "Reaching for the Moon," Holmes was frequently mentioned, Webb not at all. Seamans also suggests that "among others who were sort of captivated... by Brainerd was the President." After Holmes was introduced to Kennedy as they watched the firing of the Saturn 1 first stage at Huntsville, the press asked Holmes whether this was indeed the first time the two had met, suggesting "this is shocking that a man of your great responsibility should only be meeting the President for the first time right now." According to Seamans, "this hit Brainerd sort of in a sensitive area. He was a somewhat egotistical guy."[16]

The controversy between Holmes and NASA's top management simmered in the two months following President Kennedy's September trip as NASA's budget request for Fiscal Year 1964 was under review at the White House. On October 29, Webb wrote to Kennedy, responding to the president's question of what it would take to move the target date for the first lunar landing to 1966. Webb told the president that NASA's current target date of late 1967 "is based on a vigorous and driving program but does not represent a crash program," while "a late 1966 target date would require a crash, high-risk effort." By this time, NASA had sent BOB a request for a \$6.2 billion budget for FY1964; this represented a 68 percent increase over the agency's FY1963 budget. NASA also estimated that the first lunar landing might be targeted six months earlier if there was an immediate \$427 million supplement to the FY1963 NASA budget. To target the first landing in late 1966, a year earlier than then planned, NASA planning would have to be drastically revised and a supplementary \$900 million would have to be provided in FY1963; in addition, the NASA budget for FY1964 would have to increase to \$7.0 billion. Webb told President Kennedy that the budget and program projections were "preliminary" and not based on "detailed programmatic plans," but "we are prepared to place the manned lunar landing program on an all-out crash basis aimed at the late 1966 target date if you should decide this is in the national interest."[17]

Holmes at some point in this period had formally asked Seamans to approve a \$440 million FY1963 budget supplement, saying that such an increase would allow the first lunar landing attempt to come in late 1966. Seamans "couldn't believe" Holmes's claim that the program could be accelerated by twelve months with such a relatively modest budget increase. He denied Holmes's request; Holmes then asked for a meeting with Webb and Dryden; both also gave him a negative response. Holmes then turned to his friends at *Time* magazine, telling them that there was "an upheaval at NASA," with Holmes and Webb "locked in deadly combat" and that "one of them might have to go, and it wasn't necessarily Brainerd."[18]

The possibility of a story about this internal dispute appearing in *Time* caught President Kennedy's attention, and he asked science adviser Wiesner to meet with Hugh Sidey and Lansing Lamont, the *Time/Life* reporters preparing a story on "the lagging manned lunar program." Wiesner did so on November 16; only Lamont was able to make the meeting because Sidey's plane was grounded. Lamont reported that *Time* was being told "by NASA staff [undoubtedly Brainerd Holmes] and contractors that the lunar program is slipping for lack of funds," that "\$400 million is needed now to prevent a loss of six months," and that "Mr. Webb discounts the lunar effort and is not backing your [Kennedy's] commitment." Wiesner contradicted Lamont's conclusions, telling the reporter that "we have a hard driving program," that "we had long since passed the point where money would make a major impact on the schedule," and what was needed now was "good planning and management." Wiesner told the president that he did not think he "changed his [Lamont's] views much, though I really tried." After meeting with Lamont, Wiesner called Seamans, who told Wiesner that he also had met with the reporter and delivered the same message as had Wiesner.[19]

The White House attempt at managing the *Time* story failed. In its issue dated November 23 (which was on newsstands on November 19), the magazine reported that "the U.S. man-to-the-moon program was in earthly trouble" due to the "clashing personalities and ideas of the project's two top officials." Holmes was described in the article as a "brilliant, aggressive electrical engineer with a hard-bitten talent for ramming through tough projects," while Webb was characterized as having "a cautious eye where money is concerned." *Time* reported that Holmes believed that the lunar landing program "is already four to six months behind schedule—and the reason is that Webb is dragging his feet." Webb was reported as saying, "the moon program is important, but it's not the only important part of the space program," while Holmes argued that Apollo was "the top priority program within NASA." The article concluded that "such are the differences between Webb and Holmes that the whole program is in danger of bogging down."[20]

John Kennedy was not the type of person, or president, to ignore this public reporting of a dispute with respect to one of his high priority initiatives. He quickly called a cabinet room meeting to find out for himself what was going on.

## What is the Goal—Getting to the Moon First or Space Preeminence?

The meeting took place on November 21; it was also an occasion to review NASA's budget proposal for Fiscal Year 1964. The BOB had not yet forwarded to the president Webb's October 29 letter about the budgetary implications of accelerating the target date for the first lunar landing. Like many communications to the president from government agencies, this letter had been referred to one of the staff agencies of the executive office, in this case BOB, for review and a decision of whether it needed direct presidential attention. Kennedy may well have wondered why he had not heard from Webb after asking him about this possibility on his September tour, and that could have added to his concern about the accuracy of the *Time* article. Of course, Kennedy had also been immersed with the Cuban missile crisis and the midterm congressional elections in the interim. Budget director Bell prepared a November 13 memorandum on the NASA budget situation that incorporated the schedule and budget estimates in Webb's October letter; this memorandum was distributed to all participants in the meeting. In his memorandum, Bell identified two policy issues on which presidential guidance was needed:

1. "The pace at which the manned lunar landing should proceed, in view of the budgetary implications and other considerations," and
2. "The approach that should be taken to other space programs in the 1964 budget, i.e., should they as a matter of policy be exempted from or

subjected to the restrictive budgetary ground rules applicable in 1964 to other programs of the Government."

NASA's budget request for FY1964 was $6.2 billion, including $4.6 billion for the lunar landing program and $1.6 billion for all other NASA activities. To keep the program on an "optimum" schedule aiming at a mid-1967 landing, Bell told the president, would require a supplementary appropriation of over $400 million in 1963 and about $550 million above the estimates for FY1964 made three months earlier. NASA's recommended program aiming at a late 1967 landing would not require a FY1963 supplement, but would require the full $4.6 billion funding in FY1964. As Webb's October 29 letter had indicated, advancing the target date to late 1966 would require a $900 million supplement and "create enormous additional management problems." Bell noted that "in NASA's view and ours" such a course of action "would not appear to offer enough assurance of actually advancing the date of a successful attempt to be worth the cost and other problems involved." Bell also offered a lower cost option that would slip the landing target date to late 1968; that option would require $3.7 billion for the lunar landing program in FY1964 rather than $4.6 billion and was "significant as indicating probably about the lowest 1964 estimate under which the first actual manned lunar landing might still be expected to occur during this decade, after a realistic allowance for slippage." Bell also reported that "our understanding of the latest intelligence estimates is that there is no evidence yet that the Russians are actually developing either a larger booster... or rendezvous techniques." Thus "extreme measures to advance somewhat our own target dates may not be necessary to preserve a good possibility that we will be first." This may have been one of the first warnings to President Kennedy that the race to the Moon he thought the United States was running may not have been a race at all. But as the November 21 meeting unfolded, it became clear that Kennedy was still in a race mentality.

With respect to the other portions of the NASA budget, Bell reported that it was NASA's view that if there were any reduction in NASA's $6.2 billion request, it should be applied "at least in part to the manned lunar landing program, in order to maintain a 'balanced' total program." He added that "the Administrator and his principal assistants are fearful that the appeal and priority of the manned lunar landing program may turn NASA into a 'one program agency' with loss of leadership and standing in the scientific community at home and abroad, and inadequate provision for moving ahead with developments required for future capabilities in space." The BOB did not agree with this line of argument, suggesting that the "unique sort of national decision" that led to the lunar landing program did not "automatically endow other space objectives and programs with a special degree of urgency." The BOB suggested a $300 million cut in the "other activities" part of the NASA budget, noting that while this amount might "seem small," in the context of the overall space budget, it was "large compared to most other possibilities for adjustment in the 1964 budget."[21]

Present at the November 21 meeting in addition to President Kennedy were James Webb, Hugh Dryden, Robert Seamans, and Brainerd Holmes from NASA; David Bell, his deputy Elmer Staats, and Willis Shapley from BOB; Edward Welsh (Vice President Johnson had been invited but was out of town); and Jerome Wiesner. At some point in the meeting, President Kennedy activated the secret tape recording system that had been installed in July 1962 in the Oval Office, in the cabinet room, and on his telephone.[22] There is thus available a fascinating *verbatim* account of the portions of the meeting during which President Kennedy and James Webb got into a spirited discussion of the priority to be assigned to the lunar landing mission compared to other NASA activities.[23] Excerpts of that conversation include:

*Kennedy:* Do you think this [lunar landing] program is the top priority of the agency?

*Webb:* No sir, I do not. I think it is one of the top priority programs.

*Kennedy:* Jim, I think it is the top priority. I think we ought to have that very clear. Some of these other programs can slip six months, or nine months, and nothing strategic is going to happen... But this is important for political reasons, international political reasons. This is, whether we like it or not, in a sense a race. If we get second to the Moon, it's nice, but it's like being second any time. So if we're second by six months because we didn't give it the kind of priority [needed], then of course that would be very serious. So I think we have to take the view that this is top priority with us.

* * *

*Kennedy:* I would certainly not favor spending six or seven billion dollars to find out about space no matter how on the schedule we're doing... Why are we spending seven million dollars on getting fresh water from salt water, when we're spending seven billion dollars to find out about space? Obviously you wouldn't put it on that priority except for the defense implications. And the second point is the fact that the Soviet Union has made this a test of the system. So that's why we're doing it. So I think we've got to take the view that this is the key program... Everything we do ought really to be tied to getting on the Moon ahead of the Russians.

*Webb:* Why can't it be tied to preeminence in space?

*Kennedy:* Because, by God, we keep, we've been telling everybody we're preeminent in space for five years and nobody believes it because they [the Soviets] have the booster and the satellite... We're not going to settle the four hundred million this morning... But I do think we ought to get it, you know, really clear that the policy ought to be that this is the top priority program of the agency, and one of the two things, except for defense, the top priority of the United States government.

I think that is the position we ought to take.

Now, this may not change anything about the schedule but at least we ought to be clear, otherwise we shouldn't be spending this kind of money,

> because I'm not that interested in space. I think it's good, I think we ought to know about it, we're ready to spend reasonable amounts of money. But we're talking about these fantastic expenditures which wreck our budget and all these other domestic programs and the only justification for it in my opinion to do it in this time or fashion is because we hope to beat them and demonstrate that starting behind, as we did by a couple of years, by God, we passed them.

In the course of the November 21 meeting, Brainerd Holmes apologized for letting his differences with James Webb get into the press. He said: "I ought to add that I'm very sorry about this. I have no disagreement with Mr. Webb...I think my job is to say how fast I think we can go for what dollars."

Just before he left the meeting, President Kennedy requested a letter from NASA stating clearly the agency's position. First, Dryden drafted a relatively brief reply; then Seamans prepared a more extensive response. He took his draft to James Webb, who had stayed home from work with a severe migraine headache. (Seamans comments that "it's not surprising that one occurred at this time.") Seamans and Webb revised the letter to their and Dryden's satisfaction; it was sent to President Kennedy on November 30. At that point, final decisions on the NASA FY1964 budget had still not been made, and so the nine-page letter included a plea for approval of NASA's $6.2 billion request. The bulk of the letter supported the position that "the objective of our national space program is to become pre-eminent in all important aspects of this [space] endeavor and to conduct the program in such a manner that our emerging scientific, technological, and operational competence in space is clearly evident." The letter noted that "the manned lunar landing program provides currently a natural focus for the development of national capabilities in space and, in addition, will provide a clear demonstration to the world of our accomplishments in space." However, the letter argued, "the manned lunar landing program, although of the highest national priority, will not by itself create the pre-eminent position we seek."[24]

Because Vice President Johnson was not present at the November 21 meeting, he was separately asked for his views on the issues discussed there. Budget director Bell wrote Johnson on November 28, saying that "the President would appreciate your views" on whether the manned lunar landing "should be regarded as *the* top priority program—or as *one* of the top priority programs" and on the "desirability and feasibility of augmenting the funding for the manned lunar landing program in the present fiscal year." Johnson replied to Kennedy on December 4. He told the president that "as to the matter of relative priorities, I consider your Messages and Budget requests have made it clear that the Manned Lunar Landing Effort has the highest priority even though other projects are to be pursued vigorously." Between November 21 and his reply to Kennedy, Johnson had met with Brainerd Holmes for an hour to discuss the impact of a FY1963 budget supplement on the Apollo schedule, and had concluded that, "while I would urge any action that would have a reasonable chance of accelerating the Manned Lunar Landing project target date," he concurred with the conclusion of the NASA leadership that a "supplemental appropriation could not be made available in time to advance that date much, if any."[25]

Even with all that he had heard, President Kennedy did not easily give up on the idea that the lunar landing program could be accelerated. As Kennedy toured various nuclear facilities in New Mexico and Nevada in early December 1962, he asked Wiesner to look once more into "the possibility of speeding up the lunar landing program." Wiesner on January 10, 1963, reported to Kennedy that "approximately 100 million dollars of the previously discussed 326 million dollar supplementary could have a very important effect on the schedule." Wiesner thought that funds in this amount might be transferred from the Department of Defense budget to pay for DOD involvement in NASA's Project Gemini, the new NASA program to test out rendezvous activities in Earth orbit and to serve as a bridge between Mercury and Apollo. Wiesner told the president that funds in that amount could be used to advance the date of the first Saturn V launch by some five months, and there was some chance that this acceleration could allow an earlier attempt at the landing. Wiesner noted that "the date of the first lunar landing attempt can be accelerated *only*" if Saturn V availability were advanced. Kennedy the same day sent the Wiesner memorandum with these suggestions to Vice President Johnson, asking for his views. Johnson replied on January 18, telling Kennedy that "the people we need on the Hill tell me that the supplemental request would be inadvisable and could not be approved in time to accelerate the program." With that response, the thought of requesting supplemental funds for NASA was put to rest.[26]

Even before this January exchange of correspondence, as the final budget decisions for Fiscal Year 1964 were made by the president and BOB in December 1963, any thought of a supplemental request for FY1963 were abandoned, reluctantly on the president's part. Kennedy had once again accepted the position of NASA administrator Webb on how best to go forward. Indeed, BOB made reductions in both the lunar landing budget and the "other activities" portion of the NASA request. The president in mid-January 1963 sent a FY1964 budget proposal to the Congress requesting $5.712 billion for NASA, an almost half-billion dollar cut from what NASA had requested in September. Although the increase was not what NASA had hoped for, it still reflected a 55 percent jump over NASA's FY1963 appropriation.

## Conclusion

By the end of 1962, the White House appeared to have accepted the arguments set forth in the November 30 NASA letter arguing that the lunar landing program, though clearly a very high NASA priority, was in itself insufficient to achieve the goal of American space preeminence—a clearly leading position in all areas of space activity. Seamans noted that "whether from agreement, exhaustion, or diversion, President Kennedy gave tacit approval to NASA's programs and policies by not engaging us in further discussion on the questions of NASA's top priority." Seamans adds: "Preeminence in space on all fronts was our goal; landing men on the Moon was the top (DX) priority."[27]

During 1963, John Kennedy was no longer totally focused on how soon the United States could get to the Moon; he seemed in fact to have accepted NASA's argument that preeminence in all areas of space activity was the more appropriate goal. In addition, at least some of the president's associates, and perhaps Kennedy himself, questioned whether getting to the Moon before the Soviet Union remained a compelling national objective. Indeed, Kennedy asked, might it be both desirable and feasible to cooperate, rather than compete, with the Soviet Union in humanity's first journeys beyond Earth orbit? John Kennedy had brought with him the idea that space might be a particularly promising arena for tension-reducing U.S.-Soviet cooperation as he entered the White House in January 1961, and it had never totally disappeared from his thinking.

## Chapter 10

# Early Attempts at Space Cooperation

Theodore Sorensen recalls that "it is no secret that Kennedy would have preferred to cooperate with the Soviets" in space rather than compete with them.[1] In light of his soon-to-be-made decision to enter a space race in competition with the Soviet Union, it is worth noting that JFK's initial priority on becoming president was to make space an area for U.S.-Soviet cooperation. Kennedy came into the White House believing that science and technology could be used as tools to advance foreign policy interests and to reduce international tensions, and hoping that the habits of cooperation developed in sectors such as science and technology could spill over into areas more central to security interests. As a presidential candidate, Kennedy had said that "wherever we can find an area where Soviet and American interests permit effective cooperation, that area should be isolated and developed."

Space was one of those areas; Kennedy noted that "when the United States has at last developed rockets with larger thrust, certain aspects of the exploration of space might be handled by joint efforts; for the cost of space efforts will mount radically as we move ambitiously outward."[2] Kennedy's transition task force on space reinforced the view that space offered a promising area for cooperation. The report of the Wiesner panel said that "our space activities, particularly...in the exploration of our solar system, offer exciting possibilities for international cooperation with all the nations of the world. The very ambitious and long-range space projects would prosper if they could be carried out in an atmosphere of cooperation as projects of all mankind instead of in the present atmosphere of national competition."[3]

President Kennedy soon followed up the call in his inaugural address "to explore the stars together" with a more detailed and specific proposal. In his State of the Union address on January 30, 1961, Kennedy announced:

> This Administration intends to explore promptly all possible areas of cooperation with the Soviet Union and other nations "to invoke the wonders of science instead of its terrors." Specifically, I now invite all nations—including

> the Soviet Union—to join with us in developing a weather prediction program, in a new communications satellite program and in preparation for probing the distant planets of Mars and Venus, probes which may someday unlock the deepest secrets of the universe.
>
> Today this country is ahead in the science and technology of space, while the Soviet Union is ahead in the capacity to lift large vehicles into orbit. Both nations would help themselves as well as other nations by removing these endeavors from the bitter and wasteful competition of the Cold War.[4]

The launch of Yuri Gagarin on April 12 shifted Kennedy's attention from how to cooperate in space to how to enter, and win, the space race. Even so, the notion that cooperation was a more desirable path than competition stayed with him, and the White House in late May 1961 made one more attempt to engage the Soviet Union in a cooperative mission to the Moon, even as Kennedy announced his decision to send Americans to the lunar surface. Again in 1962, after the successful flight of John Glenn, the first American to orbit the Earth, Kennedy offered to Premier Nikita Khrushchev a range of cooperative possibilities, this time in areas other than lunar exploration. Khrushchev agreed to discuss these possibilities, and in 1962 and 1963 there were NASA-Soviet Academy of Sciences discussions that reached agreement in principle to cooperate. However, there was only modest substantive cooperative activity in subsequent months and years.

This chapter focuses on attempts during 1961 and 1962 to foster U.S.-Soviet cooperation in space, since they were of direct personal interest to President Kennedy. In a reversal from the position taken during the Eisenhower administration, where the emphasis with respect to space cooperation was on developing international arrangements and controls, Kennedy believed that direct cooperation between the two Cold War rivals was likely to make a greater contribution to an overall reduction in bilateral and thus global tensions.[5] The development of cooperative relations in space with other countries rarely rose to the presidential level for decision. NASA from its inception did develop such relations, particularly with Canada and the countries of Western Europe, but those emerging relationships are not discussed here.[6]

## The United Nations as the Venue for Space Cooperation?

The United States had in a variety of formal and informal settings tried to engage the Soviet Union in space cooperation in the 1959–1960 period, without success.[7] One possible arena for early space cooperation was the United Nations. A 1959 Soviet-sponsored General Assembly resolution had called for setting up a United Nations Committee on the Peaceful Uses of Outer Space and for convening a United Nations Conference on space matters.[8] But the Soviet Union had refused to accept the proposed structure of the new committee, and thus there had been no progress on space matters made within the United Nations by January 1961. Following up on President

Kennedy's call for space cooperation in the State of the Union Address, Secretary of State Dean Rusk in a February 2 memorandum suggested that talks be opened between the U.S. ambassador to the United Nations, twice-defeated presidential candidate Adlai Stevenson, and his Soviet counterpart, with the aim of moving forward on establishing the committee and convening the conference. These steps, said Rusk, would make the United Nations "the logical place to discuss the types of cooperative outer space proposals included in your State of the Union Message."[9]

After several weeks of discussion within the White House on how best to follow up the president's speech initiatives, McGeorge Bundy replied to Secretary Rusk on February 28. Bundy suggested that in the four weeks since Rusk had sent his memorandum to the president, "events in the United Nations, and in particular the Soviet Union's attitude toward that organization, have raised a question over here as to whether we really want to take active steps in that particular forum on this particular issue, with the Soviet Union, at this time." Bundy's feeling was "that the President would be reluctant to see us move in this direction now." Bundy noted that the White House and the State Department had in the time since Rusk's memorandum "agreed that Jerry Wiesner should be asked to take the lead in planning on the general problem of relations with the Soviet Union in this and other scientific fields."[10]

## Initial Proposals for U.S.-Soviet Space Cooperation

Bilateral U.S.-U.S.S.R. cooperation was thus the preferred alternative to cooperating through the United Nations, and active discussion of this possibility had begun soon after the president's January 30 speech. Philip Farley, special assistant to the secretary of state for atomic energy and outer space, told Secretary Rusk on February 9 that he had been meeting with Wiesner and acting NASA administrator Dryden and had found "a good deal of uncertainty and diversity of expert opinion as to what makes technical and practical sense in the three areas mentioned in the President's State of the Union message... as well as other possible areas of space cooperation." To address this situation, the White House set up a "Task Force on International Cooperation in Space" in early February. The group was composed of both non-governmental people, particularly members of and consultants to the President's Science Advisory Committee (PSAC), and officials from NASA, the Department of State, and the White House science office. The task force was charged both with identifying "the full range of possibilities for cooperative efforts" and describing "the optimum shape of possible international cooperation in outer space... on the basis of pooling or even merging of efforts in a world-wide venture... Such a description of optimum international cooperation in space activities would be an important contribution to re-examination of U.S. objectives and programs in outer space."[11] The group was chaired by Bruno Rossi, a professor of physics at MIT, who had been a member of the Wiesner transition panel on outer space.

The task force carried out its examination between February 17 and mid-March.[12] It came up with twenty-two specific proposals for U.S.-Soviet space cooperation, ranging from projects involving only data exchanges or coordination of separate projects, to intimate cooperation in ambitious projects for the human exploration of the Moon and the robotic exploration of the planets, particularly Mars. At the first meeting on February 17, one of the members of the task force, Richard Porter of General Electric, submitted a memorandum suggesting a U.S.-Soviet "Rendezvous on the Moon" project to establish an international base on the surface of the Moon, along the lines of scientific bases in the Antarctic. Porter suggested that "if agreement could be reached on this major project between the United States and the USSR, all other bilateral and multilateral cooperative projects involving the USSR would become feasible and operative"; the group found this proposal intriguing. The thought was that if the United States and the Soviet Union agreed to cooperate in lunar exploration, they would then invite other nations to join in.[13]

As the Rossi task force was completing its examination, senior government officials were also discussing U.S.-Soviet space cooperation. On March 8, Rusk met with new NASA administrator Webb and deputy administrator Dryden. Rusk told the NASA officials that there was "keen interest in the possibility of a productive approach to the Soviet Union for outer space cooperation." Webb's reaction was that, given the current uncertainty regarding the Kennedy administration's approach to space, a "decision on approaches to the Soviets was secondary to deciding what the United States wanted to do in space." Webb spoke of the opportunities for foreign relations inherent in the space program, and told Rusk that in his three weeks since becoming NASA administrator he had concluded that there was a need for acceleration of the NASA program and increased funding "by a substantial amount." Rusk in reply wondered "what the purpose was of activities in space on this major scale. Should not the objectives be clearly identified and undertaken not competitively but on behalf of the human race as a whole."[14] State Department leadership remained throughout the Apollo program a source of skepticism of the value of a unilateral large-scale space effort.

The Rossi task force submitted its final report to Wiesner on March 20. The report noted that "a cooperative enterprise in this new and bold venture would stimulate constructive thinking on the part of people all over the world and help reduce world tensions," and that "it is vitally important for the avoidance of future conflicts to establish early cooperation in such fields where unchecked competition is likely to produce a dangerous situation" such as "meteorological activities that might eventually lead to weather control, and large scale exploration of the moon and planets." The report suggested that the United States should "give preference to projects that avoid the difficulties connected with a high degree of [Soviet] involvement or else to projects that are sufficiently bold and dramatic to sweep aside these difficulties." Wiesner told Rossi that the task force had done a "superb" job in providing "the essential scientific judgment that is prerequisite for any political action that may follow."[15]

The task force report was next reviewed and revised by several of the government agencies involved in the space program. An April 4 draft report of this second review set out the political rationale for expanded U.S.-Soviet space cooperation: " The objectives are to confirm concretely the U.S. preference for a cooperative rather than competitive approach to space exploration, to contribute to reduction of Cold War tensions by demonstrating the possibility of cooperative enterprise between the U.S. and the USSR in a field of major public concern, and to achieve the substantive advantages of cooperation that in major projects would impose more of a strain on economic and manpower resources if carried out unilaterally."[16]

Interestingly, this paragraph was missing in an April 13 draft of the report, perhaps reflecting a shift in White House thinking toward a more competitive stance in space in the wake of Yuri Gagarin's April 12 orbital flight. Three categories of cooperative proposals were suggested in that draft: (1) use of existing or easily attainable ground facilities for the exchange of information and services; (2) coordination of independently launched satellite experiments; and (3) "coordination or cooperation in ambitious projects for the manned exploration of the moon and the unmanned exploration of the planets." With regard to the third category, the document suggested "as a first step in non-limited cooperative effort, the U.S. and the USSR would each undertake to place a small party (about 3) of men on the moon." Such an undertaking would have the "greatest potential for matching the President's theme that 'Both nations would help themselves as well as other nations by removing these endeavors from the bitter and wasteful competition of the Cold War.' "[17]

An early start on U.S.-Soviet space cooperation was not in the cards, however. Even as the task force began its work, President Kennedy was receiving initial indications that the Soviet Union was unlikely to be receptive to the kind of initiatives he had in mind. On February 13, he congratulated Soviet premier Nikita Khrushchev on the launching of a Soviet scientific probe to Venus. In his reply, Khrushchev took note of Kennedy's cooperative overtures in his inaugural address and State of the Union speech, but indicated that the creation of "favorable conditions" for space cooperation would require "settlement of the problem of disarmament."[18] This repeated the Soviet policy line of several years standing, linking cooperation in various areas, including space, to a U.S.-Soviet disarmament agreement, and did not auger well for the Kennedy approach of isolating areas of common U.S.-Soviet interest for cooperation even if tensions remained in the two countries' security relationship. Even so, as he met with new NASA administrator James Webb on March 22, the president stressed to Webb "his desire that we try to work out as many ideas as possible for utilization in the proposed conference with the Russians [a hoped-for Kennedy-Khrushchev summit meeting] for international cooperation. He said he hoped we would give this very high level attention."[19]

## A Last Try at U.S.-Soviet Space Cooperation before Beginning the Race to the Moon

On April 12, 1961, the Soviet Union once again achieved a spectacular space "first"—the orbiting of Soviet Air Force pilot Yuri Gagarin. As President Kennedy and his associates discussed how best to react to this newest Soviet achievement in space, notions of a cooperative initiative took a back seat. But the preference for cooperation was never far from Kennedy's mind, and even as he decided that the United States should enter a space race, he at the same time made one more attempt to engage the Soviet Union in space cooperation.

On May 16, President Kennedy received a letter from Soviet premier Nikita Khrushchev formally accepting Kennedy's February suggestion that he and Kennedy meet. Khrushchev suggested that the meeting take place in Vienna in early June. This was to be the first (and what turned out to be the only) time the two superpower leaders met face-to-face. This late acceptance of Kennedy's suggestion for such a meeting came after several rounds of communications between Moscow and Washington. The White House had believed the idea of a Kennedy-Khrushchev meeting dead after the Bay of Pigs debacle, but Khrushchev thought a meeting between the two heads of state could still be useful.[20] From Khrushchev's perspective, Kennedy and the United States were in a weakened position after the Bay of Pigs, and a summit meeting might result in compromises favoring Soviet interests. In a May 16 message to German chancellor Konrad Adenauer explaining his reasons for considering Khrushchev's proposal for a meeting, Kennedy explained that he was "faced with the problem of going ahead with this or withdrawing my previous indication of my willingness to do so." He added that he had given Khrushchev an "interim agreement" to a "get acquainted" meeting and that his "present disposition is to proceed . . . provided that there are no further untoward developments meanwhile." Kennedy expected the discussions at the meeting to be "quite general in character."[21]

In preparing for the meeting with Khrushchev, the idea of proposing U.S.-USSR space cooperation was resurrected, even as planning for setting the goal of a lunar landing before the Soviets was in its final stages. On May 18, Jerome Wiesner sent the president two papers "on possible cooperative projects in space to be explored with the Soviet Union." One of these papers was a memorandum for the president from the Department of State dated May 12 summarizing the work that had been carried out in the February–April time period; the other was the April 13 report on possible cooperative initiatives. The State Department memo listed ten possible areas of cooperation, ranging from "reciprocal ground-based support for space experiments" to "coordination or cooperation in manned exploration of the moon." It suggested that the initial approach to the Soviet Union should offer "a range of choice as to the degree and scope of cooperation they wish to embark on" and should be "unpublicized and low-pressure." The paper also suggested that "we should not exclude from our list a mention of the possibility of cooperating in the most ambitious projects related to manned exploration of the moon and investigation of the planets." This was so because such an omission would leave the United States "open to a Soviet initiative which would make them proponents of large-scale international cooperation, thus aligning them with a wide-spread sentiment... that such ventures should be undertaken cooperatively on behalf of all mankind." The paper noted that "our recent astronaut flight [the May 5 suborbital mission of Alan Shepard] and the crystallization of plans for the expansion and acceleration of our space program have served to place us in a relatively favorable posture for an approach to the Soviets." Wiesner asked Kennedy, "would you like me to do anything more about this?"[22]

Kennedy did indeed want these possibilities explored. According to one of those involved in preparing Kennedy for the summit meeting, the president did not have "any definite plan of procedure in mind." Rather, "he wanted to keep himself flexible with a minimum of fixed positions, so as to be able to explore fully any openings that might emerge" during his conversations with the Soviet leader." However, "the president did have in mind certain concrete possibilities for improving relations should the opportunity for proposing them arise. One that he considered might be particularly fruitful was cooperation in space."[23]

Beginning in mid-May, Kennedy pursued both formal and informal paths to explore Soviet interest in space cooperation.[24] As the agenda for the summit meeting with Khrushchev was being planned, the president directed Dean Rusk to raise the possibility of space cooperation, including a joint lunar mission, with Soviet foreign minister Andrei Gromyko. Gromyko's May 20 reply echoed the Soviet line—that without progress towards general disarmament, "all cooperation in the field of rocket research and any exchange of information about Soviet rocket technology is inconceivable."[25]

Kennedy did not accept this response as definitive. Beginning in the aftermath of the Bay of Pigs crisis, the president had established a secret, back-channel line of communication with the Soviet leadership. His brother, Attorney General Robert Kennedy, developed a relationship with Georgi Bolshakov, a mid-level agent of the intelligence service of the Soviet Army, the GRU, who was working under the cover of being a Soviet reporter in Washington. In the weeks before the summit, Robert Kennedy and Bolshakov met privately several times. The attorney general used the GRU agent to communicate President Kennedy's thinking on what might be accomplished at a summit meeting directly to the Kremlin, and to receive back messages from the Soviet leadership, particularly with respect to the possibility of a summit agreement on a nuclear test ban as a step toward arms control. The Soviet response, which had been approved by the Presidium, left little hope for a successful summit unless the United States was willing to address the issue of the future status of Berlin after the Soviet Union signed a peace treaty with the communist-controlled German Democratic Republic (East Germany).[26]

On May 21, following Gromyko's negative response, Robert Kennedy added an inquiry to the Kremlin about the possibility of space cooperation to his conversations with Bolshakov. There was no response to this suggestion, and the president proceeded with announcing the lunar landing project in his address to Congress on May 25.

## Kennedy and Khrushchev Meet

Space was one of four areas of scientific cooperation initially identified for possible discussion at the June 3–4 Vienna summit; the others were nuclear science, earth science, and life science. In a May 29 memorandum to the president on summit preparations, national security adviser Bundy attached "a new and much improved memorandum from Wiesner's office." This memorandum listed only four potential cooperative projects, two in space and two in nuclear science. The two space projects suggested were "use of ground facilities for cooperative experiments" and "planetary probes." Cooperation in a lunar landing program was only briefly mentioned as one of the "other possible areas for cooperative projects."[27] In transmitting the memorandum, Bundy cautioned the president that "your own proposals [on scientific cooperation] to Khrushchev should probably go no further than to express your own interest and to suggest that the matter be discussed at experts' meeting arranged by Ambassador [Llewellyn] Thompson." This approach was prudent, suggested Bundy, because "the practical process of scientific cooperation can be very difficult even with friends, and you will not want to get your own prestige hooked to specific negotiations that could be made sticky at any time by the Soviets."[28]

The Vienna summit was, in President Kennedy's words, "a very sobering two days." During their meetings, both alone with just interpreters present and with their staff, "Khrushchev had not given way before Kennedy's reasonableness, nor Kennedy before Khrushchev's intransigence."[29] The Soviet leader insisted that he would sign a peace treaty with the German Democratic Republic by the end of the year, and that the new East German government would then have the right to cut off U.S. access to Berlin. Kennedy responded that this was unacceptable, and that if necessary the United States would use force to assure its access. Khrushchev replied "force will be met with force." The president concluded his conversation with Khrushchev with the observation that "it would be a cold winter."[30]

In this grim atmosphere, there was little chance to bring up secondary topics such as space cooperation. The only opportunities for more relaxed conversation came at two luncheons for the U.S. and Soviet delegations. At the June 3 lunch hosted by President Kennedy, the talk turned to the flight of Yuri Gagarin and then to the possibility of launching a man to the Moon. None of the other proposals for scientific cooperation prepared for Kennedy's use were discussed. With respect to a lunar mission, Khrushchev said that "he was cautious because of the military aspects of such flights." Then, "in response to the President's inquiry whether the US and the USSR should go to the Moon together, Mr. Khrushchev first said no, but then said 'all right, why not?' " Reportedly, the second response was made "half-jokingly." The next

John F. Kennedy and Nikita Khrushchev at the Vienna Summit Meeting, June 3–4, 1961 (JFK Library photograph).

day it was Khrushchev's turn to host a lunch, and once again Kennedy turned the conversation to a mission to the Moon. Khrushchev commented that "he was placing certain restraints on projects for a flight to the moon." The Soviet premier noted that "such an operation" would be "very expensive" and might "weaken Soviet defenses." He added: "Of course, Soviet scientists want to go to the moon," but "the U.S. should go first because it is rich and then the Soviet Union would follow." President Kennedy once again suggested a cooperative lunar landing effort; Khrushchev retracted his casual agreement of the preceding day, noting that "cooperation in outer space would be impossible as long as there was no disarmament." This was the case because "rockets are used for both military and scientific purposes." To Kennedy's suggestion that perhaps Soviet and U.S. lunar missions could be coordinated in their timing in order to save money, Khrushchev replied "that this might be possible but noted that so far there had been few practical uses of outer space launchings. The race was costly and was primarily for prestige purposes."[31]

The reason for this overnight change of mind, even if he had been serious in his response on the preceding day, was apparently Khrushchev's consultations with his advisers. The Soviet chairman's son, Sergey Khrushchev, has suggested that in May 1961 "my father was faced with the decision of whether to accept the challenge [of a race to the Moon] and be prepared to spend billions for the sake of keeping the palms of victory, or whether he should step aside and allow his undoubtedly richer competitor to get ahead of him . . . My father was not prepared to answer this question." In addition, "the military men came out against this proposal—they wanted to protect their secrets. Korolev [Sergei Korolev, the "chief designer" of the Soviet space program] was also against it, since he did not like the idea of sharing the palms of leadership with anyone."[32] In his memoirs, Nikita Khrushchev explains his unwillingness to cooperate in space as being due to the Soviet weakness in intercontinental ballistic missiles. He noted that "we had only one good missile at the time; it was the *Semyorka* [the R-7 ICBM] . . . Had we decided to cooperate with the Americans in space research, we would have had to reveal to them the design of the booster for the *Semyorka*." He added: "We knew if we let them have a look at our rocket, they'd easily be able to copy it." Thus, "they would have learned its limitations, and from a military standpoint, it *did* have serious limitations. In short, by showing the Americans our *Semyorka*, we would have been giving away both our strength and revealing our weakness."[33]

There the discussion on space cooperation ended. During the remainder of 1961, Cold War tensions were high and the Berlin Wall was erected; the outlook for any significant space cooperation between the United States and the Soviet Union was correspondingly bleak. President Kennedy's proposal to the Congress and the nation that the United States embark on an extremely ambitious space effort, with Project Apollo as its centerpiece, received widespread political support. The idea that the lunar landing program might be a joint U.S.-Soviet undertaking appeared stillborn, and other areas of potential space cooperation remained unexplored.

## The Beginnings of U.S.-Soviet Space Cooperation

On February 20, 1962, U.S. astronaut John Glenn completed the first U.S. orbital space flight. Among the congratulations for the success received at the White House was a February 21 telegram from Nikita Khrushchev. In the message, Khrushchev suggested that "if our countries pooled their efforts—scientific, technical and material—to master the universe, this would be very beneficial for the advance of science and would be joyfully acclaimed by all peoples who would like to see scientific achievements benefit man and not be used for 'Cold War' purposes and the arms race."[34] What was surprising in this message was that Khrushchev did not link the possibility of space cooperation to prior agreement on steps toward disarmament.[35]

The White House lost no time in its reply, since it appeared to provide the opening that President Kennedy had been seeking since his first days in office. On the next day, Kennedy sent a replying telegram to Khrushchev, saying "I welcome your statement that our countries should cooperate in the exploration of space. I have long held this belief and indeed put it forth strongly in my first State of the Union message." Kennedy added: "I am instructing the appropriate officers of this Government to prepare new and concrete proposals for immediate projects of common action, and I hope that at a very early date our representatives may meet to discuss our ideas and yours in a spirit of practical cooperation."[36]

On February 23, McGeorge Bundy issued a National Security Action Memorandum (NSAM) to the secretary of state, directing him to cooperate with the NASA administrator and the special assistant to the president for science and technology to "promptly develop" the new proposals called for in the president's telegram, "together with recommendations as to the best way of opening discussion with Soviet representatives on these matters." On February 27, Bundy sent out a revised memorandum, adding the executive secretary of the National Aeronautics and Space Council to the list of those being asked to develop the cooperative proposals.[37] (There were continuing tensions over roles and responsibilities between Wiesner's science office, which reported directly to the president, and the Space Council, which reported to the vice president. Leaving mention of the Space Council out of the original draft of the National Security Action Memorandum could have been either inadvertent or intentional; it does suggest that Vice President Johnson and the Space Council were not seen as central to international space issues.) On February 23, Bundy also sent a separate memorandum to NASA administrator Webb, explaining that while the NSAM had been addressed to the secretary of state because it involved international negotiations, President Kennedy "wants you to know how much he understands the central role of your organization in this problem." Moreover, said Bundy, the president had asked him to add a "private word." Kennedy recognized "that there are lots of problems in this kind of cooperation, and he knows that you have a great head of steam in projects which we do not want to see interrupted or slowed down." However, "there is real political advantage for us if we can make clear that we are forthcoming and energetic in plans for peaceful cooperation with the Soviets in this sphere. It is even conceivable that progress on this front would have an automatic dampening effect on the Berlin crisis." For these reasons, suggested Bundy, "the President hopes that you will urge your people to go a little out of their way to find good projects."[38]

The process of drafting a letter for the president to send to Khrushchev revealed some significant differences in perspective among NASA, the Department of State, and the White House science office. According to Eugene Skolnikoff, who at the time was on Wiesner's White House staff handling international issues, NASA proposed to include only those projects that it judged both technically and politically desirable, and tended to emphasize information exchanges rather than more extensive and intimate cooperation. This was consistent with the approach that had been developed several years earlier by Hugh Dryden and NASA's international affairs head, Arnold Frutkin. NASA argued that any cooperative undertaking must have meaningful substantive merit as a necessary condition, and not be undertaken primarily for political reasons.

Only through the intervention of the White House science office was a broader range of potential projects added to the draft list of proposals. Then the State Department watered down the list, favoring the NASA approach. Jerome Wiesner was, together with President Kennedy, interested in cooperative undertakings that might produce substantial political benefits, even if their technical contributions were relatively minimal. Wiesner and his staff were often at odds with NASA over the space agency's conservative approach to space cooperation.[39]

On March 6, Secretary of State Dean Rusk forwarded to the president a draft letter for Chairman Khrushchev that contained "a range of specific proposals...in a manner designed to facilitate a positive Soviet response." The letter reflected the State Department concern that the potential contributions of other nations be recognized, but it did not link any potential bilateral discussions to the upcoming initial meeting of the UN Committee on the Peaceful Uses of Outer Space, which was scheduled for later in March, saying only that the results of the bilateral talks would be reported to the Committee. Rusk told Kennedy, "if you approve this proposed letter, we would plan to deliver it promptly...without publicity for a sufficient period of time to allow for a serious Soviet reply. We would, however, inform a few interested countries confidentially of this action."[40]

Kennedy approved the letter; it was sent to Nikita Khrushchev on March 7. It first listed some relatively modest proposals for specific areas of cooperation, including a joint weather satellite system, cooperation in space tracking, research on the Earth's magnetic field, satellite communications experiments, and space medicine. There was no mention of specific cooperation in human flights to the Moon or even in Earth orbit.[41] However, Kennedy noted in the letter that "beyond these specific projects we are prepared now to discuss broader cooperation in the still more challenging projects which must be undertaken in the exploration of outer space." Observing that "leaders of the United States space program have developed detailed plans for an orderly sequence of manned and unmanned flights for the exploration of space and the planets," the president suggested that "out of discussion of these plans, and of your own, for undertaking the tasks of this decade would undoubtedly emerge possibilities for substantive scientific and technical cooperation in manned and unmanned space investigations."[42]

Khrushchev replied to Kennedy's letter on March 20. He noted "with satisfaction" that his proposal "that our two countries unite their efforts in the conquest of space has met with the necessary understanding on the part of the Government of the United States." He found that Kennedy's message showed "that the direction of your thoughts does not differ in essence from what we conceive to be practical measures in the field of such cooperation," and asked "what then should be our starting point?" He reacted favorably to most of the U.S. proposals contained in Kennedy's letter, and added to the list of areas for possible cooperation joint robotic exploration of the Moon and the planets, search and rescue of re-entered satellites, especially spacecraft with people on board, and various legal problems associated with space activity. Like Kennedy, Khrushchev agreed that "in the future, international cooperation in the conquest of space will undoubtedly extend to even newer fields of space exploration if we can now lay a firm foundation for it." Khrushchev also noted, reinjecting the longstanding Soviet position into the interchange, "it appears obvious to me that the scale of our cooperation in the peaceful conquest of space, as well as the choice of lines along which such cooperation would seem possible is to a certain extent related to the solution of the disarmament problem... Considerably broader prospects for cooperation and uniting our scientific-technical achievements, up to and including joint construction of spacecraft for reaching other planets—the moon, Venus, Mars—will arise when agreement on disarmament has been achieved." Khrushchev agreed to an early start in discussions on cooperation, saying that "representatives of the USSR on the UN Space Committee will be given instructions to meet with representatives of the United States in order to discuss concrete questions of cooperation."[43]

With this exchange of letters, the logjam in getting started on U.S.-Soviet discussions on space cooperation had been broken. On March 27–30, 1962, toward the end of the first meeting of the Committee on the Peaceful Uses of Outer Space, there was an initial round of discussions between a U.S. delegation led by NASA deputy administrator Dryden and a Soviet delegation led by academician Anatoly Blagonravov. The State Department guidance for this first round of talks was "*to make clear* the genuine U.S. interest in cooperation" and "*to explore* the Soviet attitude toward cooperation." Dryden was instructed to "make it clear we are willing to take concrete, practical first steps in order to get started despite the problems of secrecy which particularly troubles the Soviet side."[44] The talks were to be "private with no publicity other than to acknowledge that exploratory talks had been held and would continue"; U.S. allies and other interested parties would be "informed confidentially" about the nature of the talks. Although Dryden was a senior official of NASA and thus intimately knowledgeable about U.S. space policy and programs, Blagonravov as a member of the Soviet Academy of Sciences was several steps removed from a similar position of influence and knowledge in the Soviet Union. Even after President Kennedy had designated Dryden as the head of the U.S. delegation, McGeorge Bundy had raised the possibility of identifying as the lead negotiator for future discussions a more publicly known figure, "with good standing in the Congress," so that "there would be confidence that the United States is not giving anything away, but at the same time taking a positive approach."[45] When the initial discussions, which were characterized as "amicable but inconclusive," suggested that no high-profile cooperative agreement was in the offing, this idea was abandoned, and Dryden was designated as the U.S. lead in subsequent negotiations.

While these initial meetings were in general free of cold war propaganda, on March 28 Blagonravov stated that prior to his departure from Moscow he had been instructed by his government "to convey a message from Soviet scientists to the effect that they would welcome a joint statement by US and USSR scientists restricting outer space to peaceful purposes and condemning the use of 'spy-in-the-sky' satellites." The freedom to operate reconnaissance satellites was a very high priority for President Kennedy, and Dryden quickly told Blagonravov that such a suggestion was a "legal and political question" and thus not within the scope of the current discussions. The talks identified several areas of mutual interest for further negotiations, and the delegations agreed to meet again in one or two months.[46]

At an April 20 senior-level meeting chaired by under secretary of state George McGhee to review the U.S.-Soviet interactions, Dryden noted that the "Soviets clearly prefer that cooperative arrangements with the US be developed and implemented on a step-by-step basis" and that "at the moment it appears unlikely that any significant measure of joint effort in outer space activities will develop." Both NASA administrator Webb and Space Council executive secretary Welsh agreed that such a step-by-step approach was preferable; Wiesner also thought that this would probably have to be the way to proceed, even though, reflecting the president's priorities, he would have preferred more substantial cooperative engagements. In preparing Under Secretary McGhee for the meeting, Philip Farley suggested that "in view of the past interest of the White House in a possible major political negotiation with the Soviet Union for outer space cooperation, if the step by step approach is agreed to be the most feasible one," it would be desirable "to obtain the President's concurrence."[47]

To obtain that agreement, Secretary of State Dean Rusk wrote the president on May 15, reporting that the president's senior advisers on space had agreed "that the present low-key, step-by-step approach through informal talks by scientific representatives holds the most promise of breaking through Soviet reservations and initiating cooperation." He added that "for the time being we do not think it necessary or wise to set a specific deadline by which these talks should be completed successfully or terminated."[48] In another meeting with Under Secretary McGhee on May 18, Dryden was told that President Kennedy was not interested in propaganda payoffs but rather "he had in mind real cooperation," and that Kennedy "was anxious to go just as far as the Soviets could go."[49]

From May 30 to June 8, 1962, there was a second round of meetings, this time in Geneva, between a U.S. delegation, once again headed by Dryden, and a Soviet delegation, once again headed by Blagonravov. These negotiations led to a June 8 agreement to cooperate in three areas: (1) the exchange of weather data from satellites and the eventual coordinated launching of meteorological satellites; (2) a joint effort to map the geomagnetic field of the Earth; and (3) cooperation in the experimental relay of communications.

Under secretary of state George Ball wrote the president on July 5 to inform him of the results of these discussions and to propose a future course of action. Ball told Kennedy that the agreements reached in Geneva "represent a sound way of proceeding so long as they are adhered to by the Soviet Government and are developed in such a way as not to foster an impression abroad that they represent a more significant step toward US-Soviet cooperation than they actually do or that US-USSR cooperation will in any way preempt the cooperation already being developed with other countries."

Ball's memorandum set out seven next steps in the cooperative process; these steps had emerged from a recent interagency meeting similar to that following the March talks.[50]

Bundy responded to the Ball memorandum on July 18, saying that "the President concurs in the general approach described in the report." In his cover memorandum obtaining the president's concurrence, Bundy noted that "I know that you [Kennedy] have been concerned lest Dryden make agreements that might come under political attack. I believe that these three projects are quite safe. They have been reviewed with a beady eye by CIA [Central Intelligence Agency] and Defense, and they have been reported in detail to determined and watchful Congressmen like Tiger Teague, with no criticism." In these projects, Bundy suggested, "we get as much as we give" and "neither our advanced techniques nor our cognate reconnaissance capabilities will be compromised."[51]

Thus a modest start had been made in U.S.-Soviet space cooperation. The three initial areas for cooperation were quite limited in comparison with both the hopes of early 1961 and President Kennedy's June 1961 suggestion to Nikita Khrushchev of cooperation in going to the Moon. But, as Donald Hornig, Princeton chemist and chair of the space panel of the President's Scientific Advisory Committee (and later science adviser to President Lyndon Johnson), who had been a member of the U.S. delegation to the Geneva talks, observed to Wiesner: "I believe about as much was achieved as was possible at this time. The USSR representatives were apparently anxious to conclude an agreement" but "were extremely cautious," suggesting that "we must take little steps before we can take bigger ones."[52]

This incremental approach, favored by NASA and the State Department, was not sufficient for President Kennedy. As 1963 unfolded, he once again sought Soviet cooperation in going to the Moon.

# Chapter 11

# To the Moon Together: Pursuit of an Illusion?

President Kennedy's suggestion to Nikita Khrushchev at the June 1961 Vienna summit that the United States and the Soviet Union cooperate in flights to the Moon was made privately, and was not subsequently widely reported. The 1962 discussions on space cooperation were carried out on a low-key basis, with their results being made public only after agreement had been reached. In contrast, President Kennedy's next cooperative initiative came in a most public fashion. Addressing the General Assembly of the United Nations on September 20, 1963, Kennedy said "in a field where the United States and the Soviet Union have a special capacity—in the field of space—there is room for new cooperation . . . I include among these possibilities a joint expedition to the moon." "Why," Kennedy asked, "should man's first flight to the moon be a matter of national competition? . . . Surely we should explore whether the scientists and astronauts of our two countries—indeed of all the world—cannot work together in the conquest of space, sending some day in this decade to the moon not the representatives of a single nation, but representatives of all our countries."[1]

Kennedy's proposal came as a major surprise to all but a few people who had been involved in preparing his United Nations speech or had been advised by the president of his intent. The decision to include the proposal in the president's speech was made just a day or two before September 20, although Kennedy had been mulling the idea for some time. The offer was the personal initiative of the president and a few of his closest advisers.

Responsible for drafting the UN address were presidential assistant Arthur Schlesinger, Jr. and State Department official Richard Gardner. Schlesinger suggests that as the two "canvassed the scientific and technical agencies of the government, we discovered that specific proposals of American-Soviet cooperation seemed trivial compared to the enormities of the space age." As they searched for more dramatic initiatives, "there swam into our minds the thought of merging the Russian and American expeditions to the moon."[2]

Without clearing the idea with anyone else, Schlesinger included language proposing such a cooperative lunar mission in the speech draft "to see how it sounded."[3] Schlesinger says that he "had forgotten that the President had himself suggested this to Khrushchev in Vienna in 1961," and thus was not "prepared for his quick approval."[4]

Between the September 20 speech and his assassination two months later, President Kennedy continued to hope for a positive response to his proposal and, when it seemed to come in early November, to push NASA to come up with ways of turning the proposal into reality. Given all the practical difficulties of doing so, in addition to continuing skepticism within NASA and among many in Congress about the wisdom of the proposal in the first place, he may well indeed have been in "pursuit of an illusion"—the thought that the space arena might "be used as a means to swing the US and the USSR from competition to cooperation."[5] But certainly Kennedy was not practicing what Walter McDougall has characterized as "benign hypocrisy"—a willingness to cooperate only in areas "where the United States was safely dominant." McDougall suggests that Kennedy's words about U.S.-USSR space cooperation "were just exercises at image-building."[6] The record suggests a different interpretation—that in 1963 Kennedy was quite serious in his hope that there were practical ways of making U.S. space projects, including the challenging undertaking of sending people to the Moon, an area for reducing U.S.-Soviet tensions and for developing habits of working together.

## Background to the President's Proposal

The reasons why President Kennedy chose to propose such a major step in U.S.-Soviet space cooperation were well summarized by Theodore Sorensen:

> I think the President had three objectives in space. One was to ensure its demilitarization. The second was to prevent the field to be occupied to the Russians to the exclusion of the United States. And the third was to make certain that American scientific prestige and American scientific effort were at the top. Those three goals all would have been assured in a space effort which culminated in our beating the Russians to the moon. All three of them would have been endangered had the Russians continued to outpace us in their space effort and beat us to the moon. But I believe all three of those goals would also have been assured by a joint Soviet-American venture to the moon.
>
> The difficulty was that in 1961, although the President favored the joint effort, we had comparatively few chips to offer. Obviously the Russians were well ahead of us at that time... But by 1963, our effort had accelerated considerably. There was a very real chance we were even with the Soviets in this effort. In addition, our relations with the Soviets, following the Cuban missile crisis and the test ban treaty, were much improved—so the President felt that, without harming any of those three goals, we now were in a position to ask the Soviets to join us and make it efficient and economical for both countries.[7]

### *A New "Strategy of Peace"*

In the aftermath of the Cuban missile crisis, President Kennedy sought ways of lessening the U.S.-Russian tensions and mistrust that had led to that situation. He first tried to once again engage Nikita Khrushchev in discussions on a test ban treaty, but progress toward that objective was slow. By June 1963, he was ready for a broader approach—a new "strategy of peace."[8] In a June 10 commencement address at American University in Washington, DC, Kennedy outlined his approach to a "more practical, more attainable peace," one based on "a series of concrete actions and effective agreements which are in the interest of all concerned... Both the United States and its allies, and the Soviet Union and its allies, have a mutually deep interest in a just and genuine peace... Genuine peace must be the product of many nations, the sum of many acts."[9]

President Kennedy did not mention space cooperation in this speech, an omission seen as "striking" by Harvey and Ciccoritti "in view of the great stress he had placed earlier on space as a means of bridging differences between the two countries." They suggest that "Kennedy for some time had been having second thoughts about pushing space cooperation under existing circumstances." They offer as evidence for this view Kennedy's disappointment with the results of the 1962 space cooperation agreement and the fact that "Kennedy had become more and more enthusiastic over the competitive aspects of the space endeavor." Kennedy, they suggest, "was really interested [in space cooperation] only if the Russians should prove ready to cooperate in a manner and on a scale that would involve meaningful movement toward a genuine rapprochement between the two countries. Otherwise, the US would continue with its program in strict competition with the USSR, since he considered it essential to the national interest that the US continue to develop the capabilities for the full mastery of space."[10]

A somewhat different interpretation of Kennedy's views at the time is more convincing. As suggested in Sorensen's view cited earlier, in the post–Cuban missile crisis détente atmosphere of 1963, and with the increasing costs and mounting criticisms of the lunar landing program, it is likely that President Kennedy was even more interested in U.S.-Soviet space cooperation than had previously been the case. The President in mid-1963 was actively considering resurrecting the idea. He certainly did not seem to think that he was "in pursuit of an illusion," but rather pursuing a course of action that was in the U.S. interest. Other than a call for negotiations on a nuclear test ban treaty, Kennedy did not mention any other area for potential U.S.-Soviet cooperation in his American University speech, so the fact that he did not mention space cooperation specifically is less than "striking." It was logical for him to return to a proposal for cooperation in the lunar landing program as one of the "concrete actions" needed to implement his "strategy for peace."

An August 9 Central Intelligence Agency (CIA) memorandum on "The New Phase of Soviet Policy" provided evidence that Kennedy's strategy was well conceived. On July 25 the Soviet Union agreed to sign a Limited Test

Ban Treaty, including a prohibition on tests in outer space. The CIA analysis thought that Nikita Khrushchev in the post-missile crisis period had "a vested interest in perpetuating . . . the impression that a new era in East-West relations has begun" and that "the USSR intends to sustain an atmosphere of reduced tensions for some time." These observations were certainly supportive of a proposal for enhanced space cooperation as part of President Kennedy's new approach.[11]

By mid-1963 there was increasing criticism of the lunar landing program from both sides of the political spectrum. Theodore Sorensen was asked whether by 1963 "the size of this [space] program and its rate of growth were beginning to worry the President, and that he was more eager to stress the cooperative issue because he was dubious about either the wisdom or the possibility of maintaining the rate of increase that NASA was suggesting." Sorensen replied that Kennedy "was understandably reluctant to continue that rate of increase. He wished to find ways to spend less money on the program . . . How much that motivated his offer to the Russians, though, I don't know."[12]

### *Were the Soviets Actually Racing?*

One issue as Kennedy considered resurrecting a cooperative proposal was whether the U.S.-USSR race to the Moon was real. The White House in 1963 in fact did not know whether there was an ongoing Soviet effort to send people to the Moon. A December 1962 National Intelligence Estimate regarding the Soviet space program had observed: "Our evidence as to the future course of the Soviet space program is very limited. Our estimates are therefore based largely on extrapolation from past Soviet space activities and on judgments as to likely advances in Soviet technology." The estimate went on to say that "the top Soviet leaders have not committed themselves publicly to competition with the US in achieving a manned lunar landing, and it is highly unlikely that they will do so . . . On the basis of present evidence, we cannot say definitely at this time that the Soviets aim to achieve a manned lunar landing ahead of or in close competition with the US, but we believe that the chances are better than even that this is a Soviet objective."[13]

It seems as if the president was not aware of this intelligence estimate. He asked CIA director John McCone on April 29, 1963, "Do we have very much information, and if so, what does it indicate, on the Soviet effort in space?" Kennedy that day had read an article in *The Christian Science Monitor* suggesting that there was an increased Soviet effort in space and asked McCone "What is our view on it?"[14]

A formal response to Kennedy's question did not come for several months. In a CIA analysis dated October 1, 1963, and titled "A Brief Look at the Soviet Space Program," the agency gave an even less precise estimate of Soviet capabilities and intentions than it had the prior December, saying that the Soviet space plans unquestionably include manned lunar landings... but there is no evidence that the program is proceeding on a crash basis... It is believed that the Soviets intend to compete vigorously in the early exploration of the moon and that this effort will include manned flights, although probably not early manned landings.

It is not yet possible to settle with assurance whether the Soviets are engaged in a manned lunar landing program competitive with the United States. Definitive indications of the Soviets being in such a race have not been found, but could be submerged to such an extent that they might exist without being so identified. In December 1962, it was estimated that there was a better than even chance that the Soviets had a competitive manned lunar landing program though no firm conclusion could be reached. A later review of pertinent material produced essentially the same judgment. At present there is still no firm evidence of the existence of such a program, but because of the passage of time, it is estimated that a competitive program aimed at the 1968–1970 time period is somewhat less likely than before. Though the flight testing of a new larger booster and a new manned capsule have been predicted, no firm evidence of their early introduction has as yet been noted.[15]

The uncertainty in the United States at this time about the exact character of a Soviet program to send men to the moon is in retrospect understandable, since the situation in the Soviet Union was both complex and confusing. While design work on a large booster able to carry out a manned lunar mission was already underway, those developments were not yet known to U.S. intelligence services. There was a debate within the Soviet space system over both the wisdom of a lunar mission and the assignment of responsibility for such a mission, should it be initiated. Final Soviet approval of a lunar landing mission did not come until 1964.[16]

### *A British Intervention*

An unsolicited suggestion that the Soviet Union did not in fact have a lunar landing program came from a somewhat questionable source, but was widely reported. On July 17, 1963, there were press accounts that British scientist Sir Bernard Lovell, director of the Jodrell Bank Radio Observatory, who had just returned from a visit to the Soviet Union, was saying, "a month ago I believed, like everyone else in the West, that the US-Soviet Moon race was a real struggle. Now I seriously doubt it." One NASA official deeply involved in international affairs characterized Lovell's attempt to influence the course of affairs in 1963 "by all odds the strangest chapter in US/USSR space relationships."[17]

Asked at a press conference on July 17 about whether, in light of Lovell's statement, the United States intended to continue its lunar landing program, President Kennedy replied "in the first place, we don't know what the Russians are—what their plans may be." But "there is every evidence that they are carrying on a major campaign and diverting greatly needed resources to their space effort... I think we ought to go right ahead with our own program and go to the moon before the end of the decade." Pressed on the issue, Kennedy continued, in apparent agreement with the position taken by James Webb in November 1962: "The point of the matter always has been not only of our excitement or interest in being on the moon, but the capacity to dominate space, which would be developed by a moon flight... I think we should continue and I would not be diverted by a newspaper story." Asked about the possibility of the United States cooperating with the Soviet Union in a lunar mission, Kennedy said for the first time publicly "we have said before to the Soviet Union that we would be very interested in cooperation." However, he added, " the kind of cooperative effort which would be required for the Soviet Union and the United States to go to the moon would require a breaking down of a good many barriers of suspicion and distrust and hostility which exist between the Communist world and ourselves." Kennedy concluded that he would "welcome" such cooperation, but that he "did not see it yet, unfortunately."[18]

In a July 23 letter to NASA deputy administrator Dryden, Lovell provided more details on his conversations with M.V. Keldysh, president of the Soviet Academy of Sciences. He reported that Keldysh had informed him of "the rejection (at least for the time being) of the plans for the manned lunar landing" because of several uncertainties regarding the feasibility of such a mission. Keldysh also said that "the manned project might be revived if progress in the next few years gave hope" that such an undertaking would indeed be feasible. Keldysh was reported as saying that "he believed the appropriate procedure would be to formulate the task on an international basis." More specifically, Keldysh suggested "that the time was now appropriate for scientists to formulate on an international basis (a) the reasons why it is desirable to engage in the manned lunar enterprise and (b) to draw up a list of scientific tasks which a man on the moon could deal with that which could not be solved by instruments alone."[19] As noted earlier, the Soviet Academy of Sciences had limited involvement in, and knowledge of, the Soviet space program, and particularly its human spaceflight aspects, yet Keldysh's statements were seen by the media and some politicians as authoritative.

President Kennedy was kept aware of the issues raised by Lovell's letter. The CIA told the White House that the letter was "another step in a Soviet move to internationalize manned lunar exploration." Wiesner forwarded to Kennedy a July 25 article in the *New Scientist* magazine written by Lovell about his views on the Soviet program; Wiesner highlighted the sections of the article dealing with human space flight.[20]

During August, "speculation mounted... with more and more of a tendency to move to an assumption that the USSR has in fact indicated that it wanted to cooperate rather than compete in a moon landing... There was a feeling in NASA that the state of Soviet thinking should be fully checked out," on the outside chance that "the USSR may indeed wish to inspire a slowdown or mutual accommodation in this space race." Thus, in an August 21 letter to Soviet Academy President Keldysh, Dryden offered to meet with Blagonravov "to discuss further proposals for cooperation."[21]

The two met over lunch at the United Nations in New York on September 11. Dryden reported that "Blagonravov stated that 'Lovell's statement (i.e., that there was a temporary hold in the lunar program) might be true as of today.'" Dryden told his counterpart that "it was not necessary to use Lovell as a channel to convey Soviet desires to the U.S." Blagonravov also raised "the possibility of cooperation in manned lunar exploration after instrumented landings on the moon had been made." According to Dryden, "this is a real change from previous discussions in which he had taken the point of view that there was no use in discussing cooperation in this area because of the political situation." Dryden judged "that the Russians as well as us are having discussions on the value of manned lunar landing," but that it was "dangerous" to rely only on statements coming from the Soviet Academy for an understanding of Soviet plans, since he was convinced that the Soviet lunar landing program "is a program originated and operated by the military."[22]

The reality was that neither President Kennedy, nor NASA, nor anyone else in the U.S. government knew the true state of Soviet space efforts and internal debates as of September 1963. Each participant in the decision process brought his own values and objectives to the deliberations. Thus it is somewhat ingenuous to have observed, as did one senior NASA official, that the Lovell letter and the Dryden-Blagonravov conversation "contributed to an apparently coherent and progressive picture of Soviet readiness either to abandon their own lunar program or join in a cooperative effort," and that this was "a dangerously misleading view for the credulous, the uninformed, and the wishful thinkers in official and unofficial places."[23]

## Kennedy Proposes a Joint Lunar Mission

Mid-1963 developments—improved U.S.-Soviet relations, growing criticisms of the U.S. Moon program, White House concerns about its costs, and possible signals of Soviet openness to collaboration—formed the background against which President Kennedy decided in September 1963 to include a suggestion of U.S.-Soviet cooperation in going to the Moon in his September 20 address before the United Nations General Assembly.

### *JFK Still Interested*

Whether or not Kennedy had ever given up on the idea of such cooperation during the difficult days of 1961 and 1962, the changed situation in 1963 made him again interested in actively pursuing the idea. As noted above, in his July 17 press conference, Kennedy for the first time had publicly stated his preference for a cooperative approach to lunar exploration.

The United States and the Soviet Union agreed to a Limited Test Ban Treaty on July 25, six weeks after JFK's American University speech, and the relationship between the two nuclear powers was less tense then at any time since Kennedy had come to the White House. As part of

Secretary of State Dean Rusk's agenda when he was in Moscow in early August to sign the treaty, Kennedy asked Rusk to raise the space cooperation possibility with Nikita Khrushchev. When Rusk did so, Khrushchev responded only with a quip: "Sure, I'll send a man to the moon. You bring him back."[24] Kennedy himself discussed the possibility in an August 26 meeting with Soviet ambassador Anatoly Dobrynin. At the end of a wide ranging conversation, the president "raised the question of activities in outer space." He talked about possible cooperative projects, "including going to the moon." Dobrynin found this "an interesting thought" and told Kennedy he would raise it with Khrushchev, saying that he was aware that Khrushchev was interested in "more cooperation in outer space." Kennedy told the Soviet ambassador that "if each knew the other's ambitions and plans, it might be easier to avoid all-out competition" and that "if Mr. Khrushchev thought that a cooperative effort was possible, he would be interested."[25]

On September 10, U.S. ambassador Foy Kohler visited Soviet foreign minister Gromyko in Moscow. Kohler referred to President Kennedy's August 26 conversation with Dobrynin, and asked whether the Soviet government "had given consideration to the President's broad, imaginative proposal for joint cooperation in outer space projects and if he would be prepared to discuss this subject" during his forthcoming visit to the United States to attend the United Nations General Assembly's opening sessions. Gromyko indicated that the Soviet Union "agreed in principle with the idea and he would of course be prepared to examine any specific proposals [that the] US might have in mind."[26]

Kohler reported this conversation to the president at a September 17 White House meeting. Kennedy first asked Kohler for his views on the concept of a joint lunar mission. Kohler told Kennedy that Gromyko had found the suggestion "interesting"; however, Kohler thought that the "Soviets were both intrigued and puzzled by what the president might have in mind." Thus Gromyko, while giving a "cautious welcome" to the president's idea, had asked that "we come up with some concrete suggestions." Kennedy replied that "while this was not an idea that he had considered in detail, he continued to be interested in developing it and thought it would in fact be useful, for example, and save a great deal of expense if we could come to some kind of agreement with the USSR on the problem of sending a man to the moon." Kohler repeated that he thought that "there might be some real interest in developing cooperation in this field since Khrushchev had a problem of allocation of extremely limited resources" and that made carrying out Kennedy's proposal "relatively simple."[27]

Speechwriter Arthur Schlesinger, Jr., may not have been aware of these presidential initiatives and conversations when he inserted language proposing a joint U.S.-Soviet moon mission in his draft of the UN address, although it is hard to conclude that he independently came up with the same idea. But there is no doubt that the concept had been widely discussed by President Kennedy and others between July and September 1963.

### *Kennedy Decides to Make the Offer*

The meeting with Ambassador Kohler on September 17 was apparently the final confirmation of Soviet interest Kennedy needed to decide to insert the cooperative offer into his United Nations speech. Kennedy kept a previously scheduled September 18 appointment with NASA administrator James Webb to discuss a variety of space policy and budget issues. In a memorandum to the president in advance of the meeting, national security adviser McGeorge Bundy reported that Webb had called him to say that there had been "more forthcoming noises about cooperation from Blagonravov in the UN" and that "Webb himself is quite open to an exploration of possible cooperation with the Soviets" in the lunar landing effort. Bundy added that "the obvious choice is whether to press for cooperation or to continue to use the Soviet space effort as a spur to our own," and that his "own hasty judgment is that the central question here is whether to compete or to cooperate with the Soviets in a manned lunar landing." Bundy noted that:

1 *If we compete*, we should do everything we can to unify all agencies of the United States Government in a combined space program which comes as near to our existing pledges as possible;
2 *If we cooperate*, the pressure comes off, and we can easily argue that it was our crash effort on '61 and '62 which made the Soviets ready to cooperate.

Bundy added: "I am for cooperation if it is possible, and I think we need to make a really major effort inside and outside the government to find out in fact whether in fact it can be done."[28] Bundy's preference for a cooperative approach was an important complement to Kennedy's own inclinations, given the increasing reliance that the president was placing on Bundy's views on national security and foreign policy issues.

By the time he met with Webb on September 18, Kennedy had all but finally decided to proceed with the cooperative proposal. According to Webb, "the President said that he was thinking of making another effort with respect to cooperation with the Russians, and that he might do it before the United Nations, and he said 'Are you in sufficient control to prevent my being undercut in NASA if I do that?' So in a sense he didn't ask me if he should do it; he told me he thought he should do it and wanted to do it and that he wanted some assurance from me as to whether he would be undercut at NASA."[29]

Robert Gilruth, the director of the Manned Spacecraft Center in Houston, the NASA facility with the lead role in the Moon mission, on September 17 (at which point he had no idea that the President would propose just that in three days) had "ruled out as impractical" the suggestion of a joint mission, even though the proposal "would be very interesting." The article reporting Gilruth's remarks appeared in *The New York Times* on the morning of September 18, and was probably the reason Kennedy asked

Webb at their meeting that day if he could control the NASA response to his cooperative proposal. Harvey and Ciccoritti suggest that "actually Webb had serious reservations about the enterprise, but felt that since the President was telling him and not asking him, it would be best to simply go along with the President's wishes." Webb's fear was that damage might be done the U.S. program without any real prospect of achieving anything insofar as the Russians were concerned. Webb also felt that there had not been sufficient consultation within the administration and with congressional leaders. Indeed, given the last minute insertion of the cooperative proposal into the speech, no one in the Congress had been consulted. Neither, apparently, had Vice President Johnson or at least his Space Council staff; Edward Welsh called the proposal "startling" and wondered whether "it will have any impact other than to show our willingness to cooperate and possibly to suggest further slow-downs by the Congress." The staff of NASA was also not happy to hear of the president's intent; its effect was "to cause consternation in the Space Agency because it had not been consulted on a matter so vital to its objectives and timetable."[30]

On September 19, Soviet foreign minister Gromyko in his address to the UN General Assembly suggested that following the Limited Test Ban Treaty with additional steps in relaxing global tensions was desirable; this was interpreted at the White House as a further indication that the time was ripe for a dramatic U.S. proposal on space cooperation. The cooperative proposal was incorporated in Theodore Sorensen's final draft of Kennedy's United Nations speech, prepared only on September 19. The same day, Bundy telephoned James Webb and told him that the president had decided to go ahead with the proposal. Webb "immediately telephoned directions around to the [NASA] centers to make no comment of any kind or description on this matter."[31]

Thus the stage was set. Kennedy's September 20 address was intended to set out the role of the United Nations in his strategy of peace. This was so because, he proclaimed, in the organization's development, "rests the only true alternative to war, and war appeals no longer as a rational alternative." Kennedy noted that "the clouds have lifted a little" as result of various U.S.-Soviet interactions over the preceding months, leading to a "pause in the Cold War." Such a pause, he suggested, could lead to the Soviet Union and the United States, together with their allies, finding additional areas of agreement. It was in this context that Kennedy proposed that the United States and the Soviet Union join together, so that the first people to travel to the moon "would not be representatives of a single nation, but representatives of all our countries."[32]

### *Reactions to the Proposal*

Reactions to Kennedy's proposal were quick to appear and mixed in character. *The New York Times* editorialized that the proposal showed that Kennedy was "courageous" and that he was "able and willing to seize the opportunity of the moment to exercise an imaginative political initiative." In the same edition of the newspaper, reporter John Finney noted that Kennedy's proposal had "caught many Government officials by surprise," caused "bewilderment," and was seen by many in Washington as "the first step toward pulling out of the costly 'moon race.'" The *Times* reported the next day that "Europe's Press Praises Kennedy," citing the *Manchester Guardian's* characterization of the president's proposal as a "minor master-stroke." Two weeks later, the *Times* reported that Brainerd Holmes, who had left his position in NASA as head of manned space flight in June, thought that a cooperative lunar mission would be "a very costly, very inefficient, probably a very dangerous way, to execute the program." The trade publication *Missiles and Rockets* was most negative, rejecting "such a naïve internationalist approach to the lunar project. It can only harm the U.S. effort to the benefit of the Soviet Union." The magazine characterized Kennedy's proposal as "ill conceived," potentially leading to the U.S. manned space flight program facing the prospect of "dwindling from one of the most exciting challenges ever accepted by a nation to an unimportant pawn in the Cold War to be sacrificed in the first gambit of appeasement."[33]

### Congressional Criticism

While some in Congress, such as Senator William Fulbright (D-AK), chairman of the Senate Committee on Foreign Relations, received the president's proposal positively, many members questioned whether the Kennedy was intending to back away from the commitment to a U.S. lunar landing program for which they had been willing to approve exponential budget increases in 1961 and 1962. Webb was "a little surprised" by this reaction, thinking that "a President ought to be able to put that kind of thing forward in a speech at the UN for discussion on a world-wide basis." Based on his September 18 conversation with the president, reducing his support for Apollo was not Webb's understanding of the reason for Kennedy's proposal; Webb took "Kennedy's word on the basis of full faith and credit" that his reason for suggesting cooperation was related to broader strategic issues.

Columnist Drew Pearson noted that "one of the most significant points about JFK's U.N. speech was that he bucked the wrath of the senior and sometimes wrathy moguls of Congress." Pearson expected Representative Albert Thomas to "bellow like a Texas steer at the idea of taking part of the moon project away from Houston and putting it in Moscow." Pearson recognized that the reasoning behind Kennedy's proposal was "his new strategy of pushing for peace: The belief that you have to build one success on top of another if the peace is to be won. He had scored one important international success with the test-ban treaty. And he had the alternative of sitting still and letting the favorable atmosphere which it created slowly get nibbled away by the harpies, or of proposing new dramatic moves to strengthen the foundation for peace."[34]

Pearson was correct about Albert Thomas. Thomas, who had used his position as chair of the House appropriations subcommittee in control of NASA funds to bring the Manned Spacecraft Center to the Houston area, wrote to President Kennedy on September 21. He first commended the president on his speech the previous day, saying that "it clearly sets you out as the leader of the world in international affairs." He then noted that "the press and many private individuals seized upon your offer to cooperate with the Russians in a moon shot as a weakening of your former position of a forthright and strong effort in lunar landings." Thomas asked Kennedy for "a letter clarifying your position with reference to our immediate effort in this regard."[35]

Kennedy quickly replied in a letter that stands as the clearest statement of his rationale for the cooperative proposal. He stated that "in my view an energetic continuation of our strong space effort is essential, and the need for this effort is, if anything, increased by our intent to work for increasing cooperation if the Soviet Government proves willing." He noted that "the idea of cooperation in space is not new," and that "my statement in the United Nations is a direct development of a policy long held by the United States Government." He added:

> This great national effort and this steadily stated readiness to cooperate with others are not in conflict. They are mutually supporting elements of a single policy. We do not make our space effort with the narrow purpose of national aggrandizement. We make it so that the United States may have a leading and honorable role in mankind's peaceful conquest of space. It is this great effort which permits us now to offer increased cooperation with no suspicion anywhere that we speak from weakness. And in the same way, our readiness to cooperate with others enlarges the international meaning of our own peaceful American program in space.
>
> In my judgment, then, our renewed and extended purpose of cooperation, so far from offering any excuse for slackening or weakness in our space effort, is one reason more for moving ahead with the great program to which we have been committed as a country for more than two years.
>
> So the position of the United States is clear. If cooperation is possible, we mean to cooperate, and we shall do so from a position made strong and solid by our national effort in space. If cooperation is not possible—and as realists we must plan for this contingency too—then the same strong national effort will serve all free men's interest in space, and protect us also against possible hazards to our national security. So let us press on.[36]

Even before the president's United Nations speech, the House of Representatives had reduced NASA's budget for Fiscal Year 1964 from the $5.7 billion that had been requested by the president to $5.1 billion, a cut of almost 11 percent. A post-speech attempt on the floor of the House to reduce the NASA budget by an additional $700 million was defeated by a 47 to 132 vote after four hours of acrimonious debate, but the House did approve by a 125–110 margin a resolution saying that no part of the NASA appropriation could be used for a cooperative program involving any "Communist, Communist-dominated or Communist controlled country."[37]

James Webb was able to convince the Senate Appropriations Committee to modify, but not delete, this prohibition, and the final NASA FY1963 appropriations bill included a statement that "no part of any appropriation made available to the National Aeronautics and Space Administration by this Act shall be used for the expenses of participating in a manned lunar landing to be carried out jointly by the United States and any other country without the consent of Congress."[38] This statement was incorporated into the appropriations bill over the objections of the White House.

### *No Soviet Response*

There was no immediate response to the president's proposal from Nikita Khrushchev or any other official Soviet source. The newspaper *Za Rubezhom* on September 28 suggested that Kennedy's proposal was "propaganda" and "distracts attention from joint earthly exploits directed at attaining peace and reduction of world tension." *Pravda*, the Communist Party newspaper, on October 2 reprinted without comment a column by Walter Lippman, who was widely respected in both the United States and the Soviet Union; the column had appeared in the American press on September 24. In the column, Lippman had suggested that "the main merit of the proposal" was "the opportunity it offered for the US to escape its commitment to the moon goal."[39]

As the White House waited for a Soviet reply, a report from a new source appeared that seemed to counter the idea that the Soviet Union had abandoned or postponed its lunar landing program. *The Washington Post* on October 8 reported that Leonid Sedov, characterized as the "father of the Sputnik," had said that it would be two or three years "at least" before an initial Soviet lunar landing attempt. The headline for the story read "Red Expects Moon Shot in 3 Years." Wiesner reported to the president that there was nothing in the story that "really supports the headline that was attached to it."[40] When cosmonauts Yuri Gagarin and Valentina Tereshkova (the first woman in space) visited the UN General Assembly in October, they made no direct mention of Kennedy's proposal.

In response to an October 25 question posed to Nikita Khrushchev about whether the Soviet Union had a lunar landing program planned for the not too distant future, the Soviet leader said: "It would be very interesting to take a trip to the moon. But I cannot at present say when this will be done. We are not at the present planning flight by cosmonauts to the moon . . . I have a report to the effect that the Americans want to land a man on the moon by 1970. Well, let's wish them success . . . We do not want to compete with the sending of men to the moon without careful preparation. It is clear that no benefits would be derived from such a competition."[41]

There were a number of interpretations of Khrushchev's remarks within the U.S. intelligence community. A "current intelligence memorandum" prepared by the CIA suggested that "Khrushchev's statement on a manned lunar landing suggests that at least one program bearing on defense may already have fallen victim to his new economic priorities," interpreting the Soviet premier's remarks as acknowledging the cancellation of an ongoing lunar landing program. This memorandum noted that "Khrushchev's actual remarks hardly warrant the dramatic US news agency treatment that the Soviet premier has 'withdrawn' from the moon race." A different office in the CIA on October 29 advised McGeorge Bundy that "the primary intent of Khrushchev's statement was to change the focus of the space race." This analysis noted that Khrushchev's remarks were similar to statements he had made to visiting journalists in 1961 and 1962 and to views "deliberately given to Western scientists by Soviet scientific officials earlier this year." (This presumably referred to the Soviet contacts with Bernard Lovell.) Thus, the analysis suggested, Khrushchev's statements should not be interpreted as indicating "that the Soviet leaders have taken some major decisions in recent weeks affecting the scope or pace of their lunar program." Rather, a major intent of Khrushchev's statement was a "deliberate effort to downgrade the urgency of a manned lunar landing" and thus influence "U.S. Congressional and public opinion on the question of the expenditures and pace of the U.S. lunar program," thereby "making it clear that the Soviet Union is unwilling to allow the United States to set the terms for competition in space." The head of the State Department's Bureau of Intelligence and Research told Secretary of State Rusk that in his answer at the press conference, Khrushchev "did not withdraw from the space race," "did not say that the USSR might not make the first successful moon landing," "did not accept President Kennedy's proposal," and "committed to nothing." The State Department analysis suggested that Khrushchev regarded Kennedy's offer of cooperation "as a vague one, to which he can appropriately respond in vaguely approving terms without undertaking negotiations or obligations." One of Wiesner's staff suggested that Khrushchev's statement "recognizes that the U.S. determination to send a man to the moon and back has called the bluff of their pretentions since 1957 to world technological leadership." The U.S. lunar landing decision has also "contributed in a *non-belligerent* way to imposing major strains on the Soviet economy and their ability to carry out expansionist objectives. *Our technological challenge,* along with steadfastness over Cuba exactly a year ago, *has been successful in getting them to trim their sails.*"[42]

## Was Cooperation Really Possible?

In the first weeks of November 1963 there was seemingly a real opportunity to act on President Kennedy's initiative to turn the lunar landing program into a cooperative enterprise. But was the United States ready to seize that opportunity? And was meaningful cooperation technically feasible?

In the weeks following the president's United Nations speech, the lack of a Soviet response had left U.S. officials somewhat uncertain on how to proceed. This did not put a halt, however, to thinking about the issue. On the Monday following the president's Friday address, NASA administrator Webb drafted "policy guidance for NASA staff." Webb sent a copy to the White House, where it was approved by McGeorge Bundy the same day. The only deletion Bundy made from Webb's draft was to strike a sentence saying "No one should be misled by any feeling that the President has put this forward as a political move or as a sign of weakening support for the program." Webb pointed out that the president had said only that "we should explore" whether a joint moon mission was feasible, and that "the key word here is 'explore,' and the projection of the purpose as 'joint' is a statement of how far we would be willing to go in our 'exploration' talks and examination." Webb added that "while we are putting forward to the Russians the possibility of working with them, and opening up to the world the image of a nation prepared to address itself to all the problems of cooperation in this extremely important area to which weapons systems have not yet been extended, we must continue the forward drive of the US effort." The NASA official in charge of the agency's international relations, Arnold Frutkin, noted that "to jump from the suggestion that the matter be explored to the conclusion that the President was explicitly asking to put a US spacecraft upon a Soviet booster for a lunar voyage, or vice versa, or suggesting that American and Soviet astronauts be paired off for joint trips to the moon . . . were unwarranted" conclusions, yet "they served as the straw men for the shafts and arrows directed at the 'feasibility' of the proposal."[43] In thinking about how the president's proposal might be implemented, should a positive response from the Soviet Union be received, NASA once again preferred a cautious, step-by-step approach. Webb pointed out that as a "first step" the United States and the Soviet Union could cooperate on choosing a landing site, and that "a joint effort could start with lots of things short of putting each other's men on the same space craft."[44]

Deputy under secretary of state U. Alexis Johnson wrote to Webb on October 14, noting that while the United States had not received a Soviet response to the president's proposal, "we should have as clear as possible an understanding of the broad technical aspects involved," should such a response be forthcoming. Johnson noted that "it seems doubtful that the Soviets will soon bring themselves to face up to the severe security, programmatic and political problems involved in discussing such a joint undertaking." He asked Webb to suggest "What modes of cooperation would be most useful? Which would be practicable? Which would be most advantageous from the viewpoint of our national program? Which would be most likely to evoke a constructive response from the Soviets?"[45]

The NASA response to the State Department letter noted that "the objective, as we understand it, is a substantive rather than a propaganda gain in relations with the Soviet Union, to be achieved through meaningful rather than token projects, with comparable contributions by both sides and without, at this stage, compromising our ability to pursue our own programs." After listing several of the "virtually unlimited number of specific proposals" for cooperation, NASA indicated that it would "strongly prefer" that such proposals be used only "as a second resort." Rather, "first priority should be placed instead on an escalating series of exchanges which are, in their initial stages, subject to verification and are, therefore, calculated to build a level of confidence upon which progressively significant cooperative activities may be based." NASA pointed out that even the modest cooperative activities agreed to in June 1962 had not yet been implemented by the Soviet Union, which was "seriously delinquent" in this respect.[46]

Senior White House interest in cooperation intensified at the end of October. On October 25, McGeorge Bundy requested that interested parties prepare concrete proposals for how to proceed with planning for space cooperation negotiations with the Soviet Union. A few days later, Bundy received a specially prepared estimate of Soviet intentions from the CIA. That estimate was likely similar to the October 1 CIA analysis of the Soviet space program cited earlier. Since the later analysis had been completed after President Kennedy had made his September 20 cooperative proposal, it noted that "if the Soviets are not engaged in an all-out manned landing program, it is expected that they will substitute major goals or somehow reduce the effects of the U.S. accomplishment." The CIA added that "a cooperative venture with the U.S. for a manned lunar landing would reduce the Soviet problems in this regard tremendously," since then the Soviet Union could divert its space spending toward establishing several types of orbiting stations "of enough significance to dim somewhat the luster of the Apollo program."[47]

On October 29, Wiesner gave President Kennedy a memorandum proposing a technical strategy for cooperation. He said that his proposal would "decisively dispel the doubts that have existed in the Congress and the press about the sincerity and feasibility of the proposal itself." Wiesner's idea was "a joint program in which the USSR provides unmanned exploratory and logistic support for the U.S. Apollo manned landing." Wiesner noted that "such a program would utilize the combined resources of the US and USSR in a technically practical manner and might, in view of Premier Khrushchev's statement, be politically attractive to him." He also suggested that while Apollo "would remain a purely U.S. technical program...A Russian could easily be included as a member of the landing team." Wiesner suggested to Kennedy that "It might be extremely advantageous to you to publicly offer this plan to the USSR...while the Khrushchev statement is still fresh in the mind of the public. If the proposal is accepted we will have established a practical basis for a cooperative program. If it is rejected we will have demonstrated our desire for peaceful cooperation and the sincerity of our original proposal."[48]

In the weeks following his United Nations speech, President Kennedy himself apparently did not push for taking next steps on an urgent basis, but he did maintain his interest in his proposal. On October 23, he sent a copy of *The New York Times* article of September 18 reporting on the Dryden-Blagonravov talks to James Webb, saying "I think it would be helpful to collect clippings similar to the attached showing that the Russians are interested in getting a man on the moon. This would make an additional defense for our efforts."[49] It is not clear whether the "efforts" he was referring to were the cooperative overture to the Soviet Union or to the lunar landing program itself.

In light of the negative comments about a Soviet moon program made on October 25 by Nikita Khrushchev, Kennedy was asked at an October 31 press conference: "do you think that Premier Khrushchev has actually taken the Soviet Union out of the so-called moon race, and in any case do you think that the United States should proceed as if there were a moon race?" Kennedy replied:

> I didn't read that into his statement.... I did not get any assurances that Mr. Khrushchev or the Soviet Union were out of the space race at all.
>
> * * *
>
> The fact of the matter is that the Soviets have made a very intensive effort in space, and there is every evidence that they are continuing and that they have the potential to continue. I would read Mr. Khrushchev's remarks very carefully. I think that he said before anyone went to the moon, there should be adequate preparation. We agree with that.
>
> In my opinion the space program we have is essential to the security of the United States, because as I have said many times before it is not a question of going to the moon. It is a question of having the competence to master this environment.
>
> I think that we ought to stay with our program. I think that is the best answer to Mr. Khrushchev.

Asked whether it was true that the Soviet Union had made no response to his proposal for joint moon exploration, Kennedy replied "that is correct."[50]

## Khrushchev Seems to Accept Kennedy's Offer

What was correct on October 31 changed dramatically on the following day. Nikita Khrushchev finally addressed President Kennedy's proposal in a statement at a November 1 Kremlin reception. He said that "it was with great attention that we studied President Kennedy's proposal for a joint moon project."[51] He suggested that "were a relaxation in international tension in relations between states not only reached morally, so to speak, but were it supported by practical steps in the field of disarmament, then the sphere of cooperation between states in exploring outer space could be materially expanded. We consider, with due attention to the proposal of the U.S. President, that it would be useful if the USSR and the US pooled their efforts in exploring space for scientific purposes, specifically for arranging a joint flight to the moon. Would it not be fine if a Soviet man and an American or Soviet cosmonaut and an American woman flew to the moon? Of course it would."[52]

This statement appeared to represent a decision by Nikita Khrushchev to reverse Soviet policy and to accept President Kennedy's offer of cooperation in going to the Moon. According to Khrushchev's son Sergey, this indeed is what happened in the weeks following Kennedy's United Nations speech. The Soviet leader "for the first time openly spoke about cooperating with the United States on a lunar landing project." Khrushchev was "steeling for a fight to change the military's position on the issue, certainly a difficult undertaking given the kind of secrets that would be put at risk in implementing such a joint project."[53] According to Sergey Khrushchev, there were several reasons for this shift in his father's views. In contrast to 1961, when a positive response to Kennedy's Vienna offer of cooperation was rejected because such cooperation would provide the United States a way of knowing how few long-range missiles the Soviet Union possessed, in 1963 "we had a sufficient number of the R-16 missile, and from the combined work the Americans could learn about our strength and not our weakness." Also, "he [Nikita Khrushchev] was attracted by the perspective of sharing expenses and in this way economize his own resources." Finally, "the political climate had changed after the Cuban missile crisis, and my father began to trust Kennedy more."[54]

Nikita Khrushchev's comment that he was open to cooperation accelerated the U.S. planning process; the need to create a coherent proposal from the ideas put forward in the preceding month was evident. Schlesinger of the White House and Harlan Cleveland and Richard Gardner of the State Department's International Organization Bureau visited NASA on November 5 for a briefing on NASA's planning on "the stages which might be involved in exploring whether collaboration might be possible." Schlesinger noted that NASA's plans were "procedural rather than substantive in character," focusing, as was the consistent NASA preference, on exchange of information on existing programs and plans. It was Schlesinger's impression "that NASA remains rather negative about the whole idea" of lunar collaboration. The NASA view, as expressed in its response to the October 14 U. Alexis Johnson letter, was that "eventual substantive steps would depend on the confidence established by these early procedural steps." Schlesinger suggested that to move the effort forward "an expression of Presidential interest in their progress" might be appropriate.[55]

Missing from the discussion both before and after the presidential initiative regarding cooperation in going to the Moon had been Vice President Lyndon B. Johnson and the National Aeronautics and Space Council. By this point in the Kennedy administration, the vice president often had not been included in the development of new space initiatives; even so, his seeming lack of involvement in a fundamental shift in U.S. space policy is notable. Schlesinger comments that by 1963 Johnson "had faded astonishly into the background" and that the vice president "appeared almost as a spectral presence."[56] On November 1, Space Council executive secretary Welsh had told Johnson that the significance of the Khrushchev statement was "further support to the view that the Soviets have a lunar program"; it was "neither an acceptance nor a refusal of President Kennedy's proposal for cooperation—just an expression of interest," reflecting "the standard Soviet line that they are for peace and disarmament, and that cooperation is dependent upon tensions first being relieved."[57] This memorandum suggests how far Welsh and by implication Johnson were from the main thrust of White House thinking on cooperative prospects.

The White House moved quickly on Schlesinger's suggestion of an expression of presidential interest in moving forward on planning for cooperation. On November 8, Schlesinger and National Security Council staff person Charles Johnson drafted a presidential directive in this vein, and "checked [it] in substance" with "dependable people in NASA and State," who were reported to be "enthusiastic" about such a message. NASA administrator Webb also "heartily" concurred with the draft directive. The plan was to get the president to sign the directive before his planned trip to visit the space and missile facilities at Cape Canaveral on November 16, so that Kennedy in his conversations with Webb during the trip could "be saying the same things we have put in the directive."[58]

The directive was signed by President Kennedy and issued on November 12 as National Security Action Memorandum 271, "Cooperation with the USSR on Outer Space Matters." Unlike the February 1962 memo on space cooperation, which had been addressed to the secretary of state, this directive was addressed to NASA administrator Webb, and ordered him "to assume personally the initiative and central responsibility within the Government for the development of a program of substantive cooperation with the Soviet Union in the field of outer space, including the development of specific technical proposals." The directive added that "these proposals should be developed with a view to their possible discussion with the Soviet Union as a direct outcome of my September 20 proposal for broader cooperation between the United States and the USSR in outer space, including cooperation in lunar landing programs." Kennedy asked for "an interim report on the progress of our planning by December 15."[59]

Ten days later, President John F. Kennedy was dead, felled by an assassin's bullets in Dallas. With him died the possibility of U.S.-Soviet cooperation in going to the Moon, although there was sufficient momentum behind the Kennedy initiative to keep it alive for a few more months.

## Epilogue

U.S. ambassador to the United Nations Adlai Stevenson was scheduled to make a speech to the UN Committee on the Peaceful Uses of Outer Space on November 27, and the speech was already in preparation in early November. Recognizing NASA's reluctance to move out aggressively in support of Kennedy's proposal, Schlesinger suggested to Bundy that "it might help the current State Department-NASA debate" over the seriousness of the president's offer if Bundy were to send a message to the State Department saying "I trust that Governor Stevenson's speech . . . will include an adequate follow-up of the President's moon proposal."[60]

Drafts of Stevenson's speech prepared prior to November 22 had indeed included such a reiteration. The senior State Department official on U.N. matters, Harlan Cleveland, in a lengthy November 21 memorandum had developed a reasoned justification for continuing to push the cooperative effort. He noted Nikita Khrushchev's "somewhat erratic" statements on accepting Kennedy's proposal, which, thought Cleveland, "no doubt reflect internal differences in the Soviet leadership over the desirability of cooperation with the U.S. They may also reflect financial difficulties." Further, "by postulating that it takes two to make a race . . . Khrushchev has put himself in the best position available in the circumstances—unless it can be demonstrated that it is he who is declining international cooperation." Thus, "if the U.S. were to go silent on a dialogue initiated by the President, the conclusion no doubt will be drawn that the President has given in to advocates of non-cooperation." Cleveland noted that "it is assumed that the Soviet Union has much more difficulty with the mere thought of cooperation than we do and that they will have more serious 'security problems' at any realistic level of cooperation than we will have." Thus, the United States would be "safe in shooting for the maximum amount of cooperation that the Soviets can be talked into yielding"; the United States should "egg on the Russians to cooperate in an open forum where the maximum influence of the on-looking world community can be brought to bear." Cleveland opposed integrating the two programs, so that the United States would not have to "weld a U.S. capsule on a Russian rocket, or to mate a clean-cut American astronaut with a chubby cosmonette, or to compromise the security of either state, or to make progress of one national program dependent upon the progress in the other." Rather, "the point is to put a largely symbolic umbrella over *both* national programs, plus the contributions of other countries, and to create the image of a mutually cooperative world program to put men on the moon as 'representatives of all our countries' regardless of the nationality of the first arrivals."[61]

In the aftermath of President Kennedy's assassination, Cleveland noted in a November 23 memorandum that "the entire world will be watching every Administration statement in the days ahead for hints of a change in policies enunciated by President Kennedy." Cleveland reported that after a November 23 Cabinet meeting, Ambassador Stevenson had raised with new President Lyndon B. Johnson the question of what he should say about President Kennedy's proposal. Stevenson understood Johnson's response to be that "he did not want to retreat an inch from the idea of international cooperation in the lunar program." To make sure that this was indeed the case, Cleveland suggested that President Johnson be asked to review and concur with the specific language on this issue in the final draft of Stevenson's speech. Secretary of State Rusk sent the draft speech to the White House on the next day for presidential review.[62]

The White House did change the speech draft. Originally the draft said with respect to Kennedy's proposal: "that offer stands." The revised version said: "President Johnson has instructed me to reaffirm that offer today." It also said: "If giant strides cannot be taken at once, we hope that shorter steps can. We believe that there are areas of work—short of integrating the two national programs—from which all could benefit. We should explore the opportunities for practical cooperation, beginning with small steps and hopefully leading to larger matters." Stevenson included those words in his speech, which was actually delivered on December 2. Although Webb apparently would have preferred that Stevenson not mention space in his speech, he was reported "resigned thereto and will not raise the same with the President." Frutkin was reported as "very pleased" with the revised text, which reflected the NASA approach to cooperation.[63]

According to Harvey and Ciccoritti, President Johnson "wanted to keep his options open with respect to cooperation with the Russians, but he wanted to do so in a way that would end controversy at home... He was prepared to keep faith with the Kennedy lunar offer provided that the Russians came through with something worthwhile on their side." But as the United States waited for a Soviet offer, "he wanted to get the lunar issue off the front burner. Emphasis for the moment needed to be shifted back to small first steps that were compatible with existing political realities."[64]

James Webb on January 31, 1964, transmitted to President Johnson the report that had been prepared in response to JFK's November 12 request. In his transmittal letter, Webb proposed guidelines to govern negotiations with the Soviet Union on space cooperation: "substantive rather than propaganda objectives alone; well-defined and comparable obligations for both sides; freedom to take independent action; protection of military and national security interests; opportunity for participation by friendly nations; and open dissemination of scientific results." These guidelines severely limited the scope of potential cooperation. Webb suggested that "on balance, the most realistic and constructive group of proposals which might be advanced to the Soviet Union... relates to a joint program of unmanned flight projects to support a manned lunar landing. These projects should be linked so far as possible to a step-by-step approach, ranging from exchange of data already obtained to joint planning of lunar flight missions." Webb also suggested that "no new high-level US initiative is recommended until the Soviet Union has had a further opportunity (possibly three months) to discharge its current obligations under the existing NASA-USSR Academy agreement or, in the alternative, until the Soviets respond affirmatively to the proposal you have already made in the UN."[65]

The January 31 Webb report was a return to exactly the step-by-step approach the space agency preferred; it reflected "a considerable lowering of sights for cooperation in the lunar area."[66] It certainly did not represent the kind of dramatic approach to space cooperation that had led John F. Kennedy to make such cooperation a continuing element in his strategy for reducing tensions in the U.S.-Soviet relationships.

Whatever the U.S. approach, it was made irrelevant by the lack of a formal Soviet response to President Kennedy's September 20 proposal or to its reiteration by the Johnson administration. Indeed, the Soviet Union did not fully honor even the agreements on limited cooperation that had been reached in 1962, even though some modest U.S.-Soviet cooperation did eventually take place, especially in the biomedical area. As the months passed, President Johnson turned his attention to his ambitious agenda for the Great Society. The United States was of course first to the Moon, and the Soviet Union experienced a series of failures in its lunar program. The opportunity to test whether dramatic space cooperation between the United States and the Soviet Union could serve as a counterweight to their Cold War rivalry had passed.

# Chapter 12

# Apollo under Pressure

A combination of factors—the increasing costs of Apollo, emerging Congressional opposition to those rapid increases, growing critiques of Apollo from leaders of the scientific and liberal communities, the uncertainty of whether the Soviet Union was in fact racing the United States to the Moon, and suggestions, coming primarily from the Republican opposition, that there needed to be additional emphasis on the national security uses of space—led to President Kennedy's asking several times during 1963 whether the original justifications for keeping Project Apollo on its planned schedule—and indeed, for the project itself—were still valid. In addition, the successful outcome of the Cuban missile crisis, the easing of tensions over Berlin, the signing of the Limited Test Ban Treaty, and Kennedy's overall desire to reach out to the Soviet Union with a new "strategy of peace" may have suggested to Kennedy that demonstrating U.S. technological and managerial superiority vis-à-vis the Soviet Union through a spectacular space achievement had lost some of its urgency.

President Kennedy's rationale regarding the reasons why he had decided to accelerate the U.S. space program matured in the course of 1963. In November 1962 he had disagreed with James Webb, who had argued that the goal of the acceleration was preeminence in all areas of space activity; in response, Kennedy had insisted that "everything we do ought really to be tied to getting on the Moon ahead of the Russians," and that he was "not that interested" in being the leader in other areas of space activity. Webb won his point; from mid-1963 on, Kennedy justified the fast-paced space program primarily in terms of its overall contribution to national power rather than as a race to the Moon. For example, in his July 17 press conference following reports that the Soviet Union did not in fact have a lunar landing program, Kennedy said: "The point of the matter always has been not only of our excitement or interest in being on the moon, but the capacity to dominate space, which would be demonstrated by a moon flight, I believe is essential to the United States as a leading free world power. That is why I am interested in it and that is why I think we should continue." Again in his October 31 press conference, Kennedy had said: "In my opinion the space program we have is essential to the security of the United States, because as I have said many times before it is not a question of going to the moon. It is a question of having the competence to master this environment."[1]

## Debating Space Priorities

In the immediate aftermath of President Kennedy's May 25, 1961, declaration that "we should go to the Moon," and for most of the following twenty months, there was very little Congressional or public questioning of pursuing the ambitious lunar landing goal as a high national priority. The Congress, by large bipartisan majorities and after only limited debate, approved increases in NASA appropriations of 89 percent for Fiscal Year 1962 and 101 percent for Fiscal Year 1963. Leading newspapers and other shapers of public attitudes seemed caught up in the excitement of Project Mercury and the initial steps toward the Moon. But as 1963 began, questioning of the lunar landing project began to emerge in various circles; the wisdom of the commitment to Apollo became the focus of a national discussion on the best U.S. path forward in space.

### *Growing Criticism of Project Apollo*

President Kennedy on January 17, 1963, sent to Congress a Fiscal Year 1964 budget request for NASA of $5.712 billion. *The New York Times* editorialized that "whether the $20 billion (or $40 billion) race to the moon is justified on scientific, political, or military grounds, we do not think the matter has been sufficiently explained or sufficiently debated. We hope it will be in the present Congress." In his March 21 press conference, President Kennedy was questioned about the pace of the U.S. space program as compared to that of the Soviet Union. He responded: "The United States is making, as you know, a major effort in space and will continue to do so. We are expending an enormous sum of money to make sure that the Soviet Union does not dominate space. We will continue to do it."[2]

On March 29, Congressman Charles Halleck (R-IN) released a letter from former President Dwight Eisenhower in which Eisenhower suggested that "the space program, in my opinion, is downright spongy. This is an area where we particularly need to demonstrate some common sense. Specifically, I have never believed that a spectacular dash to the moon, vastly deepening our debt, is worth the added tax burden it will eventually impose on our citizens." President Kennedy was asked about Eisenhower's comments at an April 3 press conference; he responded: "We are second in space today because we started late. It requires a large sum of money. I don't think we should look with equanimity upon the prospect that we will be second all through the sixties and possibly the seventies. We have the potential not to be. I think having made the decision last year, that we should make a major effort to be first in space. I think we should continue to do so." He added: "Now President Eisenhower—this is not a new position for him. He has disagreed with this, I know, at least a year or year and a half ago when the Congress took a different position. It is the position I think he took from the time of Sputnik on. But it is a matter on which we disagree." Kennedy added: "It may be that there is waste in the space budget. If there is waste, then I think it ought to be cut out by the Congress, and I am sure it will be. But if we are getting to the question of whether we should reconcile ourselves to a slow pace in space, I don't think so."[3]

Respected *New York Times* columnist James Reston soon suggested that "the debate on the nation's space program is getting out of hand. Some Republicans are attacking the program as if it were a vast boondoggle, and President Kennedy is defending it as if it were the Bill of Rights." Reston added that "the space program deserves a more serious response. For a large and influential sector of the scientific community of the nation... believes that the scientific objectives of the program can be achieved at a fraction of the cost by putting instruments, rather than man, on the moon." Thus scientists see the issue as "whether the immense additional cost of the man-landing should take a higher priority than using a part of the savings for other essential tasks that would invigorate the economy and create jobs." Three weeks later, Reston reported that "the debate over the Government's space budget is getting rough and threatening to create a crisis of confidence in the Administration's whole space program." Reston cited the contradictions between Vice President Johnson's claims that the space program was having a positive economic impact with statements by others in the Kennedy administration that in fact the program was taking scientists and engineers away from economically more valuable pursuits. The result, he suggested, was "a confusion of testimony that is bewildering the Congress and dragging the space program into the arena of politics."[4]

The scientific community's critique of Apollo was very visibly articulated in an April 19 editorial in the leading journal *Science* signed by its editor, Philip Abelson. Abelson suggested that "the lasting propaganda value of placing a man on the moon has been vastly overestimated. The first lunar landing will be a great occasion; subsequent boredom is inevitable." He added that "most of the interesting questions regarding the moon can be studied by electronic devices" and suggested "a re-examination of priorities is in order." Abelson's editorial received attention well beyond the scientific community; his criticism was noted in front-page articles in prominent newspapers and in an April 20 appearance on the *Today* television program.[5]

### *The White House Responds*

This growing and broadly based criticism of the lunar landing program concerned President Kennedy. At an April 19 meeting with newspaper editors, he noted with some irritation that "the space program... is now under some attack. It seems to me that this indicates a certain restlessness. This program passed unanimously last year. Now suddenly we shouldn't carry out the space program, and then maybe 6 months from now, when there is some extraordinary action in space which threatens our position, everybody will say, 'Why didn't we do more?' "[6]

The White House behind the scenes was developing its response to the criticisms of Apollo. On April 5, the day of Reston's column in the *Times*, Kennedy's national security adviser McGeorge Bundy, who had assumed a lead policy role on the space program within Kennedy's inner circle of advisers, told NASA administrator Webb that "the President would like to have an appraisal of the basic changes and modifications that we have made in the space program since this Administration came in, so that he will be fully equipped to defend our current position against what looks like an emerging attack." Bundy added that "what the President would like particularly to have is an account of the deficiencies of the space program as it was, and was projected, under the outgoing Administration. What would we have failed to do, and what are we now going to be able to do that could not have been done on the Eisenhower basis." Bundy asked for a "careful and complete answer" to the president's questions, but also gave Webb only two weeks to develop that answer.[7]

Kennedy soon also decided to use the Space Council to coordinate the process of developing the defense of his space program. In an April 9 memorandum to Vice President Johnson, Kennedy noted that "in view of recent discussions, I feel the need to obtain a clearer understanding of a number of factual and policy issues relating to the National Space Program." The president asked Johnson to lead a Space Council effort to develop responses to five questions:

1. What are the salient differences...between the NASA program projected on January 1, 1961 for the years 1962 through 1970, and the NASA program as redefined by the present Administration?
2. What specifically are the principal benefits to the national economy we can expect to accrue from the present, greatly augmented program in the following areas: scientific knowledge; industrial productivity; education, at various levels beginning with high school; and military technology?
3. What are some of the major problems likely to result from continuation of the national space program as now projected in the fields of industry, government, and education?
4. To what extent could the program be reduced, beginning with FY1964, in areas not directly affecting the Apollo program (and therefore not compromising the timetable for the first manned lunar landing)?
5. Are we taking sufficient measures to insure the maximum degree of coordination and cooperation between NASA and the Defense Department?[8]

The Space Council's Edward Welsh led the process of getting inputs from the council's member agencies. He received brief papers from staff officials at the Atomic Energy Commission (AEC) and the Department of State, a one-page letter with supporting material from Jerome Wiesner, and a seven-page response from Secretary of Defense Robert McNamara. NASA's James Webb first sent a four-page cover letter with a sixty-three page attachment; this was later reduced to a seventeen-page memo signed by Webb; it still had ten pages of attachments. Webb was as verbose in his written communications as when he spoke.

The AEC response discussed only the organization's efforts to develop nuclear power sources for space uses. The State Department's somewhat skeptical response noted that "from the viewpoint of U.S. foreign policy objectives generally, it seems quite likely that several adverse effects may develop because of the increasingly large percentage of our space effort to be devoted to the GEMINI and APOLLO programs." This was because "our program will increasingly accentuate, rather than mitigate, the gap between ourselves and other countries in space and space related activities." Another concern was "the inevitable merging of national security requirements and scientific requirements in the conduct of our program," which would make the United States "increasingly the object of skepticism on the part of the neutralist or non-allied countries with respect to the 'peaceful' or 'beneficial' objectives and character of our program." The State Department cautioned against reductions in the nonhuman space flight parts of NASA's activities, saying that such a reduction "would be particularly harmful, if it were to result in a retrenchment in our extant commitments or stated objectives with respect to cooperation in space activities with other countries."[9]

Wiesner commented that "it is my feeling that the NASA-DOD relationship should be substantially improved." In his submission, Secretary of Defense McNamara identified "$600–675 million of the NASA effort which appears to have direct or indirect value for military technology." Of that amount, "about $275–$350 million stems from the augmentation of NASA programs since January 1961." Agreeing with Jerome Weisner, McNamara told the vice president that "I am not satisfied, and I am sure that Mr. Webb is not satisfied, that we have gone far enough to eliminate all problems of duplication and waste in administration." McNamara suggested that the coordination problem was of "sufficient importance" that it required attention at the White House level. In a suggestion sure not to be welcomed by the vice president and his Space Council staff, he suggested that responsibility for monitoring coordination between NASA and DOD "be assigned to the Bureau of the Budget and to the Director of the Office of Science and Technology."[10]

NASA on May 3 not only provided to the vice president copious material responding to the president's five questions; it also submitted a draft reply to those questions. In a May 10 memorandum that "refines and reduces, in volume," the information provided on May 3, Webb suggested that "there is little evidence to indicate that significant national problems are likely to be made critical or require radical solutions as a result of the continuation of the space program." In particular, Webb argued, NASA demand for scientific and engineering personnel to carry out its program would rise only from 3 to 6 percent of total national requirements, and this increase would not unduly divert needed technical personnel from other national programs. Webb suggested that "cooperation between NASA and the DOD is good," but noted "areas for further improvement," including "greater participation by the DOD in NASA projects."[11]

A "principals only" meeting of the Space Council was held on the afternoon of May 7 to review and approve the vice president's report to the president. Welsh read the president's questions and Lyndon Johnson, reading from a seven-page draft prepared by the Space Council staff, indicated the intended replies. The draft was kept purposely relatively short compared to the inputs from the Space Council member agencies; Welsh had been reminded by the vice president that "the president does not like a lot of 'verbiage'" and that the responses to Kennedy's questions should be kept "as short and simple as possible." A few revisions and additions to the draft reply were made "after careful consideration and discussion," and the report was approved; Council members initialed the vice president's original typed copy of the reply "based on the premise that agreed upon changes would be made."[12]

Johnson delivered the Space Council's report to President Kennedy on May 13; it painted the overall situation in very positive terms. The report estimated that the augmented space program would require approximately $30 billion more in the FY1961 to FY1970 period than had been planned as the Eisenhower administration left office ($48.086 billion rather than $17.917 billion). It noted that the "basic difference between the two programs is that the plan of the previous Administration represented an effort for a second place runner and the program of the present Administration is designed to make this country the assured leader before the end of the decade." With respect to the impacts of the additional space spending, the report suggested that "the 'multiplier' of space research and development will augment our economic strength, our peaceful posture, and our standard of living." It noted that "the introduction of a vital new element into an economy always creates new problems but, otherwise, the nation's space program creates no major complications" and that "despite claims to the contrary, there is no solid evidence that research and development in industry is suffering significantly from a diversion of technical manpower to the space program." Reducing the portion of the NASA program not related to the lunar landing would "lessen the quantity and quality of benefits to the economy" and "give additional ammunition to those who criticize the major funding weight given to the lunar program on the grounds that it diverts money from other programs." With respect to NASA–DOD coordination, the report suggested that "it is inevitable that controversies will continue to arise in any field as new, as wide-ranging, and as technically complicated as space."

The Space Council report concluded:

> There is one further point to be borne in mind. The space program is not solely a question of prestige, of advancing scientific knowledge, of economic

> benefit or of military development...A much more fundamental issue is at stake—whether a dimension that well can dominate history for the next few centuries will be devoted to the social system of freedom or controlled by the social system of communism.
>
> We cannot close our eyes to what would happen if we permitted totalitarian systems to dominate the environment of earth itself. For this reason our space program has an overriding urgency that cannot be calculated solely in terms of industrial, scientific, or military development. The future of society is at stake.[13]

The Space Council report was received with some degree of skepticism by the White House staff. Wiesner discussed it with President Kennedy on May 16, telling him that "the impact of the NASA program on the nation's supply of scientists and engineers is much greater" than that indicated in the report. While NASA might require the services of only 7 percent of the total supply of U.S. scientists and engineers, it would utilize up to 30 percent of those involved in research and development, he suggested. Wiesner also noted that the report unduly minimized "current management and other differences between NASA and Defense and does not reflect the concerns and views expressed by the Secretary of Defense in his letter to the Vice President." He also pointed out that the report ignored Secretary McNamara's suggestion that "the responsibility for monitoring NASA-Defense problems in the space area be assigned to the Bureau of the Budget and to the Director of the Office of Science and Technology."[14]

The Bureau of the Budget (BOB) noted that "the statement of the 'benefits' to the national economy from the space program has the unintended effect of showing, probably accurately, how slight and intangible such benefits have been and are likely to be in the future." Also, "the question of whether the space program is having detrimental effects on the supply or activities of scientific and technical manpower is still not clearly answered either way." The BOB suggested that while "the report does gloss over current problems and differences between the Department of Defense and NASA," because "the issues involved are either petty or very complex...we would see no useful purpose in presenting any of them to the President at this time." Finally, BOB suggested that "while the point might be overstated" in the concluding paragraphs of the report, "we are inclined to agree with the conclusion that the fundamental justification at this time for a large-scale space program lies...in the fundamental unacceptability of a situation in which the Russians continue space activities on a large scale and we do not."[15]

The individual on McGeorge Bundy's staff with primary responsibility for tracking space issues was Charles E. Johnson. When Bundy received a copy of the Space Council report, he asked Johnson "do you think well of this?" Johnson replied to Bundy that while the "purple prose" in the report's conclusion left him "quite unimpressed," his primary concern was that the report might be released as a public statement of the administration's position on the space program. He saw "no need or urgency for such a statement," believing that public debate could go on "without getting the

President more firmly signed on to a hard position with respect to the space program." He suggested that "the 'Cold War' aspects should not be magnified," since "there is already too much religion in the space program." The NSC staffer suggested that "the Administration's ability to maneuver should be retained so we can adjust to developments." He added, in what turned out to be a perceptive comment, that "it is possible that the Soviets may not be engaging in a race (the intelligence is not conclusive) or may wish to join in a cooperative space program as an alternative to an expensive national program that strains their economy."[16]

Although President Kennedy requested that an unclassified version of the Space Council report be prepared, it was not publicly released or referred to in the continuing debate over space priorities. While Kennedy himself remained publicly committed to proceeding with the accelerated space program he had endorsed in May 1961, the questioning of the program's values and implementation by his senior staff suggested not only that was there a public debate over the proper goals and pace of the space program, but also that a similar debate was beginning to take place inside the top levels of the Kennedy administration.

## The Debate Continues

The criticism of Project Apollo took on a more partisan tone as the Senate Republican Policy Committee on May 10 released a report suggesting that there were other important national problems that "should, perhaps, be examined side by side with the moon shot program." The report suggested that "the question is not, then, whether man will ultimately reach the moon and beyond. The question is, rather, how shall it be done, and whether other aspects of human needs should be bypassed or overlooked in one spasmodic effort to achieve a lunar landing at once." It suggested that "a cold, careful examination is past due." The report was distributed to all Republican senators; it concluded that "for momentary transcendence over the Soviet Union we have pledged our wealth, national talent, and our honor" and suggested that "a decision must be made as to whether Project Apollo (the moon program) is vital to our national security or merely an excursion, however interesting, into space research...If our vital security is not at stake, a less ambitious program may be logical and desirable." A month later, at a breakfast meeting with Republican congressmen, former President Eisenhower made a widely reported comment that spending $40 billion to beat the Soviet Union to the Moon was "nuts."[17]

The Kennedy administration in May began an intensified effort to respond to the critics of its space program. NASA administrator Webb added to a previously scheduled speech the declaration: "At the earliest appropriate stage in the program scientists will be included on Apollo missions." Vice President Johnson in a May 11 speech responded to criticism that the costs of Apollo would undermine the strength of the dollar as an international currency, saying that "we are not told what would happen to the dollar—or to America—if space were defaulted to the Communists." He added: "The question is what kind of philosophy, democratic or Communist, will dominate outer space?... I, for one, don't want to go to bed by the light of a Communist moon."

On May 26, in an effort coordinated by NASA, "eight scientists, three of them Nobel laureates and most of them in academic positions, spoke out... in support of the United States program of landing men on the moon." *Life* magazine in a May 17 editorial added its support to the Moon program, suggesting that the United States could "abdicate its national greatness by not doing enough... The U.S. commitment to space seems a natural undertaking for the American people, who are a venturesome lot." A June 3 editorial in *Aviation Week and Space Technology* suggested that "gradually, the point that the manned lunar landing Apollo program is simply the best possible focal point [for] development of a broad capability in space technology" is "emerging from the verbal pyrotechnics of the current debate."[18]

Arguments for and against proceeding with Project Apollo were aired at June 10–11 hearings of the Senate Committee on Aeronautical and Space Sciences; ten scientists testified and other interested parties submitted written statements. There was general agreement in the hearings that the deadline set for the first lunar landing was probably conducive to waste, and that many national problems deserved equal attention; there was no agreement that the American science enterprise was being distorted by so much attention to space. The strongest protest against the program was a written statement submitted to the committee by Warren Weaver, vice president of the Alfred P. Sloan Foundation; it listed many other desirable uses for $30 billion of federal spending, which Weaver projected as Apollo's ultimate cost. Thirty billion dollars, Weaver said, would give every teacher in the U.S. a 10 percent annual raise for 10 years; give $10 million each to 200 small colleges; provide 7-year scholarships at $4,000 per year to produce 50,000 new Ph.D. scientists and engineers; give $200 million each to 10 new medical schools; build and endow complete universities for 53 underdeveloped nations; create 3 more Rockefeller Foundations; and leave $100 million over "for a program of informing the public about science."[19]

## Another Round of Presidential Questions

As he began during the summer to think again about suggesting to the Soviet Union that sending men to the Moon become a cooperative undertaking, President Kennedy was faced not only with the lingering doubts regarding whether Russia in fact was intending to go to the Moon, but also questions regarding the possible hostile purposes of the Soviet space program. The August 1963 issue of the widely read *Reader's Digest* featured an article headlined "We're Running the Wrong Race with Russia!" that asked "are we suffering from moon madness?" and suggested that "the over-publicized 'race' to get a man on our faraway neighbor has obscured an imminent threat to our security—Soviet strides toward military conquest of the space just over our heads."[20]

Not surprisingly, this article caught President Kennedy's attention. On July 22 he sent a memorandum to Robert McNamara and James Webb, noting "the lead article in the *Reader's Digest* this month states that the Soviet Union is making a major effort to dominate space while we are indifferent to this threat. I wonder if you could have some people analyze this and give me a response to it." A week later, he wrote a similar memorandum to Vice President Johnson: "The attack on the moon program continues and seems to be intensifying. Note *Reader's Digest* lead article this month." Kennedy asked the vice president to develop answers to two sets of questions: (1) "Did the previous administration have a moon program? What was its time schedule? How much were they going to spend on it?" and (2) "How much of our present peaceful space program can be militarily useful? How much of our capability for our moon program is also necessary for military control in space?" Kennedy added: "I would be interested in any other thoughts that you may have on the large amounts of money we are spending on this program and how it can be justified."[21]

In his response to this second Space Council review, Webb suggested that "all" of the civilian space program "can be directly or indirectly militarily useful." An important justification for the sums being spent on the NASA program, said Webb, was to develop "the power to operate in space" and "as insurance against surprise and as the building of the necessary underlying capacity" for an accelerated military space program, should the United States decide that such a speed-up was needed. The Department of Defense reply to Kennedy's questions was signed by deputy secretary Roswell Gilpatric. He told the president that "the article is based for the most part on Soviet propaganda statements, faulty and greatly exaggerated interpretation of technical data, quotes by U.S. authorities taken out of context or distorted, excerpts from Air Force magazine articles, and the author's personal opinions and unsupported statements." Gilpatric added: "At the same time, he [the author] deliberately ignores or is strangely uninformed about our ongoing military space program."[22]

A rapidly convened meeting of the Space Council on July 31 discussed the appropriate reply to the president's questions. Vice President Johnson noted that "we had entered a very tricky period," and that there seemed to be a "political basis" for much of the criticism of the lunar landing program, with "more trouble to be expected as we get closer to [election year] 1964." Johnson suggested that "we are facing a Congress where a majority is for the program, but there is a very vocal minority." The group discussed the language to be included in the response to the president, and agreed to have Edward Welsh draft that response, which took the form of a one-page letter signed by Johnson that told the president that "there was no Administration moon program until your message to Congress in 1961." Johnson, agreeing with Webb's argument, added that "all of the scientific and engineering ability in space has direct or indirect [military] value" and that "the space program is expensive, but it can be justified as a solid investment which will give ample returns in security, prestige, knowledge, and material benefits."[23]

Webb on August 9 sent to the White House a separate response to Kennedy's original July 22 memorandum, noting that NASA had also "received from the Vice President a number of questions which we understand he is answering." This somewhat disingenuous comment, since Webb had participated in the July 31 Space Council meeting, was indicative of the preference on Webb's part to report directly to the president rather than working through the Space Council. Webb associated himself with the views in Gilpatric's July 31 memo to the president and added that Apollo would require extensive operations in near-earth orbit and that "75–80% of the cost of the Apollo program will be devoted to the development of a capability for conducting near-earth orbital operations which could form a basis for any military systems we may require." Webb noted that "the *Reader's Digest* article ignores the fact that these basic resources—large launch vehicles, advanced spacecraft, extensive and complex ground facilities—are vitally important resources for future military missions as well as in fulfillment of the NASA program." Webb's belief in the military value of NASA's activities was not shared by Secretary of Defense Robert McNamara. Webb observed that McNamara "was unwilling to stand up and be counted for the [NASA] program." He told President Kennedy later in 1963 that "the Secretary of Defense will not want to support the program as having substantial military value."[24]

John F. Kennedy's late July 1963 questioning of the justifications for continuing to spend large amounts of money to get to the Moon before the Soviets came at the same time as very public discussion of the suggestion that the Soviet Union in fact did not have a lunar landing program. At the end of August, Kennedy in a conversation with Soviet ambassador Anatoly Dobrynin "raised the question of activities in outer space, pointing out that these are very expensive." "If outer space was not to be used for military purposes," thought Kennedy, "then it became largely a question of scientific prestige, and even this was not very important, as accomplishments in this field were usually only three-day wonders."[25] This was certainly a rather different attitude toward Project Apollo than what Kennedy had been saying publicly, and may well have reflected his emerging doubts about proceeding with the lunar landing program at its planned pace and increasing costs.

## Congress Cuts the NASA Budget

The $5.712 billion Fiscal Year 1964 budget request for NASA sent to the Congress in January 1963 was almost $500 million less than what NASA had requested from the White House the previous September, but still represented a 55 percent increase over NASA's appropriation for Fiscal Year 1963. As the Congressional examination of the NASA budget request began in February and March 1963, *Aviation Week and Space Technology* speculated that NASA would be faced with "a sizeable budget cut—up to a half billion dollars—unless a new Soviet space spectacular changes the attitude of an economy-minded Congress."[26] This forecast proved prophetic; by the time that the Congress completed work on the NASA appropriation on December 10, the agency's approved budget was $5.1 billion, a reduction of $612 million, almost 11 percent less than what had been requested.

While the president's September 20, 1963, United Nations proposal to turn lunar exploration into a cooperative undertaking was viewed with dismay by NASA and its congressional advocates, the reality was that most of the reductions in the NASA budget, particularly by the House of Representatives, predated the cooperative proposal. In June and July, the House Committee on Science and Astronautics cut a total of $475 million from the NASA budget, and during floor debate an additional $34 million was taken out; the House on August 1 approved a NASA FY1964 authorization of $5.203 billion. NASA fared somewhat better in the Senate Committee on Aeronautical and Space Sciences, but still took a $201 million reduction; the full Senate on August 8 authorized a $5.511 NASA budget. On August 28, after a conference committee had compromised on the differences between the two bills, the Congress approved a $5.351 billion NASA authorization; this amount was almost $400 million less than the president had requested and $850 million less than what NASA the previous fall had thought needed to keep Apollo on schedule. The White House made no public statements in support of reversing the cuts in the NASA budget, although science adviser Wiesner in an August 2 memorandum to President Kennedy did note that the House cuts in robotic missions intended as precursors to human missions to the Moon would make it difficult to ascertain lunar surface characteristics, an understanding critical to a successful lunar landing. Wiesner recommended to the president that the White House inform the chairman of the Senate Space Committee, Clinton Anderson (D-NM) (Robert Kerr had died on January 1, 1963 and was replaced as committee chair by Anderson) about the importance of the robotic missions "with the request that funds deleted by the House Committee be reinstated." This message apparently reached Senator Anderson, and funds for the robotic Lunar Orbiter and Surveyor missions were included in the Senate version of the authorization bill.[27]

NASA's hope that there would be no further cuts in its budget proved illusory. Authorization bills set the upper limit on the funding for a particular federal agency; the actual funds available are contained in the congressional appropriation for the agency. Hearings on the NASA FY1964 appropriation began in the House of Representatives on August 19. Webb in his testimony urged the Appropriation Subcommittee on Independent Agencies, which had jurisdiction over NASA and which was chaired by space program supporter Albert Thomas, to approve the full amount that Congress would soon authorize. The members of the subcommittee were not swayed by Webb's plea; *The New York Times* on September 19 (the day before President Kennedy's address to the United Nations) reported that the subcommittee members were "contemplating a cut in the space budget of more than $700 million, which would make it virtually impossible to fulfill the Presidential objective of achieving a manned lunar landing by the end of the decade." The *Times* also reported that "administration officials are working frantically behind the scenes to ward off such an unexpectedly large cut." On September 24, in the aftermath of President Kennedy's United Nations speech, the House Subcommittee approved a $5.1 billion NASA appropriation. The full Appropriations Committee confirmed the $5.1 billion budget on October 7, leading James Webb to say that NASA could not achieve a lunar landing before 1970 at that budget level. Even so, the House of Representatives approved the $5.1 billion NASA appropriation on October 10.

NASA's expectation at this point was that the Senate Appropriations Committee, which in the past had been a strong NASA supporter, would restore the $250 million that the House appropriation had cut from the NASA authorization level. However, taking the White House and NASA "somewhat by surprise," the Senate committee on November 13 approved a NASA FY1964 budget of $5.19 billion, only $90 million above the House level. As the appropriations bill was being debated on the Senate floor, Senator J. William Fulbright (D-AK) proposed an additional 10 percent cut in the NASA budget. The Senate rejected this proposal, but did accept an amendment from Senator William Proxmire (D-WI) to reduce the budget to the House level of $5.1 billion. With no difference in the budget level approved by the House and the Senate, there was a real prospect of missing the "end of the decade" target date for the first lunar landing.[28] Only a bit more than two years after Apollo was begun, the Congress was beginning to sour on providing the resources needed to meet the program's end-of-the-decade goal.

## Project Apollo in Management and Schedule Trouble

Congressional budget cuts and widespread criticism were not the only threats to Apollo's success during 1963. The relationship between James Webb and "Apollo czar" Brainerd Holmes never recovered from their differences in the final months of 1962 with respect to requesting additional funding to try to move forward the date of the initial lunar landing attempt. It became increasingly clear in the following months that Webb and Holmes could not work together effectively. As the accomplishments of Project Mercury were being celebrated by various ceremonies and receptions in Washington on May 21, 1963, Holmes became incensed that he was not mentioned at any point during the day; he called Robert Seamans, complaining that "there is absolutely no excuse for the lack of recognition" and that Webb "hates me." Seamans later commented that "to say he was upset is to put it very mildly," and that Holmes's reaction that day "was really the start of the sequence of events that led to his leaving." During a reception that evening at Webb's home, Holmes and Webb got into a public argument. In a series of meetings a few days later, first with Seamans, then with Seamans and Dryden, and finally with Seamans, Dryden, and Webb, Holmes was asked to resign. On June 12 he announced that he would be leaving NASA within the next few months to return to industry.[29]

NASA sought the president's assistance in quickly finding a replacement for Holmes. On June 11, Webb sought JFK's help in recruiting to the NASA position Ruben Mettler, president of Space Technology Laboratories, an organization providing systems engineering support for the Air Force ICBM and space programs. Webb told the president that Mettler had "exactly the qualifications and the experience necessary...and has the complete confidence of men like Secretary McNamara and Dr. Wiesner." Webb suggested that the president could assist the recruitment effort by joining McNamara and Webb in signing a letter to the chairman of the Board of the Thompson-Ramo-Wooldridge Company, the parent company of Space Technology Laboratories, requesting Mettler's services and indicating that "we all will be working together in this program and that we all want and need him and are presenting the request in the form of a national draft."[30] It is not clear whether such a letter was ever sent.

At any rate, NASA was not able to convince Mettler to leave his West Coast position, and so turned to one of his senior associates at the Space Technologies Laboratories, George Mueller, as Holmes's successor. As he formally joined NASA on September 1, 1963, Mueller was greeted by a front-page article in *The New York Times* headlined "Manned Test Flight Lags 9 Months in Moon Project" and saying that such a delay "has led some space officials to question whether it will be possible to achieve the Administration's objective of landing men on the moon by the end of the decade." *Newsweek* in its September 23 issue reported that "the Apollo man-on-the-moon program is almost a year behind its original timetable—and almost certainly will not meet the target set by Mr. Kennedy." The magazine suggested that "the crux of the delay is threefold—money, machines, and men," and suggested that there was "lagging morale and confusion inside NASA."[31]

Soon after assuming his position at NASA, George Mueller asked two senior NASA engineers to conduct a quick and discreet inquiry into the state of the Apollo program. On September 28, the two reported to Mueller that "if funding constraints...prevail," the "lunar landing cannot be attained within the decade at acceptable risk," and that the "first attempt to land men on the moon is likely about late 1971." Mueller showed this report to Robert Seamans, who directed that it not be distributed, much less publicized; there are reports that he told Mueller to destroy the report since it was so at variance with what NASA was saying publicly, but at least some copies were retained. On the basis of this report and his own experience, by the end of October Mueller mandated a dramatic change in the Apollo schedule, known as "all up" testing; this required that all parts of the Saturn V launch vehicle be tested together, rather than separate tests for each launcher stage. This critical management decision made feasible getting to the Moon by the end of the decade.[32]

Whether NASA's problems with the Apollo schedule were known to the White House is not clear from the written record. Given John Kennedy's avid reading of the general media, it is probable that he noticed the *Times* and *Newsweek* stories. The program's troubles in maintaining its schedule are likely to have played a role in a major White House review of the nation's civilian and national security space programs that was just beginning in early October 1963.

## Conclusion

Certainly if the Soviet Union had responded positively to Kennedy's September 20, 1963, offer to cooperate in sending people to the Moon, there could have been profound changes in the character of the Apollo program. But even if such cooperation were not to have materialized, there is strongly suggestive evidence that Kennedy's advisers, if not the president himself, were thinking about significant changes in the national space program in the October–November 1963 period. Those changes might well have included relaxing the schedule aimed at an initial lunar landing by late 1967, or even abandoning the Moon goal altogether. *The New York Times* noted as NASA celebrated its fifth birthday in early October that "technically, politically, financially, the space agency was in trouble...After five years of seemingly unlimited growth, the agency had suddenly and unexpectedly found its future ambitions and growth questioned by segments of the scientific community it had tried so hard to patronize and by a Congress that had always seemed so open-handed and enthusiastic."[33] That questioning extended to John Kennedy's inner circle, and it was very uncertain in the fall of 1963 whether the White House would maintain the lunar landing program on its planned course.

# Chapter 13

# Were Changes in the Wind?

Responding both to President Kennedy's concern over the increasing costs of the U.S. space effort and criticisms such as those in the August *Reader's Digest* that there was too much emphasis on the lunar landing program at the expense of space efforts more directly relevant to national security, the White House in late September 1963 initiated a sweeping review of the U.S. civilian and military space programs and the balance between them. Representative Olin Teague a few days after Kennedy's September 20 United Nations speech had written to President Kennedy, saying that he was "very anxious to know whether this national goal [being first to the Moon] was being abandoned or changed" and that he was "disappointed" at the suggestion of cooperation in the undertaking with the Soviet Union.[1] National security adviser McGeorge Bundy replied on October 4 that he and White House congressional liaison Larry O'Brien had discussed Teague's letter with the president. Bundy told Teague that Kennedy asked Bundy to contact Teague with "an interim answer to the important question which you raise" regarding the national security implications of the cooperative proposal. Bundy told the congressman that "the relation between national security and the space program is very clear and important in the President's judgment, and he is currently engaged in a major review of the relative roles of different agencies... We can assure you that there will be new expressions of the Administration's point of view."[2]

What precisely was meant by the tantalizing term "new expressions" was not specified. But apparently there were some people advising President Kennedy that it was not necessary to continue the fast-paced effort to reach the Moon by the end of the decade; for example, secretary of state Dean Rusk in an October 3 meeting with the president suggested taking 15–20 years to reach the lunar surface. Others argued that more emphasis should be placed on human flights in Earth orbit carried out under Department of Defense auspices. NASA's Webb saw Kennedy's United Nations speech as "a slight withdrawal of support" for Apollo, a "slight testing of the sentiment as to whether the program could stand without his strong support." Webb saw the speech as reflecting a "feeling that this was just the beginning of a group around him [Kennedy] who wanted to withdraw support." Who the members of this "group" were was not clear to Webb; he suggested in a 1969 interview that "I don't know whether it meant Schlesinger and Sorensen or whether it meant the disarmament [and] arms control [advocates] or whether it meant Mr. McNamara. I would simply say those around him."[3]

## NASA-DOD Review

In this context, on October 5, 1963, senior officials from NASA and Department of Defense (DOD), including James Webb and Robert McNamara, met at the Pentagon with McGeorge Bundy. Bundy told them that the White House wanted a comprehensive review of the space program and that as part of that review, the two agencies would be asked several questions:

1. "What are the minimum elements of the space program essential to the lunar landing program?"
2. "What are the minimum elements of the space program essential to clearly specified military requirements?"
3. "What are the minimum elements of the space program essential to user requirements in areas common to military and commercial users (e.g. communications satellites, weather satellites, etc.)?"
4. "What are the elements essential to or desirable for scientific objectives in space?"

Bundy informed the officials that the NASA and DOD efforts in response to this set of questions would not only be used to rationalize the overall national space program to balance civilian and national security space efforts, to minimize duplication of effort, and thus to contain costs. They would also be inputs into a second review in which a task force headed by the director of the Bureau of the Budget (BOB) would "determine, in light of the FY1964 and FY1965 budget pressures, what should be set forth as the goals of the U.S. space program and what the nature and pace of the program should be."[4] The continuing commitment to a lunar landing before the end of the decade was very much a part of this second review.

In response to this directive, NASA and DOD in October and November first conducted a series of separate studies on the various elements of their individual programs. Then, during November and December, they carried out five joint studies in the areas of launch vehicles, manned Earth orbital activities, communications satellites, geodetic, mapping, and weather satellite programs, and various ground facilities. These studies were not completed until January 1964, and with one important exception discussed below, did not lead to major changes in the already planned NASA and DOD space efforts.[5]

As the White House–mandated review was underway, Nikita Khrushchev on October 25 made a statement reported by *The New York Times* with the headline "Soviet Bars Race with U.S. to Land Men on the Moon." A few days later, well-informed *Times* reporter John Finney suggested "for months the Administration had been trying to back away from the idea of a lunar race." Even so, suggested Finney, for the United States "the question is not whether to go to the moon or not," but rather "the pace at which the lunar expedition should be pursued," particularly, as Khrushchev had seemed to suggest, if the Soviet Union was not engaged in a competitive lunar effort. Finney's conclusion was that "the United States will not abandon the lunar expedition, but it will be pursued with less competitive zeal and at a more leisurely pace."[6]

It seems clear that indeed there was White House consideration being given at this point in time to carrying out Apollo at "a more leisurely pace." In preparation for an October 31 presidential press conference, Charles Johnson of the National Security Council staff had suggested to McGeorge Bundy that there was "some merit in trying to unhitch ourselves from the idea of going to the moon in this decade as a hard proposition and focusing public attention on the critical period 1966–1967 when we will know if we have achieved adequate booster power" with the first launch of the Saturn V rocket. Bundy appears to have been sympathetic to such a suggestion; in September, he had told Kennedy that "the obvious choice [with respect to the future of the lunar landing program] is whether to press for cooperation or to continue to use the Soviet space effort as a spur to our own," and that his preference was for cooperation, since "*If we cooperate*, the pressure comes off, and we can easily argue that it was our crash effort on '61 and '62 which made the Soviets ready to cooperate."[7]

Charles Johnson's suggestion of focusing on demonstrating the superior booster power of the Saturn V launcher rather than achieving a lunar landing by the end of the decade was in line with the thinking of some top people in NASA. The NASA assistant administrator for public affairs, Julian Scheer, sent to the White House both an initial statement issued by the space agency the day after Khrushchev's October 25 remarks and a fuller statement reflecting "thoughts developed by the Administrator, Associate Administrator and the Assistant Administrator for Public Affairs." The October 26 statement said: "We will continue to conduct our own program according to our own needs" and the U.S. program "has a brake and a throttle." The fuller statement noted that "as a practical matter, the time for a decision on whether to speed up or slow down a space program such as the program we have developed and is now underway cannot be made at this time." This was because "technology comes from passing certain critical points. One of these is booster power and we will not pass this critical point until 1965–66 when we should equal or surpass the Russian booster launch vehicle power."[8]

NASA was concerned, however, that the White House not make a premature statement suggesting significant changes in the end-of-the-decade goal, even if that possibility was under active consideration. On October 30, Seamans, who was acting NASA administrator because both James Webb and Hugh Dryden were absent,[9] tried to reach Bundy to voice NASA's concerns with respect to what the president might say at his October 31 press conference. He was unsuccessful in contacting Bundy and so relayed NASA's concerns to budget official Willis Shapley. He told Shapley that "it seems to NASA extremely important that the President not indicate at this time that there is any vacillation in the executive branch with respect to the manned lunar program." He added that both NASA and the White House would find themselves in an "extremely awkward position... if it were to be indicated by the President or any other official administration spokesman that the current objectives of the manned lunar landing program are likely to be relaxed or abandoned" before the ongoing White House reviews were completed. Seamans told Shapley that "NASA's present strong recommendation is that we should keep going" with the planned program "at this time," so the country would "not lose the benefit of the near-term objectives (e.g., launch vehicle development, further manned space flight experience with Gemini)... even if it is subsequently decided not to press on to a manned lunar landing attempt as now planned."[10]

These various activities and press reports in preparation for the president's news conference strongly suggest that there was at the end of October 1963 serious consideration being given within the White House to significant changes in Project Apollo, and that these developments were seen by NASA as threatening the integrity of its efforts. NASA on October 30 also told the White House that it "did not want to suggest any new lines for consideration because NASA is committed to a policy of maximum effort directed towards a lunar landing in this decade." The space agency reported to the White House that during the preceding 48 hours it "had received important expressions of support for the present program and timetable—some of this support is from unexpected sources. There is reason to believe that there is a reaction in the country of 'don't quit when you are ahead.'"[11] (Who these sources may have been is not clear from the historical record.)

President Kennedy was very likely aware of the arguments among his associates for and against slowing down the pace of the space buildup. At his October 31 press conference, Kennedy said, when questioned about Khrushchev's statement that the Soviet Union did not have a lunar landing program, "I think that we ought to stay with our program. I think that is the best answer to Mr. Khrushchev."[12]

## A Final Kennedy Visit to the Apollo Launch Site

On November 16, 1963, John Kennedy made the short flight to Cape Canaveral from his family home in Palm Beach, Florida for an inspection tour of progress being made by NASA in the Gemini and Apollo programs. He also took a helicopter to a Navy ship offshore to witness the launch of a submarine-based Polaris missile. The president's visit was seen as "an effort to focus attention on the nation's space program" as the Congress made final decisions on the NASA FY1964 budget.

Kennedy was first briefed on the Gemini program by astronauts Gus Grissom and Gordon Cooper. He then had a short presentation on Apollo by George Mueller in the launch control center at Launch Complex 37; a Saturn 1 booster was sitting on that launch pad for a planned December launch attempt. As his party left the control center, the president lagged behind to inspect the models of the various launch vehicles being used by NASA, ranging from the small Redstone booster that had been used for the suborbital launches of Alan Shepard and Gus Grissom to the mighty Saturn V that would be used to send astronauts to the Moon. When he was assured that the models were all to the same scale, Kennedy used words like "amazing" and "fantastic." Robert Seamans, who accompanied Kennedy throughout the visit, suggests that the President "maybe for the first time, began to realize the dimensions of these projects."

President Kennedy is briefed on Apollo plans by associate administrator for manned space flight George Mueller on November 16, 1963. In the first row (l–r) are: manned space flight official George Low; director of NASA's Launch Operations Center Kurt Debus; NASA associate administrator Robert Seamans; NASA administrator James Webb; the president; NASA deputy administrator Hugh Dryden; director of Marshall Spaceflight Center Wernher von Braun; commander of the Air Force Missile Test Center at Cape Canaveral Major General Leighton Davis; and Senator George Smathers (D-FL). In front of the president is a model of the massive Vehicle Assembly Building within which the Saturn V moon rocket, also shown, would be assembled before being transported to the launch pad (NASA photograph).

Kennedy then walked to the vicinity of the Saturn 1 booster, where he was briefed on the rocket's dimensions and capabilities by Wernher von Braun (see cover image). Before leaving the launch pad, and much to the discomfort of his Secret Service detail, Kennedy walked over and stood directly underneath the rocket. At that moment, he asked whether the upcoming launch "will be the largest payload that man has ever put in orbit." When told that this was indeed the case, Kennedy responded: "That is very, very significant." He recognized that with the upcoming launch the United States would finally surpass the Soviet Union in lifting capacity, a goal that he had pursued from his first presidential decisions on space. The party then flew by helicopter over the Saturn V launch facilities under construction at Launch Complex 39 on the adjoining Merritt Island; Kennedy had been shown a model of the complex during his earlier briefing by Mueller.

As he returned to the mainland after witnessing the Polaris launch, Kennedy said to Seamans, "I'm not sure that I have the facts really straight" with respect to the launch capability of the Saturn 1. "Will you tell me about it again?" Seamans responded that "the usable payload is 19,000 pounds, but we'll actually have 38,000 pounds up there in orbit." The rocket's lift-off thrust would be 1.5 million pounds. Kennedy then asked, "what is the Soviet capability?" and Seamans told him that it was approximately 15,000 pounds of usable payload and that the Soviet booster had only a lift-off thrust of 800,000 pounds. Kennedy once again said: "That's very important. Now, be sure that the press understands this." As he was preparing to return to Palm Beach, he turned to Seamans and said: "Now, you won't forget, will you, to do this?" He asked Seamans to get on the press plane to emphasize that the United States would soon close the weight-lifting gap in space. Seamans was successful in his presidentially assigned mission. *The New York Times* reported the next day that Kennedy was "enthralled by the sight of the Saturn 1 vehicle, which is expected to make space history next month" by putting the United States ahead of the Soviet Union in the weight placed in orbit.[13]

Six days later, on the short flight from Fort Worth to Dallas, John F. Kennedy told Representative Olin Teague that "he wanted to go to the Cape for the Saturn launch in December. He thought the space program needed a boost and he wanted to help."[14]

## Apollo to Go Forward as Planned

What might have happened to Project Apollo if John F. Kennedy had not been assassinated in Dallas on November 22, 1963, is an unanswerable question. Historian Roger Launius has suggested that "had Kennedy served two full terms, it is quite easy to envision a point... in which he might have decided that the international situation that sparked the announcement of a lunar landing 'by the end of the decade' had passed and he could have safely turned off the landing clock." Such a thought could well have been the president's mind as he worried during the summer and fall of 1963 about the increasing criticisms and costs of Apollo and sought Soviet cooperation in a lunar mission; it was certainly one of the options being considered at the White House senior staff level. However, backing off of being first to the Moon did not seem to be on Kennedy's mind in November 1963. In remarks he planned to deliver in Dallas on November 22, Kennedy would have said "the United States of America has no intention of finishing second in space. This effort is expensive—but it pays its own way for freedom and for America."[15]

As already noted, during October and November, NASA and DOD carried out independent reviews in response to the October 5 White House questions; they then began to prepare joint reviews on several key questions, such as what should be the Earth orbital human space flight effort and what should be the country's launch vehicle program. The pace of the lunar landing program was not an issue in the joint NASA-DOD effort. Rather, the focus with respect to human space flight was the balance between NASA and DOD in Earth-orbital spaceflight activities. The Department of Defense at this point was becoming increasingly interested in the military and intelligence potentials of humans in orbit, and there were suggestions that DOD at some point might take over from NASA the leading role in the Gemini project. NASA during this period examined what was "the minimum manned earth-orbit program required to support a manned lunar landing"; the space agency wanted to make sure that there were enough Gemini flights to satisfy its requirements in support of Apollo before any consideration was given to the potential transfer of the project to Air Force control. The NASA study operated under the assumption that the White House objective in the joint program review was "to find ways and means of reducing the projected cost of the total (NASA and DOD) national space program without abandoning the objective of the manned lunar landing, while at the same time increasing the responsiveness of the program to military needs."[16]

The pace of the lunar landing program *was* an issue in the BOB review of the national space program, which was taking place in parallel with the NASA-DOD effort. In late November, a report summarizing BOB's conclusions was drafted by the BOB staff, primarily Willis Shapley, in consultation with senior representatives of NASA and the Department of Defense. This report had two sections, one dealing with the "Manned Lunar Landing Program" and the other with "Military Space Objectives," with a particular focus on the future of a separate Air Force effort in human space flight.

A November 13 draft of report of this "Special Space Review" contained a clear statement of the goal of the lunar landing program: "to attempt to achieve a manned lunar landing and return by the end of the decade" with "principal purposes" of (1) "demonstrating an important space achievement ahead of the USSR"; (2) "serving as a focus for technological developments necessary for other space objectives and having potential significance for national defense"; and (3) "acquiring useful scientific and other data to the extent feasible." (This statement reinforces the reality that science was never the primary goal of going to the Moon.) The conclusion of the draft review was that "after examining the pros and cons and the fiscal implications of possible alternatives . . . the goal as stated above should be adhered to at this time, with due recognition of the problems involved and of the possibility that it may be necessary to change the objective at some future date."[17]

The next draft of the report was dated November 20. It expanded on the analysis of why it was prudent to stay with the lunar landing program as planned, saying that "the feasible alternatives available for major changes . . . are quite limited." First, "there were generally good and sufficient reasons" for setting the lunar landing goal which were "still valid today." In addition, "previous administration policy statements, testimony, and 'commitments' tend to limit the flexibility for major changes." Also, "significant losses in time, money, and other disruptions are likely to result from major changes at this time."[18]

The final draft of the Special Space Review was dated November 29, one week after President Kennedy's assassination. This draft best represents the analysis and conclusions that would have been presented to the president for final decision if he had lived. The draft noted that its contents had been prepared "by Bureau of the Budget staff in consultation with senior representatives of NASA and the Department of Defense." The report first asked: "Should consideration be given at this time to backing off from the manned lunar landing goal?" Three alternative actions were offered:

1. "Adhere to the present goal."
2. "Decide now to abandon current work directly related to the manned lunar landing objective but to continue development of the large launch vehicle (Saturn V) so that it will be available for future space programs."
3. "Decide now to abandon both current work toward the manned lunar landing objective and the development of the Saturn V large launch vehicle."

In support of alternative 1, the report suggested that "in the absence of clear and compelling external circumstances a change in present policies and commitments would involve an unacceptable 'loss of face' both domestically and internationally" and that cancelling Apollo would not "in fact reduce criticism of the total magnitude of the budget or increase support for other meritorious programs to which the funds might be applied." The arguments in support of alternative 2 included "doubts that Congress will provide adequate support for the manned lunar landing program in 1965 and succeeding years" and "the apparent absence of a competitive USSR manned lunar landing program at this time." Arguments in support of alternative 3 included "that an adequate continuing space program can be built around the use of the Saturn IB (and perhaps Titan III) launch vehicle."

The report also raised the possibility of deciding "that the [manned lunar landing] program should be geared to a schedule slipping the first manned lunar landing attempts one or two years later than now planned to the very end of the decade (i.e., end of CY 1969 or 1970, depending on the definition of 'decade')." The review pointed out that "some slippage in present schedules is recognized as inevitable, so that eliminating the present margin...would be tantamount to and generally recognized as an admission that the achievement of the goal has been deferred beyond the end of the decade."

Another issue addressed was "should our posture on the manned lunar landing program attribute a greater degree of military significance to the program?" The review concluded that "the facts of the situation justify the position that the launch vehicle, spacecraft, facilities, and general technology being developed by NASA... do have important future military significance." However, "overplaying this point could have the effects of undercutting the general peaceful image of the program." The review concluded with respect to a joint effort with the Soviet Union that "in the present situation we must necessarily take the posture that we are prepared to enter into any constructive agreement which will not jeopardize our national security interests and which will not delay or jeopardize the success of our MLL [manned lunar landing] program."

After it spelled out these various options, the BOB report concluded that "in the absence of clear changes in the present technical or international situations, the only basis for backing off from the MLL objective at this time would be an overriding *fiscal* decision." (This conclusion had first appeared in the November 20 draft of the report.) That decision might be either "that budget totals in FY1965 or succeeding years are unacceptable and should be reduced by adjusting the space program" or "that within present budgetary totals an adjustment should be made shifting funds from space to other programs." The BOB analysis was that the lunar landing program could indeed be accommodated within the projected FY1965 and subsequent budget levels, and thus that there was also no basis on fiscal grounds for recommending significant changes in the program's character or pace.[19]

The bottom line of the 1963 Special Space Review was that there was no reason for "backing off" the lunar landing goal. It is very unlikely that either President Kennedy's top advisers or the president himself would have countermanded this conclusion, had Kennedy lived to consider that choice.

When James Webb, Robert McNamara, Jerome Wiesner, Edward Welsh, new budget director Kermit Gordon, and their top associates met on November 30 to consider the draft Special Space Review, the possibility of changes in Apollo was not even discussed; the meeting focused on the second part of the report dealing with the Air Force program of human space flight. The group decided to cancel the Air Force DynaSoar program and replace it with the Manned Orbital Laboratory program, a small outpost combining the Gemini spacecraft and an attached module with room to experiment with various military and intelligence payloads. With respect to Apollo, Secretary of Defense McNamara was insistent that there was no military justification for a lunar landing effort, but the group agreed that broader national security considerations were part of the rationale for sending Americans to the

Moon. Without even reaching the level of a presidential decision, Apollo had survived an intense and wide-ranging review with its basic character intact.[20]

### Kennedy's Final Words on Space

The recommendation to continue Apollo on its current path would most likely have been welcomed by the president. As the BOB review was underway, John F. Kennedy repeatedly made clear his view that the United States should continue its effort to assume the leading position in space. Kennedy's excitement during his November 16 visit to Cape Canaveral in recognizing that the upcoming Saturn 1 launch would give the United States the weight-lifting lead in space reflected this determination; he referred to that soon-to-be-realized achievement several times in remarks on November 21 and the morning of November 22 as he moved forward with his tragic Texas tour.

Perhaps Kennedy's attitude on the space program on the last full day of his life are best reflected in remarks he made at the dedication of an aerospace medicine facility in San Antonio on November 21:

> I think the United States should be a leader [in space]. A country as rich and powerful as this which bears so many burdens and responsibilities, which has so many opportunities, should be second to none...We have a long way to go. Many weeks and months and years of long, tedious work lie ahead. There will be setbacks and frustrations and disappointments. There will be, as there always are, pressures in this country to do less in this area as in so many others, and temptations to do something else that is perhaps easier...This space effort must go on. The conquest of space must and will go ahead. That much we know. That much we can say with confidence and conviction.
>
> Frank O'Connor, the Irish writer, tells in one of his books how, as a boy, he and his friends would make their way across the countryside, and when they came to an orchard wall that seemed too high and too doubtful to try and too difficult to permit their voyage to continue, they took off their hats and tossed them over the wall—and then they had no choice but to follow them.
>
> This Nation has tossed its cap over the wall of space, and we have no choice but to follow it.[21]

# Chapter 14

# John F. Kennedy and the Race to the Moon

The assassination of President John F. Kennedy on November 22, 1963, had, of course, many consequences. One of them was turning the U.S. space program, and particularly the lunar landing effort, into a memorial to the fallen president. There was essentially no chance that the new president, Lyndon B. Johnson, would modify the goal set by President Kennedy in 1961, a goal that Johnson had himself so strongly recommended. To reinforce his commitment to President Kennedy's space legacy, less than a week after the assassination Johnson announced that Cape Canaveral would be renamed Cape Kennedy and that the space launch facilities located there would be called the John F. Kennedy Space Center.[1]

In the more than five-and-a-half years between Kennedy's death and the July 20, 1969 landing of Apollo 11 astronauts Neil Armstrong and Buzz Aldrin on the Moon, dedication to Kennedy's commitment to achieving that feat "before this decade is out" sustained the program through delays and difficult times, including the death of three Apollo 1 astronauts in a launch pad accident on January 27, 1967. When the Apollo 11 command module *Columbia* returned to Earth, landing in the Pacific Ocean at dawn on July 24, 1969, a large video screen in Apollo Mission Control in Houston displayed these words:

> **I believe that this nation should commit itself to achieving the goal, before this decade is out, of landing a man on the moon and returning him safely to earth.**
>
> **John F. Kennedy to Congress, May 25, 1961**

Above the image of the Apollo 11 mission patch on another screen appeared:

> **Task accomplished**
> **July 1969**

A half century has passed since President Kennedy decided to send Americans to the Moon, and almost forty years since the last two Apollo astronauts walked on the lunar surface. As noted in the prologue, historian Arthur Schlesinger, Jr. some years ago suggested that "the 20th Century will be remembered, when all else is forgotten, as the century when man burst his terrestrial bounds."[2] While the broadest historical significance of the initial journeys to Moon may indeed take centuries to fully appreciate, it is certainly possible to evaluate the impacts of the lunar landing program to date and of John F. Kennedy's role in initiating the effort and continuing to support it until the day of his death.[3] This chapter contains my assessment of what Kennedy's commitment to the race to the Moon tells us about how John F. Kennedy carried out his duties as President of the United States; asks whether such a presidentially directed large-scale undertaking can serve as a model for other such efforts; and evaluates the several impacts of Project Apollo. I carry out this last evaluation in terms of how well Apollo served the objectives sought by President Kennedy in sending Americans to the lunar

Apollo 11 astronaut Buzz Aldrin and the U.S. flag on the lunar surface, July 20, 1969 (NASA photograph).

surface, in terms of its impact on the evolution of the U.S. space program since the end of Project Apollo, and in terms of how humanity's first journeys beyond the immediate vicinity of their home planet will be viewed in the long sweep of history.

## Understanding Kennedy's Commitment to Apollo

In the American public memory, John F. Kennedy stands as one of the most successful and important of U.S. presidents. This public image, however, is not universally shared by scholars of the presidency and of American government. A half century after John Kennedy entered the White House, they disagree on how best to evaluate the Kennedy presidency. Some portray Kennedy as "a worldly, perceptive, strong, and judicious leader exuding confidence and charisma, deeply affected by the early crises of his administration, recognizing the rapid changes taking place in the world, and responding with a New Frontier of foreign policy initiatives." Others have portrayed Kennedy as "a shallow, cynical, passionless and vainglorious politician, a traditional Cold Warrior, a weak and vulnerable president not always in control of his own foreign policy." A more nuanced assessment is that President Kennedy was a "complex figure whose personality embraced elements of both images."[4]

I believe that the narrative in the preceding chapters supports this last view, but also suggests that in the case of Kennedy's commitment to the race to the Moon, it is the more positive of the two general portrayals that best describes his choices and behavior. In deciding to go to the Moon, and then reiterating that choice several times after extensive White House reviews, Kennedy demonstrated with respect to space a steadiness of purpose and a clear understanding of the arguments for and against implementing his choice. He had the flexibility to pursue a cooperative path if it were open to him, but his judgment that space leadership was in the U.S. national interest made him determined to compete if competition was necessary. Kennedy as he announced his decision to go to the Moon warned the American public and their congressional representatives that the undertaking would be "a heavy burden, and there is no sense in agreeing or desiring that the United States take an affirmative position in outer space, unless we are prepared to do the work and bear the burdens to make it successful." As his science adviser Jerome Wiesner commented, "I think he became convinced that space was the symbol of the twentieth century. It was a decision he made cold bloodedly. He thought it was good for the country."[5]

The decision to go to the Moon was a choice that reflected particularly American characteristics, such as the assumption that the U.S. democratic system of government was superior to all alternatives, that the United States was rightfully the exemplar for other nations, and that meeting challenges to the U.S. position as the leading world power justified the use of extensive national resources to achieve success.[6] Not only the security of the United States was seen at stake; the decision reflected an almost messianic, expansive drive, one resulting in a sense of destiny and mission, which has for a long time been part of the American world view. The validity of this assumption of American exceptionalism is, of course, open to challenge, but that is not my point. Rather, I conclude that it was this perspective that justified in the minds of President Kennedy and many of his key advisers the decision to begin, as Kennedy said in his speech announcing the decision, what they knew would be an expensive and difficult "great new American enterprise" aimed at winning the battle between "freedom and tyranny" for the "minds of men everywhere who are attempting to make a determination of which road they should take."

President Dwight Eisenhower had come to a different judgment of the importance of space achievement (or rather its lack of importance) in terms of preserving U.S. global leadership, which he saw as being based more on a sound defense, fiscal soundness, and social stability. John Kennedy, with his much more activist approach to government, had an opposing view. Kennedy was not at all a visionary in the sense of having a belief in the value of future space exploration; rather, his vision was that space capability would be an essential element of future national power, and thus that the United States should not by default allow the Soviet Union to have a monopoly of large-scale capabilities to operate in "this new ocean." I believe that this was a wise judgment, one from which the United States has benefitted over the past half century. Perhaps the technical capabilities developed for Apollo were in fact too large and too expensive for subsequent regular use, but the principle that the United States should be the leading spacefaring nation has served the country well.

As Walter McDougall observed, "perhaps Apollo could not be justified, but, by God, we could not *not* do it." Even the fiscally conservative Bureau of the Budget (BOB) agreed, commenting in a 1963 analysis that "we are inclined to agree with the conclusion that the fundamental justification at this time for a large-scale space program lies...in the unacceptability of a situation in which the Russians continue space activities on a large scale and we do not."

## *A Rational Choice?*

In my 1970 book *The Decision to Go to the Moon*, I portrayed Kennedy's 1961 decision to enter a space race with the Soviet Union as closely resembling the rational choice model of decision-making, in which a decision-maker identifies a desirable goal to be achieved or a problem to be addressed, assesses various options for achieving that objective, and selects the option with the best ratio of benefits to costs. It is important to note that what makes this decision process "rational" is the purposeful evaluation of alternatives to achieve a stated goal and the choice of the alternative that embodies the best relationship between benefits and costs; the goal itself is a matter of judgment, and must be evaluated on the quality of that judgment. My reconstruction of the decision process in April and May 1961 suggested that

John Kennedy made the following judgments, each of which could be open to debate:

- Kennedy defined the U.S. national interest as requiring this country to be superior to any rival in every aspect of national power.
- This conviction reflected a Cold War interpretation of the international situation in which there was a zero-sum contest for global power and influence conducted between two sharply opposed social and political systems, one led by the United States and the other by the Soviet Union, with an uncommitted "Third World," and perhaps even more developed countries, deciding with which system it was better to associate.
- National prestige, Kennedy thought, was an important element of national power. As image-conscious as he was, Kennedy judged that what other nations thought about American power and resolve to use it was as important, if not more important, than the reality of that power. Kennedy once wondered aloud "What is prestige? Is it the shadow of power or the substance of power?" He concluded that prestige was a real factor in acquiring and exercising national power.[7]
- Kennedy's own analysis, the answers he got from the many people he queried in the weeks following the April 12 launch of Yuri Gagarin, his assessment of the national and international reaction to that feat, and the advice he received from people like Lyndon Johnson and James Webb convinced the president that dramatic space achievements were closely tied to national prestige and thus "part of the battle along the fluid front of the cold war." In addition, Kennedy judged that the potential contributions of space capabilities to military power justified a significant investment in developing those capabilities, albeit through a peaceful, civilian-led effort.

Once these judgments were made, the choice of sending Americans to the Moon emerged from a rapid but searching assessment of what space activity would best achieve a dramatic space "first" before the Soviet Union, thereby both enhancing U.S. prestige and serving as the focal point for the development of various space capabilities. It was this decision process that can best be characterized as rational. For example, veteran budget official Willis Shapley, who had been observing national security policy choices since he joined the BOB in 1942, commented that "after having been through quite a few major decisions, there was never a major decision like this made with the same degree of eyes-open, knowing-what-you're getting-in-for" character. Science adviser Jerome Wiesner agreed, saying that he and Kennedy "talked a lot about do we *have* to do this. He said to me, 'Well, it's your fault. If you had a scientific spectacular on this earth that would be more useful—say desalting the ocean—or something that is just as dramatic and convincing as space, then we would do it.' We talked about a lot of things where we could make a dramatic demonstration—like nation building—and the answer was that there were so many military overtones as well as other things to the space program that you couldn't make another choice." Wiesner added that "if Kennedy could have opted out of a big space program without hurting the country in his judgment, he would have." Also, "these rockets were a surrogate for military power. He had no real options. We couldn't quit the space race, and we couldn't condemn ourselves to be second." *Time/Life* reporter Hugh Sidey suggests that the Moon project "was a classic Kennedy challenge. If it hadn't been started, he might have invented it all, since it combined all those elements of intelligence, courage, and teamwork that so intrigued John Kennedy." Some years later, the admiring Sidey added that in deciding to go to the Moon, Kennedy "heard the poets. He was beyond politics and dollars."[8]

The final words on why he decided to go to the Moon belong to President Kennedy himself. We have in the tape recording of his November 21, 1962, meeting with his space and budgetary advisers an uncensored record of his thinking on the reasons behind his commitment. Then he said:

- "This is important for political reasons, international political reasons. This is, whether we like it or not, in a sense a race."
- "I would certainly not favor spending six or seven billion dollars to find out about space no matter how on the schedule we're doing... Why are we spending seven million dollars on getting fresh water from salt water, when we're spending seven billion dollars to find out about space? Obviously you wouldn't put it on that priority except for the defense implications. And the second point is the fact that the Soviet Union has made this a test of the system. So that's why we're doing it. So I think we've got to take the view that this is the key program. Everything we do ought really to be tied to getting on the Moon ahead of the Russians."
- " We're talking about these fantastic expenditures which wreck our budget and all these other domestic programs and the only justification for it in my opinion to do it in this time or fashion is because we hope to beat them and demonstrate that starting behind, as we did by a couple of years, by God, we passed them."
- " I'm not that interested in space."

The public rhetoric of President Kennedy, particularly the memorable September 1962 speech at Rice University, created the impression that Kennedy was motivated in his support of Apollo by other reasons, and particularly by a long range vision of space exploration. One clear conclusion of this study is that Kennedy was not a space visionary. Rather, he was a pragmatic decision-maker who came to the conclusion that "whatever mankind must undertake, free men must fully share."

### *Commitment Reviewed and Reiterated*

It is important to realize that Kennedy's decision to go to the Moon was not made once and for all time in April and May 1961. By mid-1961, Kennedy began questioning the costs associated with Apollo, and several times in

1962 and again, more intensely, in 1963 there were in-depth reviews of Apollo's cost and schedule, asking each time whether the benefits of going ahead as planned justified the very high costs involved. In 1963, Kennedy saw an opportunity to cooperate with the Soviet Union in going to the Moon as a means of reducing U.S. costs while achieving other important strategic objectives; if the Soviet Union had responded positively, it certainly would have changed the character of Project Apollo.

There was thus not a single decision to aim at a lunar landing, but rather a series of decisions, each time with alternative paths being considered and each time with the resulting choice being to proceed with the program to land Americans on the Moon "before this decade is out," either as a unilateral undertaking or cooperatively. Only at the very end of the Kennedy administration was serious consideration given to slipping the end of the decade schedule, and even then the decision made was to reject such slippage and to stay with the planned schedule.

Kennedy's consistently reiterated commitment to Apollo can be best understood in terms of how he carried out his presidency overall. Theodore Sorensen, as he prepared Columbia University lectures which were later published as his book *Decision-Making in the White House*, asked national security adviser McGeorge Bundy for suggestions on what to say. Bundy replied that "the modes of Presidential decision are enormously varied," that "decisions are made through the ceaseless process by which, if an administration is lively, recommendations and proposals are ground forward," and that in a sense "the entire presidential existence is . . . a process of decision." Viewing JFK's commitment to Apollo in these terms is particularly useful. Bundy suggested that "the president's larger policies: an open door to Moscow, an open door to all underdog Americans, an open door to intelligence and hope, honor to bravery, equal sense of past and future, gallantry to beauty, and pride in politics" were "colors of a permanent palette" and were reflected "in the small as well as the large decisions, drawn from in a hundred ways."[9] Policies in the space arena were indeed a reflection of Kennedy's broader objectives as president. As Sorensen suggests, reflecting the multiple facets of Kennedy's space strategy:

> I think the President had three objectives in space. One was to ensure its demilitarization. The second was to prevent the field to be occupied to the Russians to the exclusion of the United States. And the third was to make certain that American scientific prestige and American scientific effort were at the top. Those three goals all would have been assured in a space effort which culminated in our beating the Russians to the moon. All three of them would have been endangered had the Russians continued to outpace us in their space effort and beat us to the moon. But I believe all three of those goals would also have been assured by a joint Soviet-American venture to the moon.
>
> The difficulty was that in 1961, although the President favored the joint effort, we had comparatively few chips to offer. Obviously the Russians were well ahead of us at that time . . . But by 1963, our effort had accelerated considerably. There was a very real chance we were even with the Soviets in this effort. In

addition, our relations with the Soviets, following the Cuban missile crisis and the test ban treaty, were much improved—so the President felt that, without harming any of those three goals, we now were in a position to ask the Soviets to join us and make it efficient and economical for both countries.

President Kennedy himself explained the subtlety of his space strategy as he wrote Congressman Albert Thomas in the aftermath of his September 20, 1963, proposal at the United Nations that the journey to the Moon become a cooperative undertaking: "This great national effort and this steadily stated readiness to cooperate with others are not in conflict. They are mutually supporting elements of a single policy." Kennedy added: "If cooperation is possible, we mean to cooperate, and we shall do so from a position made strong and solid by our national effort in space. If cooperation is not possible—and as realists we must plan for this contingency too—then the same strong national effort will serve all free men's interest in space, and protect us also against possible hazards to our national security."

One analyst of the Kennedy presidency correctly comments that "there would have been no race to the moon without the Cold War; the space program became as much a part of that conflict as Cuba, Berlin, and Laos."[10] Whatever President Kennedy, Vice President Johnson, and NASA administrator Webb said about the purposes of Project Apollo in their public rhetoric, from the time that Kennedy asked Johnson to identify "a space program that promises dramatic results in which we could win," it was well understood within the government that the primary objective of Apollo was winning a Cold War–inspired competition to be first to the Moon. To those more focused on the totality of the U.S. space program than was John Kennedy, it was also clear from 1961 on that a program aimed at sending Americans to the Moon could serve as a focal point for the development of space capabilities of strategic value for the United States. By 1963, President Kennedy had seemingly also embraced that view. The November 1963 "Special Space Report" recommending proceeding with Apollo on its then-planned schedule clearly stated that "principal purposes" of the lunar landing program were (1) "demonstrating an important space achievement ahead of the USSR"; (2) "serving as a focus for technological developments necessary for other space objectives and having potential significance for national defense"; and (3) "acquiring useful scientific and other data to the extent feasible." These were the reasons John F. Kennedy decided in 1961 to go to the Moon, and they remained the objectives of Apollo at the time of his death.

This stability in the actual reasons for the lunar race served as the political foundation for White House decisions to allocate the massive resources required for Apollo's success, even after Kennedy's assassination. It is perhaps his willingness to stay the course in the face of increasing criticisms of the path in space that he had chosen that most indicates the quality and strength of John F. Kennedy's original decision to go to the Moon.

### *Other Explanations*

Historian Roger Launius has been somewhat critical of explaining Kennedy's space decisions as the result of "an exceptionally deliberate, reasonable, judicious, and logical process." He finds such an explanation as overly "neat and tidy." Launius suggests that the strength of the rational choice model "is its emphasis on Kennedy's Apollo decision as a politically pragmatic one that solved a number of significant problems," and that its weakness is "its unwavering belief that individuals—and especially groups of individuals—logically assess situations and respond with totally reasonable consensus actions." He adds that "since virtually nothing is done solely on a rational basis this is a difficult conclusion to accept." Launius also wonders whether Kennedy's attraction to the race to the Moon was a reflection of his "quintessential masculinity."[11]

I agree; of course other considerations than logic were involved in decisions related to the race to the Moon. The question is whether a rational approach was the predominant influence on policy choice in the 1961–1963 period, even as politics and personalities also played a part. I believe the preceding narrative suggests that this indeed was the case. A rational decision process can address both solving current problems as well as finding a way to achieve longer-term goals. Certainly the immediate stimulus to the decision to go to the Moon was the threat to U.S. global leadership posed by the world's reaction to Soviet space successes at the same time as the United States looked weak in its conduct of the Bay of Pigs fiasco. Kennedy's desire to regain his personal prestige and his administration's momentum were also problems addressed by the Apollo choice. Finding a way at the same time to move away from current problems and to pursue a worthy goal is an optimum policy-making objective, and Kennedy's space strategy was well-crafted to achieve this outcome. John Kennedy shared with others in his family an intensely competitive personality, and that characteristic certainly influenced the way he interpreted the U.S.-Soviet space relationship. He constantly used references to "a race," the need for "winning" and being "first" in both his public and private comments on space.[12] As Manned Spacecraft Center director Robert Gilruth commented, "he was a young man; he didn't have all the wisdom he would have had. If he'd been older, he probably would never have done it." It was a combination of Kennedy's youthful faith in the future, his fundamentally competitive personality, and his broader conception of the national interest that made him willing to accept the costs and risks of the lunar enterprise.

In 1964, political scientist Vernon van Dyke suggested that Kennedy's need to restore national pride, which van Dyke characterized as "a need for national achievement and national morale" and as "gratification stemming from actual or confidently anticipated achievement," was the basic motivation for the decision to initiate the U.S. lunar landing program. John Kennedy came to the White House believing that by the force of his personality combined with forward-looking government actions he could "get this country moving again"; the combined shocks of the world reaction to the Gagarin flight and the Bay of Pigs fiasco challenged this belief. While restoring national (and perhaps personal) morale was indeed one of President Kennedy's goals, it seems to me that he saw pride in American society and its achievements not primarily in domestic terms but more as an element of U.S. "soft power"—the ability of the United States to "obtain the outcomes it wants in world politics because other countries want to follow it, admiring its values, emulating its example, aspiring to its level of prosperity and openness."[13] In Kennedy's thinking about Project Apollo, both pride and power were elements of a policy initiative aimed primarily at influencing other nations of the world. As he said in his May 25, 1961, speech announcing the decision to go to the Moon, "no single space project in this period will be more impressive to mankind."

Surprisingly, most historians of the Kennedy presidency give only passing attention to JFK's space choices, even though Kennedy himself once characterized his decision to initiate Project Apollo as "among the most important decisions that will be made during my incumbency in the office of the Presidency."[14] An exception is the work of Michael Beschloss, who in 1997 characterized Kennedy's lunar landing decision in a way that fits well with the more negative general assessment of Kennedy as president quoted earlier. Beschloss suggests that Kennedy "could easily tolerate the Gagarin success," but after the collapse of the Bay of Pigs invasion, "he was desperately in need of something that would divert the attention of the public and identify him with a cause that would unify them behind his administration." According to Beschloss, Kennedy's April 20, 1961, memorandum to Vice President Johnson asking him to lead a review of the space program was "redolent of presidential panic." Kennedy's May 25 speech announcing the acceleration of the U.S. space effort asked for "the most open-ended commitment ever made in peacetime . . . and represented a high moment of the imperial presidency." Beschloss argues that the proposed commitment was "a measure of Kennedy's aversion to long-range planning and his tendency to be rattled by momentary crises," and that "Kennedy's desire for a quick, theatrical reversal of his new administration's flagging position, especially just before a summit with Khrushchev, is a more potent explanation of his Apollo decision than any other." Beschloss concludes that "Kennedy's political objectives were essentially achieved by the presidential decision to go to the Moon, and he did not necessarily think much about the long-term consequences."[15]

I believe the record of how the lunar landing decision was made gives only modest support to Beschloss's analysis. Words like "desperately" and "panic" do not seem to me to describe Kennedy's state of mind as he considered whether to use a "space achievement which promises dramatic results" as a tool of his foreign policy. Both in the weeks before the Gagarin flight, especially during the March 22–23 review of the NASA budget, and during his own inquiries as Lyndon Johnson's space review was underway, Kennedy heard a wide variety of views on the value of a prestige-oriented space effort. Beschloss suggests that it was the Bay of Pigs failure that convinced Kennedy to move forward with a space initiative. But on April 14, before the invasion began, Kennedy met with his space advisors and commented that "there's nothing more important" than getting the United States into a leading position in space. Kennedy's final approval of the acceleration of the space effort came on May 10; the summit meeting with Khrushchev was not finally set until a week later. Even then, Kennedy sent out feelers regarding a possible agreement at the summit meeting on U.S.-Soviet cooperation in going to the Moon; this is inconsistent with assertion that the need for a "quick, theatrical reversal" of Kennedy administration fortunes before the summit was a key factor in Kennedy's space decisions.

In summary then, I conclude that President Kennedy's commitment to a lunar landing program as the centerpiece of an effort to establish U.S. space leadership was the result of thoughtful consideration, particularly given that it was reiterated a number of times between May 1961 and November 1963. The commitment was publicly embellished with rhetorical flourishes, but at its core was a Cold War–driven but rational policy choice.

The commitment also reflected values deeply embedded in the national psyche. When I wrote *The Decision to Go to the Moon* over forty years ago, my analysis of that decision reflected what Launius has correctly characterized as "a fundamentally liberal perspective on U.S. politics and society" and a celebration of "the use of federal power for public good." I suggested then that the Apollo decision reflected assumptions at the core of Western liberal philosophy. That man can do whatever he chooses, given only the will to do it and the techniques and resources required, is a belief that reflects motivations and characteristics basic to Western and particularly American civilization—a will to action, confidence in man's mastery over nature, and a sense of mission. Specific decisions on what a government should do are made by its leaders, and ideally reflect a lasting conception of the national interest rather than more parochial concerns or the specifics of their character. Through such decisions, the values and aspirations of a society can then be expressed through state action. John Kennedy embraced this activist perspective; in his much-respected June 1963 commencement address at American University, he suggested that "our problems are man-made—therefore, they can be solved by man. And man can be as big as he wants. No problem of human destiny is beyond human beings. Man's reason and spirit have often solved the seemingly unsolvable."[16]

I would today revise my 1970 assessment, but only somewhat. The liberal perspective— that it is appropriate for the Federal government to undertake large-scale programs aimed at the public good—has been embraced by American presidents such as Woodrow Wilson, notably Franklin D. Roosevelt with his New Deal initiatives, John F. Kennedy and, after Kennedy, by Lyndon Johnson and Barack Obama, as well as by the more progressive elements of the U.S. political community. Other presidents and the more conservative elements among U.S. intellectuals, media, and most probably the majority of the general public are, in contrast, skeptical of both the appropriateness and the feasibility of large-scale government programs aimed at societal improvement. So the proposal to focus massive government resources on a lunar landing effort in fact reflected only one of the two dominant strains in American political thought, the one that sees government steering of U.S. society as legitimate.

Conservative thinking as it applies to the commitment to Apollo was best articulated by historian Walter McDougall in his 1985 prize-winning study... *the Heavens and the Earth.* McDougall suggests that Kennedy's proposal that the United States send Americans to the Moon "amounted to a plea that Americans, while retaining their free institutions, bow to a far more pervasive mobilization by government, in the name of progress." The lunar landing decision was part of JFK's assumption that some areas of "private behavior, when they involved the common security and well-being of the country" should be "susceptible to political control," expressed through a "growing technocratic mentality." He suggests that Project Apollo and the other initiatives proposed by Kennedy in his first months in office resulted in "an American-style mobilization that was one step away" from the Soviet approach to a planned society. To McDougall, "the commitment to go to the moon did more than accelerate existing trends in space. It served as the bridge over which technocratic methods passed from the military to the civilian realm."[17]

Some justification for McDougall's concern about the impulses behind the lunar landing decision can be found in the language used in the May 8, 1961, report signed by James Webb and Robert McNamara that recommended setting a voyage to the Moon as a national goal. In portions of the report embodying themes first suggested by McNamara assistant John Rubel, the report argued that the diffusion of U.S. research and development efforts during the 1950s, especially in the national security sector, had had "a strong adverse effect on our capacity to do a good job in space." While the report did not suggest "that we apply Soviet type restrictions and controls upon the exercise of personal liberty and freedom of choice... we must create mechanisms to lay out and insist on achievement." This call for concentration of effort was also found in Wernher von Braun's April 29, 1961, letter to Vice President Johnson responding to the questions President Kennedy had asked in his April 20 memorandum. Von Braun concluded his letter by noting that "in the space race we are competing with a determined opponent whose peacetime economy is on a wartime footing... I do not believe we can win this race unless we take some measures which thus far have been considered acceptable only in times of a national emergency."

Overall, however, McDougall's analysis is derived more from his overall conservative perspective than from the facts of the situation in 1961–1963. While those charged with implementing the lunar mission individually went to extraordinary lengths to achieve success, neither John Kennedy's nor James Webb's management approach called for strong centralized control. The policy and budget decisions that steered Apollo in its early years were made through the normal decision-making process, not in a war-time or Soviet style. It was not the decision to go to the Moon that "militarized" civilian decision-making and led to such initiatives as President Johnson's Great Society and James Webb's attempts to use the space program as an instrument of change with respect to the U.S. educational and research systems. Those impulses stretched back to the activist presidency of Franklin D. Roosevelt. Both Johnson and Webb were committed New Dealers who used their positions in government to take actions that in their view would be for the common good.

The debate over the appropriate role of the federal government in undertaking large-scale efforts on behalf of the U.S. citizenry is a continuing one, and a full discussion of that role is well beyond the scope of this study. What *can* be discussed, however, is what lessons can be drawn from the Apollo experience, and particularly from the way it was initiated by President John Kennedy, should there be a desire to begin another very expensive multiyear government initiative.

## If We Can Put a Man on the Moon...

Project Apollo became the twentieth-century archetype of a successful, large-scale, government-led program. As peacetime engineering endeavors sponsored by the government, only the construction of the Panama Canal between 1904 and 1914 and the construction of the Interstate Highway System over several decades beginning in the 1950s rivaled Apollo in terms of the scope and difficulty of the task and the scale of human and financial resources required. The success of Apollo has also led to the cliché, "if we can put a man on the Moon, why can't we...?" In their 2009 book titled with that cliché, Eggers and O'Leary suggest that "democratic governments can achieve great things only if they meet two requirements: wisely choosing which policies to pursue and then executing those policies."[18]

I believe that this study demonstrates that Project Apollo met both of these requirements for success. Eggers and O'Leary attribute much of the successful execution of the lunar landing program to the leadership of NASA administrator James Webb. I suggest that many others, both within and outside of NASA, should share credit for that implementation success, including particularly John Kennedy.[19] President Kennedy gave Webb a great deal of freedom to manage NASA as Webb saw fit. A number of times between 1961 and 1963 Kennedy heard from others, often science adviser Jerome Wiesner or budget director David Bell, who questioned or disagreed with the path chosen by Webb. In particular, Wiesner waged a vigorous campaign to overturn NASA's choice of the lunar orbit rendezvous approach for carrying out the landing mission. Brainerd Holmes let it be known to Kennedy that Webb opposed his suggestion that the schedule for the first lunar landing be accelerated; Kennedy shared Holmes's desire for the earliest possible landing. Even the Mercury astronauts took their plea for an additional flight in the Mercury program directly to President Kennedy. In every instance, Kennedy deferred to Webb as the individual responsible for carrying out the space program and thus the person who should make these decisions. Kennedy's style as chief executive was to seek as much information as possible in formulating his policy choices, but once a decision was made, Kennedy seldom intervened in its execution.

In my 1970 book, I suggested that "the experience of the lunar landing decision can be generalized to tell us how to proceed toward other 'great new American enterprises.'" I set out in that book four conditions that seemed to me to be requirements for making a wise decision regarding an ambitious future objective:

1. The objective sought must be known to be feasible, with a high degree of probability, at the time the decision to seek it is made.
2. The objective must have been the subject of sufficient political debate so that the groups interested in it and opposed to it can be identified, their positions and relative strengths evaluated, and potential sources of support have time to develop.
3. Some dramatic "occasion for decision," such as a crisis resulting from an external or domestic challenge, must occur to create an environment in which the objective and the policies to achieve it become politically feasible.
4. There must be in leadership positions in the political system individuals whose personalities and political philosophies support the initiation of new large-scale government activities aimed at long-term payoffs and who have the political skill to choose the situations in which such activities can be initiated successfully.

Even writing in 1970, I recognized that the first of these conditions was very limiting, and would not work when the end desired required both technological breakthroughs and significant changes in deep-seated behavior patterns. However, I thought that "finding objectives with high social utility which could be achieved by a specific time using technologies, either physical or social, which are based on existing knowledge is not difficult."[20] Forty years later, I find these comments either remarkably optimistic or remarkably naïve, probably both. What was unique about going to the Moon is that it required no major technological innovations and no changes in human behavior, just mastery over nature using the scientific and technological knowledge available in 1961. There are very few, if any, other potential objectives for government action that have these characteristics.

The reality is that attempts to implement other large-scale nondefense programs over the past forty years have never been successful, in the space sector or in the broader national arena. Both President George H. W. Bush in 1989 and President George W. Bush in 2004 set out ambitious visions for the future of space exploration, but neither of those visions became reality; the political and budgetary support needed for success were notably missing. More recent attempts to re-create a space race mentality by positing that China was intending to send humans to the Moon before a U.S. return have fallen flat. In 2010, President Barack Obama proposed a dramatic move away from the Apollo approach to space exploration, stressing the development of new enabling technologies and widespread international collaboration; he also declared that the Moon would not be the first destination as humans traveled beyond Earth orbit. This proposal was met with skepticism and political controversy; as I write these words, its fate is still unclear. In the nonspace sector, there have been few opportunities for large-scale government programs that do not require for their success a combination of technological innovation and significant changes in human behavior. The attempts to declare a "War on Cancer," for example, required not only research breakthroughs but also changing the smoking habits of millions of Americans. Attempts to move toward U.S. "energy independence" run afoul both limited research and development spending and the complex ties between non-U.S. energy suppliers and the U.S. financial and government sectors. Providing adequate health care for all Americans turns out to be primarily a political, not merely a technical, challenge. Managing global environmental change has both high technical uncertainties and challenging social inertia to overcome. And so on.

Given this situation, I am now inclined to accept an alternative explanation that I rejected forty years ago: that the lunar landing decision and the efforts that turned it in into reality were unique occurrences, a once-in-a-generation, or much longer, phenomenon in which a heterogeneous mixture of factors almost coincidentally converged to create a national commitment and enough momentum to support that commitment through to its fulfillment. If this is indeed the case, then there is little to learn from the decision to go to the Moon relevant to twenty-first century choices. This would make the lament "if we can put a man on the moon, why can't we . . . ?" almost devoid of useful meaning except to suggest the possibility that governments can succeed in major undertakings, given the right set of circumstances. Other approaches to carrying out large-scale government programs will have to be developed; the Apollo experience has little to teach us beyond its status as a lasting symbol of a great American achievement.

## Apollo's Impacts

Indeed, it may be the symbolic character of America's voyages to the Moon that is the most important heritage of the Apollo program. Certainly the image of the Earth rising over the barren lunar surface taken by Apollo 8 astronaut Bill Anders on Christmas Eve 1968 and of Apollo 11's Buzz Aldrin standing next to the American flag at "Tranquility Base" have become iconic, communicating to subsequent generations that the United States did years ago achieve something unique in human experience, the first steps off the home planet.

### *Achieving JFK's Purposes*

John Kennedy chose to go to the Moon as a means of restoring the U.S. prestige that he judged had been lost during the Eisenhower administration.

In the shorter term, he also wanted to counteract the prestige loss caused by the conjunction of the Soviet success with the flight of Yuri Gagarin and the U.S. failure at the Bay of Pigs. Kennedy in 1961 conceptualized prestige in a way well described by British diplomatic historian F. S. Oliver thirty years earlier:

> What prestige is, it would be hard to describe precisely, It may be nothing more substantial than an effect produced upon the international imagination—in other words, an illusion. It is, however, far from being a mere bubble of vanity; for the nation that possesses great prestige is thereby enabled to have its way, and to bring things to pass which it could never hope to achieve by its own forces. Prestige draws material benefits in its train. Political wisdom will never despise it.[21]

In terms of both shorter-term and more lasting impacts on U.S. international prestige and the associated national pride, Apollo was a substantial success. Within months of JFK's clarion call, NASA and U.S. industry were mobilized in a high-profile pursuit of the lunar landing goal. By declaring that the United States intended to take a leading position in space, and by then taking the steps to turn that declaration into practice, Kennedy effectively undercut the unilateral Soviet space advantage in dramatic space achievements well before any comparable U.S. success. The successive achievements of Projects Mercury and Gemini, and most notably the February 1962 first U.S. orbital flight of John Glenn, became initial steps in JFK's lunar quest and thus made the U.S. space program of the 1960s a source of international prestige and national pride. The psychological and political advantages of early Soviet space successes were quickly and effectively countered.

The success of Apollo 11 in July 1969 and five subsequent missions to the lunar surface (Apollo 13, of course, had a major failure on the way to the Moon and did not complete its landing mission) cemented the international perception of the United States as a country committed to peaceful space achievements "for all mankind." Americans who were abroad at the time of the first Moon landings, U.S. diplomats, and Apollo astronauts returning from post-mission international tours all attested to an immense flow of admiration for the country that could accomplish such a feat. Senior State Department officer U. Alexis Johnson reported that "There is no question that the success of Apollo 11 mission did more to bolster prestige abroad than any single event since the termination of the Pacific War in 1945." Johnson added a qualification, noting that "no one could hope or expect that the euphoric burst of enthusiasm felt by most of the world toward our country…could be long maintained—nor has it been." According to Johnson, "we are left, however, with a very substantial residue of admiration and prestige. While benefits are impossible to measure in quantitative terms, these gains should be of very real value with respect to our posture in the world and our relations abroad for many years to come."[22] John Kennedy could not have hoped for a better report on the success of his 1961 lunar landing decision.

One analyst of the Kennedy presidency has suggested that JFK's "aggressive, militaristic, confrontational attitude" toward the Soviet Union made the world a riskier place during Kennedy's brief time in office, as he "waged Cold War."[23] Extending this judgment to the race to the Moon seems unjustified. By choosing a Cold War competitive arena that did not involve military or direct political confrontation, Kennedy channeled one dimension of the U.S.-Soviet rivalry into what some have described as "the moral equivalent of war," rather than armed confrontation.

As it turned out, however, during the Kennedy administration the United States was racing only itself to the Moon. We now know that while by 1963 the Soviet Union had begun to develop a large space rocket capable of sending a cosmonaut to the Moon, it had not yet decided to use it for lunar missions. It was not until spring 1964 that Central Intelligence Agency (CIA) analysts identified activity at the Soviet launch site in central Asia as a launch complex for a very large rocket. Until then, the CIA in its intelligence estimates had basically assumed without supporting hard evidence that the Soviet Union was pursuing a lunar landing program, both because that was a logical extension of past Soviet space activities and because the United States had identified a lunar landing as the appropriate goal for human space flight and intelligence analysts reasoned that the Soviet would follow the same course.[24]

Kennedy was aware in 1961 that his decision to go to the Moon was being made without knowledge of Soviet space intentions; he decided that the prestige benefits of the lunar landing program required that the United States be first to the Moon, whether or not the Soviet Union was in the race. But he continually referred to U.S.-Soviet competition in going to the Moon in his public statements defending his decision. This was certainly the politically expedient thing to do; it would have been far more difficult to maintain political support for Apollo if the threat of Soviet competition had been absent. Several of President Kennedy's advisers in 1962 and 1963 alerted Kennedy to the lack of evidence in support of Soviet lunar intentions, and Kennedy was quite aware of the mid-1963 claims by Bernard Lovell that the Soviets did not have a lunar landing program. By then, he seems to have accepted James Webb's argument that the U.S. lunar program was an extremely valuable focal point for developing overall U.S. space capability, and that it should proceed, cooperatively if possible but unilaterally if not, even if the Soviet Union did not have a similar program.

Both during the Kennedy administration and during the rest of the 1960s (and even until today), critics have argued that the Apollo program was an unfortunate reflection of misplaced U.S. priorities. President Kennedy was aware of these criticisms, and in 1963 worked to prepare answers to the program's doubters. Apollo came to culmination at a time when the United States was experiencing urban riots, civil rights conflicts, political assassinations, and a seemingly pointless war in Southeast Asia. Kennedy cannot be faulted for not anticipating the domestic and international upheavals of the 1960s that changed the social context in which the lunar landings actually took place. In starting Apollo, Kennedy gave more weight to the situation in 1961 than to the longer-term situation in which the landings would actually take place. From his perspective in spring 1961, Apollo looked like the right thing to do.

All in all, then, an evaluation of Project Apollo in terms of the objectives that led John Kennedy to initiate and sustain it must be positive. Although it is impossible clearly to separate the positive impacts of Apollo from the many negatives of the decade of the 1960s, if not for the achievements of the U.S. space program at the end of the decade there would be little positive for Americans to remember from that time.

### *Apollo's Impact on the U.S. Space Program*

By contrast, the impact of Apollo on the evolution of the U.S. space program has on balance been negative. Apollo turned out to be a dead end undertaking in terms of human travel beyond the immediate vicinity of this planet; no human has left Earth orbit since the last Apollo mission in December 1972. Writing in 1970, I suggested that the capabilities developed for Apollo would have "broad and significant impacts on human existence in the decades to come." Like many others close to the space program, I was caught up in the excitement of the initial lunar landings, and could not conceive of the possibility that having served its political purposes, Apollo and whatever human exploration efforts might follow it would so rapidly be brought to a close.

What happened, however, was that most of the Apollo hardware and associated capabilities, particularly the magnificent but very expensive Saturn V launcher, quickly became museum exhibits to remind us, soon after the fact, of what once was had been done. Commenting on this reality in 1989, Walter McDougall lamented the fate of Apollo: "a brilliant creation, carrying tremendous emotional baggage for the nation, achieved so quickly through such skilled and dedicated teamwork, only to be discarded, dismembered, or disinherited." Columnist Charles Krauthammer at the time of the fortieth anniversary of the Apollo 11 mission in 2009 deplored the fact that humans have not returned to the Moon since the last Apollo mission: "On it are exactly 12 sets of human footprints—untouched, unchanged, abandoned. For the first time in history, the Moon is not just a mystery and a muse, but a nightly rebuke. A vigorous young president once summoned us to this new frontier, calling the voyage 'the most hazardous and dangerous and greatest adventure on which man has ever embarked.' And so we did it. We came. We saw. Then we retreated."[25]

This rapid retreat should not have come as a surprise to careful observers. By being first to the Moon, the United States achieved the goal that had provided the sustainable momentum that powered Apollo; after Apollo 11, that momentum very rapidly dissipated, and there was no other compelling rationale to continue. In 1969 and 1970, even as the initial lunar landing missions were taking place, the White House canceled the final three planned trips to the Moon. President Richard Nixon had no stomach for what NASA proposed—a major post-Apollo program aimed at building a large space station in preparation for eventual (in the 1980s!) human missions to Mars. Instead, Nixon decreed, "we must think of them [space activities] as part of a continuing process...and not as a series of separate leaps, each requiring a massive concentration of energy. Space expenditures must take their proper place within a rigorous system of national priorities...What we do in space from here on in must become a normal and regular part of our national life and must therefore be planned in conjunction with all of the other undertakings which are important to us."[26] Nixon's policy view quickly reduced the post-Apollo space budget to less than $3.5 billion per year, a budget allocation one-quarter of what it had been at the peak of Apollo. There were in the 1960s proposals, called the Apollo Applications Program, to use Apollo hardware for a variety of Earth orbit and deep space missions. Only one of those missions, the Skylab space station, ever came to fruition; its May 1973 launch was the last use of the Saturn V. The booster's production line had been shut down in 1970. The 1975 Apollo-Soyuz Test Program mission was the last use of an Apollo spacecraft and the Saturn 1B launch vehicle. With the 1972 decision to begin the shuttle program, followed in 1984 with the related decision to develop a space station, the United States basically started over in human space flight, limiting itself to orbital activities in the near vicinity of Earth.

The policy and technical decisions not to build on the hardware developed for Apollo for follow-on space activities were inextricably linked to the character of President John Kennedy's deadline for getting to the Moon—"before this decade is out." By setting a firm deadline for the first lunar landing, Kennedy put NASA in the position of finding a technical approach to Apollo that gave the best chance of meeting that deadline. This in turn led to the development of the Saturn V launcher, the choice of the lunar orbit rendezvous approach for getting to the Moon, and the design of the Apollo spacecraft optimized for landing on the Moon. Perceptive observer Richard Lewis in 1968 spoke of the "Kennedy effect," noting that

> the political decision to send men to the moon also led to unexpected results in the development of space technology...It has determined the priorities, the engineering designs, and the scientific objectives of the space program in this decade, and it is quite likely to control future space work for the remainder of this century. This unforeseen result might be called the Kennedy effect. While its intent at the beginning was to enlarge American competence in space, its implementation has built a Procrustean bed and the American space program has been severely mutilated to fit it.[27]

The consequences of selecting the lunar orbit approach to the Moon landing were of concern to Kennedy's science adviser Jerome Wiesner as he opposed the LOR choice in 1962. President Kennedy in his determination to be first to the Moon overruled Wiesner, a decision, as Lewis noted, with profound consequences for the space program. NASA during the second half of the 1960s became what James Webb had feared, a one-program agency; given the budget constraints of the period, there was no money available for major new starts on alternative programs.

The "Kennedy effect" went well beyond rockets and spacecraft. The Apollo program created in NASA an organization oriented in the public and political eye toward human space flight and toward developing large-scale systems to achieving challenging goals. It created from Texas to Florida the institutional and facility base for such undertakings. With the White House rejection of ambitious post-Apollo space goals, NASA entered a four-decade identity crisis from which it has yet to emerge. Repetitive operation of the space shuttle and the extended process of developing an Earth-orbiting space station have not been satisfying substitutes for another Apollo-like undertaking. NASA has never totally adjusted to a lower priority in the overall scheme of national affairs; rather, as the Columbia Accident Investigation Board observed in its 2003 report, NASA became "an organization straining to do too much with too little."[28] All of this is an unfortunate heritage of John Kennedy's race to the Moon.

Yale University organizational sociologist Gary Brewer in 1989 observed that NASA during the Apollo program came close to being "a perfect place"—the best organization that human beings could create to accomplish a particular goal. But, suggests Brewer, "perfect places do not last for long." NASA "perfected itself in the reality of Apollo, but that success is past and the lessons from it are now obsolete." The NASA of 1989, according to Brewer, was "no longer a perfect place" and was "deeply troubled." He added:

> The innocent clarity of purpose, the relatively easy and economically painless public consent, and the technical confidence [of Apollo]... are gone and will probably never occur again. Trying to recreate those by-gone moments by sloganeering, frightening, or appealing to mankind's mystical needs for exploration and conquest seems somehow futile considering all that has happened since Jack Kennedy set the nation on course to the Moon.

Brewer's comments of more than two decades ago might usefully be applied to the twenty-first century NASA and its supportive space community, which still struggle to maintain the approach to human space flight developed during the Mercury, Gemini, and Apollo programs. It is well beyond the scope of this study to discuss the future of the U.S. space exploration program; the point to make here is that the conditions that made Apollo possible and the NASA of the 1960s a "perfect place" were unique and will not reoccur. I agree with Brewer's conclusion that NASA needs "new ways of thinking, new people, and new means to come to terms and cope with social, economic, and political environments as challenging and harsh as deep space itself."[29]

### *Apollo and History*

The set of judgments that led President John F. Kennedy to decide to send Americans to the Moon combined lasting characteristics of the American people, a conviction of American exceptionalism and a mission derived from that conviction, the geopolitical situation of early 1961, and the individual values and style that Kennedy brought to the White House. Apollo was a product of a particular moment in time. Apollo is also a piece of lasting human history. Its most important significance may well be simply that it happened. Humans did travel to and explore another celestial body. Apollo will forever be a milestone in human experience, and particularly in the history of human exploration and perhaps eventual expansion. Because the first steps on the Moon were seen simultaneously in every part of the globe (with a few exceptions such as the Soviet Union), Apollo 11 was the first great exploratory voyage that was a shared human experience—what historian Daniel Boorstin called "public discovery."[30] John Kennedy's name will forever be linked with those first steps. Like other ventures into unknown territory, Apollo may not have followed the best route nor have been motivated by the same concerns that will stimulate future space exploration. But without someone going first, there can be no followers. In this sense, the Apollo astronauts were true pioneers.

The iconic "Earthrise" picture taken on Christmas Eve 1968 by Apollo 8 astronaut Bill Anders as he and his crewmates became the first humans to orbit the Moon and to look back at their home planet from 240,000 miles away. (NASA photograph).

Leaving the Earth gave the Apollo astronauts the unique opportunity to look back at Earth and to share what they saw. The Apollo 8 "Earthrise" picture is surely one of the iconic images of the twentieth century. It allowed us, as poet Archibald McLeish noted at the time, "to see earth as it truly is, small and blue and beautiful in that eternal silence where it floats" and "to see ourselves as riders on the earth together, brothers on that bright loveliness in the eternal cold—brothers who truly know that they are brothers."[31] That perception alone cannot justify the costs of going to the Moon, but it stands as a major benefit from going there, one that has influenced human behavior in many ways.

I hope that sometime in the future—if not in the coming decades then in the coming centuries—humans will once again choose to venture beyond the immediate vicinity of Earth. I believe that the urge to explore—to see what is over the next hill—is a fundamental attribute of at least some human cultures. Michael Collins, the Apollo 11 astronaut who remained in orbit as Armstrong and Aldrin experienced being on the Moon, has commented that the lasting justification for human space flight is "leaving"—going away from Earth to some distant destination. As future voyages of exploration are planned, I also hope that the United States chooses to be in the vanguard of a cooperative exploration effort involving countries from around the globe. There are two things I judge as certain, whenever those voyages take place. One is that they will not be like Apollo, a grand but costly unilateral effort racing against a firm deadline to reach a distant and challenging goal. The other is that President Kennedy's name will be evoked as humans once again begin to travel away from Earth. As he said in September 1962, "We set sail on this new sea because there is new knowledge to be gained, and new rights to be won, and they must be won and used for the progress of all people." John F. Kennedy, like the astronauts who traveled to the Moon during Apollo, was a true space pioneer.

# Notes

## Abbreviations

| | |
|---|---|
| JFKL | John F. Kennedy Presidential Library, Boston, Massachusetts |
| LBJL | Lyndon B. Johnson Presidential Library, Austin, Texas |
| NHRC | NASA Historical Reference Collection, NASA Headquarters, Washington, DC |
| NSF | National Security Files, JFKL |
| POF | President's Office Files, JFKL |
| WHCF | White House Central Files, JFKL |

## Preface

1. Roger D. Launius, "Interpreting the Moon Landings: Project Apollo and the Historians," *History and Technology,* Vol. 22 (September 2006), 225–255.
2. The bibliography accompanying the Launius essay cited in the previous note includes much of the thoughtful literature on Project Apollo.

## Prologue

1. Many people think that Kennedy announced his decision to go to the Moon in the speech at Rice University in Houston, Texas on September 12, 1962. That address was indeed a dramatic reaffirmation of JFK's decision and the most articulate presentation of its rationale, but it followed the actual announcement by almost sixteen months.
2. Hugh Sidey, *John F. Kennedy, President* (New York: Atheneum, 1964), 98.
3. The quotes are taken from my book *The Decision to Go to the Moon: Project Apollo and the National Interest* (Cambridge, MA; The MIT Press, 1970). With the permission of The MIT Press, this study draws heavily, and in some instances incorporates passages *verbatim,* from the contents of the earlier book, which has long been out of print.
4. William D. Eggers and John O'Leary, *If We Can Put a Man on the Moon…Getting Big Things Done in Government* (Boston, MA: Harvard Business Press, 2009), xi.

5. For Apollo costs, see http://history.nasa.gov/Apollomon/Apollo.html. For Panama Canal costs, see http://www.pancanal.com/eng/history/history/end.html. Inflation costs were calculated using the Bureau of Labor Statistics calculator found at http://data.bls.gov/cgi-bin/cpicalc.pl. For the cost of the Interstate Highway System, see http://www.fhwa.dot.gov/interstate/faq.htm#question6. The highway costs were paid through a special user tax, not general government revenues, and thus are not directly comparable to the funding appropriated through annual congressional actions for Apollo. Since the highway funds were provided over many years, it is not possible to inflate the total cost into 2010 dollars.
6. Schlesinger is quoted in *Congressional Record*, 94th Congress, 2nd Session, September 30, 1976, p. H11946.

## Chapter 1 Before the White House

1. The point that Joseph Kennedy expected his eldest living son to run for public office is made in almost every biography of John F. Kennedy and his family. The quote is taken from Vincent Bzdek, *The Kennedy Legacy: Jack, Bobby and Ted and a Family Dream Fulfilled* (New York: Palgrave Macmillan, 2009), 53. The assessment of Kennedy as a congressman is from James N. Giglio, "John F. Kennedy and the Nation," in James N. Giglio and Stephen G. Rabe, *Debating the Kennedy Presidency* (Lanham, MD: Rowman & Littlefield Publishers, 2003), 99. The description of Kennedy is by Norman Podhoretz, editor of *Commentary* magazine, quoted in Deborah Hart Strober and Gerald S. Strober, *The Kennedy Presidency: An Oral History of an Era* (Washington, DC: Brassey's, Inc., 1993), 33.
2. Charles Murray and Catherine Bly Cox, *Apollo: Race to the Moon* (New York: Simon & Schuster, 1989), 61.
3. Memorandum on John Kennedy's voting record on space issues from Walter (Jenkins), who was Lyndon Johnson's top aide during his Vice Presidential years, to the Vice President, September 28, 1962, Vice Presidential Papers, 1962 Subject Files, Box 181, LBJL. Geoffrey Perret in *Jack: A Life Like No Other* (New York: Random House, 2001) suggests (312) that Kennedy voted against establishing NASA. This does not seem to be correct; the final bill to create NASA was approved by the Senate on July 15, 1958, by a voice vote.
4. Speech of John F. Kennedy, "United States Military and Diplomatic Policies—Preparing for the Gap," August 14, 1958, reprinted from the Congressional Record, Pre-Presidential Files, Box 901, JFKL.
5. Giglio and Rabe, *Debating the Kennedy Presidency*, 9.
6. Letter from William Everdell to Senator John Kennedy, February 6, 1960, and Letter from John F. Kennedy to William Everdell, February 16, 1960, Pre-Presidential Files, Box 747, JFKL. Who actually composed this response for Kennedy's signature is unknown; Kennedy may have dictated it himself. Kennedy's close aide Theodore Sorensen noted that "Senator Kennedy signed very little of the correspondence he approved for his signature and dictated even less of it. Staff members composed letters in accordance with his thinking... On the other hand, he sometimes answered mail not worthy of his time." Theodore C. Sorensen, *Kennedy* (New York: Harper & Row, 1965), 57.
7. Oral history interview of Theodore C. Sorenson, March 26, 1964, JFKL.

8. Address of Senator John F. Kennedy Accepting the Democratic Party Nomination for the Presidency of the United States, July 15, 1960, http://www.jfklibrary.org/Historical+Resources/Archives/Reference+Desk/Speeches/JFK/JFK+Pre-Pres/1960/Address+of+Senator+John+F.+Kennedy+Accepting+the+Democratic+Party+Nomination+for+the+Presidency+of+t.htm. As a footnote to this history, the author was present at the Los Angeles Coliseum as Kennedy made his acceptance speech.
9. Democratic Party Platform of 1960, July 11, 1960, http://www.presidency.ucsb.edu/ws/index.php?pid=29602
10. Theodore H. White, *The Making of the President, 1960* (New York: Atheneum Publishers, 1961), 272; Sorensen, *Kennedy*, 117–118; Arthur M. Schlesinger, Jr., *A Thousand Days: John F. Kennedy in the White House* (Boston, MA: Houghton Mifflin, 1965), 15, 69.
11. Memorandum from Senator [John] Kennedy to Archibald Cox, September 2, 1960, NHRC, Folder 012492. It is worth noting that President Eisenhower had approved the CORONA reconnaissance satellite program in February 1958 and that CORONA flew its first successful mission in August 1960, before JFK's question. The CORONA program was very highly classified, and clearly Senator Kennedy was not aware of its existence. For information on the CORONA intelligence satellite program, which was not declassified until 1995, see Dwayne A. Day, John M. Logsdon, and Brian Latell, *Eye in the Sky: The Story of the CORONA Spy Satellites* (Washington, DC: Smithsonian Institution Press, 1998)
12. Ralph E. Lapp in consultation with Trevor Gardner and Frank McClure, and coordinated with Samuel K. Allison, Harrison Brown, Ernest C. Pollard, Richard B. Roberts, and Harold C. Urey, "Position Paper on Space Research," September 7, 1960, 3, 11, 13–15, NHRC, Folder 012492. For an example of Lapp's views, see Ralph E. Lapp, *The New Priesthood: The Scientific Elite and the Uses of Power* (New York: Harper & Row, 1965)
13. "Briefing Paper on Space," undated, in Pre-Presidential Files, Box 993A, JFKL. In what seems to be Kennedy's handwriting, the notation "Library" appears on the first page of the paper, indicating that it may have been read by the candidate.
14. Interview of Edward C. Welsh by the author, August 14, 1967, cited in John M. Logsdon, *The Decision to Go to the Moon: Project Apollo and the National Interest* (Cambridge, MA: The MIT Press, 1970). 65. The quote is from Edwin Diamond, *The Rise and Fall of the Space Age* (Garden City, NY: Doubleday & Company, 1964), 31.
15. *Missiles and Rockets*, October 3, 1960, 10 and October 10, 1960, 12–13.
16. Speech in Portland, Oregon, September 7, 1960, http://www.jfklink.com/speeches/jfk/sept60/jfk070960_portland03.html
17. Sorensen, *Kennedy*, 178; White, *Making of the President*, 348–349.
18. This speech is quoted in Vernon van Dyke, *Pride and Power: The Rationale of the Space Program* (Urbana: University of Illinois Press, 1964), 23.
19. A draft campaign speech on space can be found in NHRC, Folder 012492.
20. *The Washington Post*, October 29, 1960, p. A1, cited in Logsdon, *Decision*, 67.
21. Lyndon B. Johnson, "The Record in Space" (campaign paper released October 31, 1960), 1–2, quoted in Ibid, 66.

22. Memorandum for the President [Eisenhower] from Allen Dulles, August 3, 1960, and Sorensen, *Kennedy*, 176.
23. Neil Sheehan, *A Fiery Peace in the Cold War: Bernard Schriever and the Ultimate Weapon* (New York: Random House, 2009), 362.
24. Sorensen, *Kennedy*, 610–613 and Theodore C. Sorensen, *Counselor: A Life at the Edge of History* (New York: HarperCollins, 2008), 189.

## Chapter 2 Making the Transition

1. Sorensen, *Kennedy*, 227.
2. Richard E. Neustadt, *Presidential Power: The Politics of Leadership* (New York: Wiley, 1960).
3. Richard E. Neustadt, "Organizing the Transition," September 15, 1960, Sorenson Papers, Box 120, JFKL. The quoted passage is on 7. Also, Schlesinger, 122–123 and Sorensen, *Kennedy*, 229.
4. Richard E. Neustadt, "Staffing the President-Elect," October 30, 1960, 3–4, Pre-Presidential Files, Box 1072, JFKL. This staffing approach, which to a large degree Kennedy adopted, was strongly criticized by another eminent scholar of the presidency, James McGregor Burns, in his *Running Alone: Presidential Leadership—JFK to Bush II: Why It Has Failed and How We Can Fix It* (New York: Basic Books, 2006).
5. Arthur Schlesinger, Jr., *A Thousand Days*, 123–124. On the transition, see also Sorensen, *Kennedy*, 228–240 and Sorensen, *Counselor*, 198–202.
6. For a full account of Sorensen's involvement with John F. Kennedy, see his two books cited above. For Salinger, see Pierre Salinger, *With Kennedy* (Garden City, NY: Doubleday & Company, 1966). O'Donnell collaborated with David Powers, another member of the "Irish mafia," and Joseph McCarthy to write *Johnny, We Hardly Knew Ye'* (Boston, MA: Little Brown, 1972). For Schlesinger's assessment of Sorensen, see *A Thousand Days*, 208.
7. Salinger, *With Kennedy*, 64–65.
8. Sorensen, *Counselor*, 211, 237–238, 334; Sorensen, *Kennedy*, 263.
9. Sorenson, *Counselor*, 199, 203; author's interview with Theodore C. Sorensen, May 28, 2009.
10. Sorensen, *Kennedy*, 254; Schlesinger, *A Thousand Days*, 127.
11. "Pre-Inaugural Task Forces Unprecedented in History," *Congressional Quarterly Weekly Report*, April 7, 1961, 620.
12. This account of the space issues facing President-elect Kennedy is by design very selective. For more comprehensive accounts of the early years of the U.S. space program, see John M. Logsdon, *The Decision to Go to the Moon: Project Apollo and the National Interest* (Cambridge, MA: The MIT Press, 1970); Walter A. McDougall,*...the Heavens and the Earth: A Political History of the Space Age* (New York: Basic Books, 1985); James R. Killian, Jr., *A Scientist in the White House* (Cambridge, MA: MIT Press, 1977); George B. Kistiakowsky, *A Scientist at the White House: The Private Diary of President Eisenhower's Special Assistant for Science and Technology* (Cambridge, MA: Harvard University Press, 1976); Vernon van Dyke, *Pride and Power: The Rationale of the Space Program* (Urbana: University of Illinois Press, 1964); Robert A. Divine, *The Sputnik Challenge: Eisenhower's Response to the Soviet Satellite* (New York: Oxford University Press, 1993); Robert L. Rosholt, *An Administrative History of NASA, 1958–1963* (Washington, DC: Government

Printing Office, 1966); and David Callahan and Fred I Greenstein, "The Reluctant Racer: Eisenhower and U.S. Space Policy" in Roger D. Launius and Howard E. McCurdy, eds., *Spaceflight and the Myth of Presidential Leadership* (Urbana: University of Illinois Press, 1997).

13. Logsdon, *Decision*, 51–52. For a comprehensive and even-handed biography of Wernher von Braun, see Michael Neufeld, *Von Braun: Dreamer of Space, Engineer of War* (New York: Alfred P. Knopf, 2007).
14. White is quoted in David N. Spires, *Beyond Horizons: A Half Century of Air Force Space Leadership*, Revised Edition (Colorado Springs, CO: Air University Press, 1998), 54.
15. Logsdon, *Decision*, 44–48.
16. James R. Killian, Jr., special assistant for science and technology; Percival Brundage, director, Bureau of the Budget; and Nelson A. Rockefeller, chairman, President's Advisory Committee on Government Organization, Memorandum for the President, "Organization for Civil Space Programs," March 5, 1958, in John M. Logsdon *et al.*, eds., *Exploring the Unknown: Selected Documents in the History of the U.S. Civil Space Program,* Volume I: Organizing for Exploration, NASA SP-4407 (Washington, DC: Government Printing Office, 1995), 638.
17. Alison Griffith, *The National Aeronautics and Space Act: A Study of the Development of Public Policy* (Washington, DC: Public Affairs Press, 1962), 14.
18. For the text of this Act, see Logsdon et al., *Exploring the Unknown*, Vol. I, 334–345.
19. Roger D. Launius, "Introduction," in the *The Birth of NASA: The Diary of T. Keith Glennan* , NASA SP-4105 (Washington, DC: Government Printing Office, 1993), xxi. Glennan dictated his observations on the issues and personalities involved in the start up years of NASA. He intended the diary only for his children, but NASA prevailed on him to have it published, and the result is fascinating reading.
20. Killian, *Sputnik, Scientists, and Eisenhower,* 139.
21. T. Keith Glennan, Oral History Interview, February 20, 1987, Glennan-Webb-Seamans Project for Research in Space History, National Air and Space Museum.
22. Loyd Swenson, Jr., James M. Grimwood, and Charles C. Alexander, *This New Ocean: A History of Project Mercury,* NASA SP-4201 (Washington, DC: Government Printing Office, 1966), 102.
23. See Roger E. Bilstein, *Stages to Saturn: A Technological History of the Apollo/Saturn Launch Vehicles*, NASA SP-4206 (Washington, DC: Government Printing Office, 1980), chapter 2, for the origins of the Saturn vehicles, and chapter 4 for the origins of the F-1 engine.
24. NASA, Office of Program Planning and Evaluation, "The Long Range Plan of the National Aeronautics and Space Administration," December 16, 1959, in Logsdon et al., *Exploring the Unknown*, Volume I, 403–405.
25. Spires, *Beyond Horizons,* 87–88.
26. John Finney, "Air Force Seeks Top Role in Space," *The New York Times*, December 11, 1960, 68.
27. Glen P. Wilson, "How the Space Act Came to Be," Appendix A to NASA History Division, *Legislative Origins of the National Aeronautics and Space*

*Act of 1958,* Monographs in Aerospace History, Number 8, 1998, 62; Killian, *Sputnik, Scientists, and Eisenhower*, 137–138; and Divine, *Sputnik Challenge*, 147–148.

28. Wilton B. Persons, Deputy Assistant to the President, "Memorandum for the Record," July 7, 1958. www.eisenhower.archives.gov/Research/Digital_Documents/NASA/15.pdf.
29. Divine, *Sputnik Challenge*, 148–149.
30. Glennan, *Birth of NASA,* 25, 27, 29, 38, 46–47.
31. Ibid, 170–171.
32. *U.S. Congressional Record*, 86th Cong., 2d sess., August 31, 1960, 18509.
33. For a fascinating account of the origins and evolution of the U.S. ICBM program, see Neil Sheehan, *A Fiery Peace in the Cold War: Bernard Schriever and the Ultimate Weapon* (New York: Random House, 2009)
34. The first U.S. attempt to launch a satellite called Vanguard, which was the approved U.S. IGY entry, failed on December 6, 1957. Vanguard was even lighter than Explorer 1, weighing only a little over 3 pounds.
35. Bilstein, *Stages to Saturn*, 50.
36. See Glennan, *Birth of NASA*, especially 273 and 282, for a discussion of the negotiations with respect to the FY1962 budget.
37. "Minutes of Meeting of Research Steering Committee on Manned Space Flight," May 25–26, 1959, in John M. Logsdon, ed., with Roger D. Launius, *Exploring the Unknown: Selected Documents in the History of the U.S. Civil Space Program,* Volume VII: Human Spaceflight: Mercury, Gemini, and Apollo, NASA SP-2008-4407 (Washington, DC: Government Printing Office, 2008), 447–448.
38. James Killian, Speech to MIT Club of New York, December 13, 1960, reprinted in *Science*, January 6, 1961, 24–25, as cited in Logsdon, *Decision*, 20.
39. "To Race or Not to Race (A Discussion Paper)" October 16, 1959, Box 15, Records of the Office of the Special Assistant for Science and Technology, Dwight D. Eisenhower Library, Abilene, Kansas. Although Kistiakowsky's name does not appear on the paper, he is identified as its author in President's Science Advisory Committee, "Major Actions of the President's Science Advisory Committee, November 1957—January 1961, January 13, 1961, in Box 1072, Pre-Presidential Files, JFKL.
40. Kistiakowsky, *A Scientist at the White House,* 409.
41. The panel's skeptical view may also have been influenced by two recent failures in the Mercury program. Orbital flights during Project Mercury would be launched by a converted Atlas ICBM. The first suborbital test of the Mercury-Atlas combination on July 29, 1960, was a complete failure. Then on November 21, 1960, a test launch of a Mercury capsule atop a von Braun—developed Redstone rocket rose only a few inches off the launch pad before settling back to the ground. The Mercury-Redstone combination was to launch the first few U.S. astronauts on brief suborbital flights before NASA would commit to sending an American into orbit.
42. President's Science Advisory Committee, "Report of the Ad Hoc Panel on Man-in-Space," December 16, 1960, in Logsdon *et al.*, *Exploring the Unknown*, Vol. I, 408–412.
43. Kistakowsky, *Scientist in the White House, 409;* Glennan, *Birth of NASA,* 292. Seamans was the "privileged source" source quoted in Logsdon, *Decision*, 35. Also, Robert C. Seamans, Jr., *Project Apollo: The Tough Decisions*, NASA

Monographs in Aerospace History No. 31, SP-2005-4537 (Washington, DC: Government Printing Office, 2005), 7–8.

44. Glennan, *Birth of NASA,* 297.
45. George M. Low, "Manned Lunar Landing Program," October 17, 1960 in Logsdon with Launius, *Exploring the Unknown*, Volume VII, 457.
46. Seamans, *Project Apollo*, 7.
47. Ibid, 9.
48. Interview with Elmer Staats, August 30, 1967, cited in Logsdon, *Decision*, 67. Richard E. Neustadt, "Preliminary Check-List of Organizational Issues," December 17, 1960, Sorensen Papers, Box 120, JFKL.
49. W.H. Lawrence ,"Kennedy Assigns Johnson to Head Two Major Units," *The New York Times*, December 21, 1960, 1.
50. Ken Belieu, Memorandum for Senator Johnson, "Governmental Organization for Space Activities," December 17, 1960, Vice Presidential Papers, Security Files, Box 17, LBJL.
51. Randall B. Woods, *LBJ: Architect of American Ambition* (New York: Simon & Schuster, 2006), 336.
52. Interview of Lyndon B. Johnson by Walter Cronkite, July 5, 1969 (aired July 21, 1969). A transcript of the interview is in the NHRC, Folder 012530.
53. Ken Belieu, Memorandum for Senator Johnson, "Space Problems," December 22, 1960, Vice-Presidential Security Files, Box 17, LBJL.
54. Richard E. Neustadt, Memorandum for Senator Kennedy, "Memo on Space Problems for you to use with Lyndon Johnson," December 23, 1960, JFKL.
55. Interview with Senator Clinton Anderson, August 18, 1967, in Logsdon, *Decision*, 68.
56. William Lawrence, "Kennedy Confers on Plans to Spur Space Research," *The New York Times,* December 27, 1960, 1.
57. Ken Belieu, Memorandum for Senator Johnson, "Governmental Organization for Space Activities," December 17, 1960, Vice-Presidential Security Files, Box 17, LBJL.
58. Seamans, *Project Apollo*, 9.
59. "Report to the President-Elect of the Ad Hoc Committee on Space," January 10, 1961, Pre-Presidential Papers, Box 1072, JFKL.
60. This last factor, "international cooperation," was added to the list provided in a 1958 PSAC report to President Eisenhower, "Introduction to Outer Space." The order in which other goals were listed was also shifted from that report to move "national prestige" to the top of the list. The task force report also covered the U.S. ballistic missile program; that portion of the report will not be discussed here.
61. *The Washington Post,* January 11, 1961, A2; *The New York Times*, January 26, 1961, 10.
62. Michael Bechloss, "Kennedy and the Decision to Go to the Moon," in Launius and McCurdy, *Spaceflight and the Myth of Presidential Leadership,* 54. A February 2, 1961, memorandum from C. Berg of the Bureau of the Budget notes that Wiesner had informed the NASA staff that he "disassociates" himself from the content of his task force's report. NHRC, Folder 012457.
63. Glennan, *Birth of NASA*, 273.
64. Ibid, 300.

65. Ibid, 307. It is not clear that Dryden was ever told of this message from Clifford. On the day that he was nominated to be the next NASA administrator, James Webb wrote to Vice President Johnson, saying that "Dryden has not been officially requested to serve as Acting Administrator." Dryden remembers that he "had submitted my resignation" but that "no acknowledgement had been received. Therefore I continued as Acting Administrator until the new officials could be appointed." Memorandum to the Vice President from James E. Webb, January 30, 1961, NHRC, Folder 12288 and Hugh Dryden, Memorandum for Eugene Emme for NASA Historical Files, September 17, 1965, NHRC, Folder 012505.
66. Glennan, *Birth of NASA*, 310.

## Chapter 3 Getting Started

1. Kennedy's inaugural address can be found at http://www.presidency.ucsb.edu/ws/index.php?pid=8032&st=&st1=. This web site contains the public papers of the presidents beginning with George Washington in 1789.
2. Sorensen, *Counselor,* 223.
3. Thomas C. Reeves, *A Question of Character: A Life of John F. Kennedy* (New York: The Free Press, 1991), 310.
4. Richard E. Neustadt, "Organizing the Transition," September 15, 1960, 1, Sorenson Papers, Box 120, JFKL; Schlesinger, *A Thousand Days,* 214.
5. For a discussion of these issues during the early months of the Kennedy administration, see Schlesinger, *A Thousand Days*, 298–318 and Richard Reeves, *President Kennedy: Profile of Power* (New York: Simon & Schuster, 1993), 58–75.
6. This account is based on Schlesinger, *A Thousand Days*, 320–334 and Roger Hilsman, *To Move a Nation: The Politics of Foreign Policy in the Administration of John F. Kennedy* (Garden City, NY: Doubleday & Company, 1967), 130–131.
7. In a speech to the National Press Club on March 16, 1967, Lyndon Johnson said that he had interviewed nineteen people; in a press conference after the Apollo 8 flight in December 1968, he put the number at twenty-eight. The transcript of the December 27, 1968, press conference can be found at http://www.presidency.ucsb.edu/ws/index.php?pid=29300&st=&st1=. In his careful biography of James Webb, W. Henry Lambright also uses the number nineteen. W. Henry Lambright, *Powering Apollo: James E. Webb of NASA* (Baltimore, MD: Johns Hopkins University Press, 1995), 82.
8. Ken Belieu, Memorandum to Senator Johnson, December 21, 1960, and Memorandum to Senator Johnson, "Suggestions for Head of NASA," December 22, 1960, NHRC, Folder 12288.
9. Unsigned memorandum dated January 25, 1961, Vice Presidential Papers, 1961 Subject Files, Box 116, LBJL.
10. Interview of Lyndon B. Johnson by Walter Cronkite, July 5, 1969 (aired July 21, 1969). A transcript of the interview is in the NHRC, Folder 012530.
11. Ken Belieu, Memorandum for the Vice President, January 23, 1961, and Memorandum for Bill Moyers, January 25, 1961, NHRC, Folder 12288.
12. Transcript of President Kennedy's News Conference, January 25, 1961, http://www.presidency.ucsb.edu/ws/index.php?pid=8533&st=&st1=.

13. Author's interview with Jerome Wiesner, September 11, 1967, cited in Logsdon, *Decision*, 83.
14. See Jay Holmes, *The Race to the Moon* (Philadelphia, PA: J. B. Lippincott Company, 1962), 189–192 for a discussion of this dispute.
15. Piers Bizony, *The Man Who Ran the Moon: James E. Webb, NASA, and the Secret History of Project Apollo* (New York: Thunder's Mouth Press, 2006), 15. This journalistic and undocumented biography of James Webb is an occasionally useful complement to the more scholarly biography by Lambright cited earlier. See also Murray and Cox, *Apollo: Race to the Moon*, 70. For Wiesner's comments, see interview of Jerome Wiesner by W. Henry Lambright, November 15, 1990, NHRC, Folder 7106.
16. Interviews with James Webb by the author, December 15, 1967, and August 12, 1969, cited in Logsdon, *Decision*, 84. This account is taken from Logsdon, *Decision*, 82–85 and Lambright, *Powering Apollo*, 82–85. See also James E. Webb Oral History Interview 1, April 29, 1969, Internet Copy, LBJL. In some accounts, Webb asked Lyndon Johnson, not President Kennedy, whether he was being hired to implement a predetermined policy.
17. Webb's words are in the foreword to Robert Rosholt, *An Administrative History of NASA, 1958–1963*, NASA SP-4101 (Washington, DC: Government Printing Office, 1966), iv.
18. Memorandum from Military Division, Bureau of the Budget, to the Director, "Meeting with Mr. James Webb," February 15, 1961, NHRC, Folder 012457.
19. Lambright, *Powering Apollo*, 87 and Robert C. Seamans, Jr., *Aiming at Targets*, NASA SP-4106 (Washington, DC: Government Printing Office, 1996), 82.
20. Bizony, *The Man Who Ran the Moon*, 18; Theodore C. Sorensen oral history interview, March 26, 1964, JFKL; Interview of Jerome Wiesner by W. Henry Lambright, November 15, 1990, NHRC, Folder 7106; Murray and Cox, *Apollo*, 70; Robert Kennedy, interview with journalist John Bartow Martin, in Edwin O. Guthman and Jeffrey Shulman, eds., *Robert Kennedy in His Own Words: The Unpublished Recollections of the Kennedy Years* (New York: Bantam Books, 1988), 340–341.
21. John Finney, "U.S. Space Setbacks: Failure of Moon Shot Raises Problem of Capability of Atlas Missile for the Job," *The New York Times*, December 16, 1960, 19.
22. This account is drawn from Logsdon, *Decision*, 86 and Lambright, *Powering Apollo*, 89–90.
23. *Missiles and Rockets*, March 27, 1961, 13.
24. Interview with privileged source (who was Robert Seamans) cited in Logsdon, *Decision*, 86.
25. Roswell Gilpatric, telephone interview with W. Henry Lambright, December 17, 1991. Lambright's notes from the interview can be found in NHRC, Folder 7106.
26. James E. Webb, "Memorandum for the Record," February 24, 1961, and letter to David Bell, February 27, 1961, NHRC, Folder 012504.
27. Lambright, *Powering Apollo*, 90–91 and Logsdon, *Decision*, 79.
28. Department of Defense Directive 5160.32, "Development of Space Systems," March 6, 1961.

29. "Report of the Air Force Space Study Committee, March 20, 1961, NSF, Box 307, JFKL. See also Spires, *Beyond Horizons,* 91 for a discussion of the report.
30. Letter from Representative Overton Brooks to President John F. Kennedy, March 9, 1961 in NHRC.
31. Letter from President John F. Kennedy to Representative Overton Brooks, March 23, 1961 in NHRC.
32. Hugh Dryden, Memorandum to NASA Historian Eugene Emme for NASA Historical Files, "Eisenhower-Kennedy Transition," September 27, 1965, NHRC, Folder 012505.
33. Letter to author from George Low, cited in Logsdon, *Decision*, 80.
34. Interview with Donald Hornig, October 11, 1967, cited in Logsdon, *Decision*, 80.
35. "g force" is a measure of the pull of gravity acting on an object. At the Earth's surface, it is 1 g. This force increases as an object is accelerated or decelerated, as during a space lift-off or reentry.
36. Neufeld, *Von Braun,* 358–359.
37. Letter to author from George Low, cited in Logsdon, *Decision*, 80. See also Swenson, Grimwood, and Alexander, *This New Ocean,* 310–326 for a discussion of these events.
38. Letter from Jerome B. Wiesner to Hugh Dryden, March 7, 1961, NHRC, Folder 012506. Oral history interview with Hugh Dryden, March 26, 1964, JFKL.
39. "Report of the Ad Hoc Mercury Panel," April 12, 1961, in Logsdon with Launius, eds., *Exploring the Unknown*, Volume VII, 177–192.
40. Ibid. See also Swenson, Grimwood, and Alexander, *This New Ocean*, 330, 347–349, and author's interviews with Donald Hornig and Jerome Wiesner in *Decision.*
41. The instructions given to Low's group are discussed in Seamans, *Project Apollo*, 8.
42. George M. Low, "A Plan for a Manned Lunar Landing," February 7, 1961. Excerpts of the report are reprinted in Logsdon with Launius, *Exploring the Unknown*, Vol. VII, 458–471.
43. Homer Newell, Memorandum for the Files, "Notes on the Space Science Board Meeting," February 10–11, 1961, in NHRC, Folder 012504. Newell in 1961 was NASA's top scientist and its liaison to the Space Science Board.
44. The Board's statement is cited in Logsdon, *Decision*, 87–88.
45. AAAS Committee on Science in the Promotion of Human Welfare, "The Integrity of Science," *American Scientist* 53 (June 1965), 184–85.

## Chapter 4 First Decisions

1. The work of this task force is discussed in chapter 10.
2. Memorandum from Bill Moyers to the Vice President, undated, with attached draft Executive Order, White House Famous Names Files, John F. Kennedy, Box 5, LBJL.
3. Letter from John F. Kennedy to the Vice President, January 28, 1961, White House Famous Names Files, Box 5, LBJL.
4. Woods, *LBJ*, 381.

5. Bureau of the Budget, Status Memorandum, "Reorganize the National Aeronautics and Space Council," March 23, 1961, attached to Richard E. Neustadt, Note for Ed Welsh, "Title for the Space Council," April 4, 1961, National Archives and Record Administration, National Aeronautics and Space Council, Record Group 220, Box 19.
6. Letter from Lyndon B. Johnson to the President, February 14, 1961, Vice-Presidential Papers, 1961 Subject Files, Box 117, LBJL.
7. Richard E. Neustadt, Memorandum for Mr. Bill Moyers, "The Space Council," February 28, 1961, with attached "Memorandum on Organizing the Space Council," February 27, 1961, WHCF, Box 114, JFKL; Letter from J. A. Van Allen to Lyndon B. Johnson, February 4, 1961, Vice-Presidential Papers, 1961 Subject Files, Box 116, LBJL; "The National Aeronautics and Space Council," March 1, 1961, NHRC, Folder 12288. Although the Bureau of the Budget is not identified as the source of this document, it follows the format then used for BOB staff memoranda and thus is likely to be a BOB document.
8. David E. Bell, Memorandum for the Vice President, March 6, 1961, NHRC, Folder 12288.
9. Holmes, *Race to the Moon,* 195–196. Oral History Interview of Dr. Edward C. Welsh, May 16, 1964, JFKL.
10. *The National Aeronautics and Space Council during the Tenure of Lyndon B. Johnson as Vice President and during His Administration as President (January 1961—January 1968)*, undated but 1968, 4–5, LBJL. This was one of the many similar organizational histories prepared during the final year of the Johnson administration.
11. Richard E. Neustadt, Note for Ed Welsh, "Title for the Space Council," April 4, 1961, National Archives and Record Administration, National Aeronautics and Space Council, Record Group 220, Box 19.
12. Note from E.B. Staats to Dave [Bell], April 7, 1961, 3:25 p.m., NHRC, Folder 012506. Elmer Staats was the Deputy Director of the Bureau of the Budget and a holdover from the Eisenhower administration.
13. *The National Aeronautics and Space Council during the Tenure of Lyndon B. Johnson as Vice President and during His Administration as President (January 1961—January 1968)*, 8–9.
14. http://www.presidency.ucsb.edu/ws/index.php?pid=8086&st=&st1=.
15. Overton Brooks, "Attitude of Committee on Science and Astronautics Relative to the National Space Program," Memorandum for White House Conference, February 13, 1961, POF, Box 82, JFKL.
16. Memorandum from Richard Hirsch to Mr. [McGeorge] Bundy and Mr. [Walt] Rostow, "Possible Military Implications of Soviet Venus Direction Probe," February 16, 1961, NSF, Box 307, JFKL. No author, "Possible Questions and Answers on the Soviet Venus—Probe for the President's Press Conference on Wed. Feb. 15," NSF Files, Box 307, JFKL.
17. J.B. Wiesner, Memorandum for the President, February 20, 1961, NSF, Box 307, JFKL.
18. Letter from James E. Webb to the Director, Bureau of the Budget, March 17, 1961, NHRC, Folder 012504.
19. Author's interview with David Bell, October 4, 1967, cited in *Decision*, 94. Bell had been a junior budget examiner when James Webb had been the BOB director during the Truman administration.

20. E.C. Welsh, Memorandum for the Vice President, "Proposed Increase in FY 1962 Budget for NASA," March 22, 1961, Vice Presidential Security Files, Box 15, LBJL.
21. Agenda for NASA-BOB Conference with the President, March 22, 1961, NSF, Box 282, JFKL.
22. James E. Webb, "Administrator's Presentation to the President," March 21, 1961. NSF, Box 282, JFKL.
23. Seamans, *Project Apollo,* 13. See also Robert Seamans, Oral History Interview, November 2, 1987, Glennan-Webb-Seamans Project for Research in Space History, National Air and Space Museum, for Seamans's account of this meeting.
24. Letter from James Webb to T. Keith Glennan, March 30, 1961, NHRC.
25. David Bell, Memorandum for the President, "National Aeronautics and Space Administration budget problem," undated but March 23, 1961, National Aeronautics and Space Council Papers, Box 3, JFKL.
26. Robert Seamans, Memorandum for the Administrator, March 23, 1961, NHRC, Folder 012504.
27. Memorandum from Associate Administrator Robert Seamans to the Administrator, "Recommended Increases in FY1962 Funding for Launch Vehicles and Manned Space Exploration," March 23, 1961, NHRC.
28. Oral History Interview of Edward C. Welsh, May 16, 1964, JFKL.
29. Memorandum from Colonel [Howard] Burris to the Vice President, "Views of Secretary McNamara on Acceleration of Booster Capabilities and Manned Exploration of Space," March 23, 1961, NHRC, Folder 12288. Burris was a military assistant to the Vice President.
30. Author's interview with Robert Seamans, cited in Logsdon, *Decision*, 99.
31. Author's interview with James Webb, December 15, 1967, and August 12, 1969, cited in Logsdon, *Decision*, 99.
32. Author's interview with David Bell, cited in Logsdon, *Decision*, 100.
33. Logsdon, *Decision*, 91.

## Chapter 5 "There's Nothing More Important"

1. Sidey, *John F. Kennedy, President*, 92–93 and Andrew Hatcher, "Memorandum for the President," April 10, 1961,WHCF, Box 654, JFKL.
2. Memorandum from Edward R. Murrow to McGeorge Bundy, "Recommended U.S. Reaction to Soviet Manned Space Shot Failure," April 3, 1961, NSF, Box 307, JFKL. In a handwritten note on Murrow's memo, Bundy said "tell him I agree."
3. For details of the Gagarin flight, see Asif A. Siddiqi, *Challenge to Apollo: the Soviet Union and the Space Race, 1945–1974,* NASA SP-2000-4408 (Washington, DC: Government Printing Office, 2000), chapter 7. For information on the decoding of the television transmissions from the Soviet spacecraft, see Sven Grahn, "TV from Vostok" at http://www.svengrahn.pp.se/trackind/TVostok/TVostok.htm.
4. Sidey, *John F. Kennedy, President*, 94–95 and John F. Kennedy, "Statement by the President on the Orbiting of a Soviet Astronaut," April 12, 1961, http://www.presidency.ucsb.edu/ws/?pid=8053. The Soviet statement is reported in Swenson, Grimwood, and Alexander, *This New Ocean,* 332.
5. Sorensen, *Counselor*, 334 and Sorensen, *Kennedy*, 525.

6. John F. Kennedy, "The President's News Conference of April 12, 1961," http://www.presidency.ucsb.edu/ws/index.php?pid=8055.
7. Logsdon, *Decision*, 101 and Harry Schwartz, "Soviet Feat Aids Propaganda Aim," *The New York Times*, April 13, 1961, 16.
8. Logsdon, *Decision*, 102.
9. Memorandum from Donald Wilson, Deputy Director, USIA to McGeorge Bundy, April 21, 1961 with attached report "Initial World Reaction to Soviet 'Man in Space,'" Report R-17-621, April 21, 1961. Quoted material is on 1. NSF, Box 307, JFKL.
10. *The Washington Post*, April 13, 1961, A18.
11. Harry Schwartz, "Moscow: Flight is Taken as Another Sign That Communism is the Conquering Wave," *The New York Times*, April 16, 1961, E3.
12. Hanson Baldwin, "Flaw in Space Policy: U.S. is Said to Lack Sense of Urgency in Drive for New Scientific Conquests," *The New York Times*, April 17, 1961, 5.
13. James R. Kerr, "Congressmen as Overseers: Surveillance of the Space Program" (unpublished Ph.D. dissertation, Stanford University, 1963), 402, cited in Logsdon, *Decision*, 103.
14. These quotations are taken from Logsdon, *Decision*, 103.
15. Ibid, 104 and Seamans, *Project Apollo*, 16.
16. *The Washington Post* story is quoted in House Committee on Science and Astronautics, *Toward the Endless Frontier: History of the Committee on Science and Technology, 1959–1979* (Washington, DC: Government Printing Office, 1980), 86. See also *The New York Times*, April 15, 1961, 3; Seamans, *Project Apollo*, 17; and James E. Webb, "Memorandum for Mr. Kenneth O'Donnell, The White House," April 21, 1961, WHCF, Box 652, JFKL.
17. Sorensen, *Counselor*, 335.
18. Memorandum from Hugh Sidey to Pierre Salinger, "Questions for the President on Space," April 14, 1961, Sorensen Files, Box 38, JFKL.
19. Jerome Wiesner, Draft Memorandum for the President, "Review of the Space Program," April 14, 1961, Wiesner Files, Box 8, JFKL. It is not clear if the memorandum was actually sent to the president, but the draft does show the state of Wiesner's thinking in the immediate aftermath of the Gagarin flight.
20. Jerome B. Wiesner, Memorandum for the President, April 14, 1961, Wiesner Files, Box 2, JFKL.
21. Glenn T. Seaborg with the assistance of Benjamin S. Loeb, *Kennedy, Khrushchev, and the Test Ban* (Berkeley, CA: University of California Press, 1981), 31.
22. This account of the April 14 presidential meeting is drawn from several sources, particularly Sidey, *John F. Kennedy, President*, 100–103. Another source is a letter dated December 6, 1971 that Sidey wrote to fellow veteran journalist Robert Sherrod, who was working on a book on the space program, describing the April 14 meeting. That letter is in Sherrod's files in the NHRC. In addition, see Sorensen, *Counselor*, 334–336 and Edward Welsh, Confidential Memorandum to the Vice President, "Discussion of Space Program Friday Evening, April 14," April 18, 1961, NHRC, Folder 012506.
23. Interview with Theodore Sorensen cited in Logsdon, *Decision*, 107.
24. Interview with Jerome Wiesner cited in Ibid.

25. *Life*, April 21, 1961, 26–27. Kennedy's interaction with Sidey is a revealing example of how JFK used the press to further his interests. Thinking that Hanson Balwin's article of April 17 cited above may have been prompted by the same kind of background briefing, I wrote Baldwin in 1967. He replied that "there was not then…any direct communication between President Kennedy and me about this or any other topic…The Kennedys expected too much tailoring of writing to make any such relationship palatable to me…I was not a writer admired by the administration." Hanson Baldwin, Letter to the author, September 21, 1967.
26. This is the title of a book by Trumbull Higgins, *The Perfect Failure: Kennedy, Eisenhower, and the CIA at the Bay of* Pigs (New York: W. W. Norton, 1989). Every history of the Kennedy presidency contains a detailed analysis of the Bay of Pigs failure, the decisions that led to it, and its aftermath. They will not be repeated here.
27. Sorensen, *Counselor*, 316–317; Robert Dallek, *An Unfinished Life: John F. Kennedy, 1917–1963* (Boston, MA: Little Brown and Company, 2003), 367. The Robert Kennedy quote is from an interview with Walt Rostow, at the time of the Bay of Pigs on McGeorge Bundy's national security staff, in Strober and Strober, *Kennedy Presidency,* 349.
28. Interviews with Edward Welsh, Jerome Wiesner, McGeorge Bundy, and Theodore Sorensen cited in Logsdon, *Decision*, 111–112.
29. Interview of Lyndon B. Johnson by Walter Cronkite, July 5, 1969 (aired July 21, 1969). A transcript of the interview is in NHRC, Folder 012530.
30. John F. Kennedy, "The President's News Conference of April 21, 1961," http://www.presidency.ucsb.edu/ws/index.php?pid=8077 .

## Chapter 6 Space Plans Reviewed

1. The first quote is drawn from an interview of Jerome Wiesner by W. Henry Lambright, November 15, 1990, NHRC, Folder 7106; the second, longer, quote, from an earlier interview of Wiesner by the author on September 11, 1967, cited in Logsdon, *Decision,* 110–111. The consistency between Wiesner's assessment of Kennedy's behavior in 1967 and in 1990 is striking.
2. On the "Johnson system," see Rowland Evans and Robert Novak, *Lyndon B. Johnson: The Exercise of Power* (New York: New American Library, 1966), 88–118. On the organization of the review, see Holmes, *Americans on the Moon*, 199.
3. T.C.S. (Theodore C. Sorensen), "On Space for the National Security Council," April 22, 1961, NHRC, Folder 012506.
4. WWR (Walt Whitman Rostow), Memorandum to the President, "The Hundred Days," April 19, 1961, Sorensen Papers, Box 38, JFKL.
5. Memorandum, presumably to James Webb, from Franklyn Phillips, April 21, 1961, NHRC.
6. NASA Presentation to the Vice President, April 22, 1961, NHRC, Folder 012504.
7. Robert McNamara, Memorandum for the Vice President, "Brief Analysis of Department of Defense Space Program Efforts," with attached "Resume of Existing Programs," April 21, 1961, Vice Presidential Security Files, Box 17, LBJL.

8. Letter from George J. Feldman and Charles S. Sheldon II to the Vice President, April 24, 1961, with attached report, Vice-Presidential Security Files, Box 17, LBJL, and what appears to be a transcript of Johnson's orders to his staff, undated but either April 24 or April 25, 1961, Vice Presidential Papers, 1961 Subject Files, Box 116, LBJL.
9. This account of the meeting is drawn from the author's interview with Edward Welsh cited in Logsdon, *Decision*, 114 and a second interview with Welsh conducted by NASA Historian Eugene Emme, NASA Assistant Historian William Putnam, and LBJ Library Oral History specialist David McComb on February 20, 1969. This latter interview can be found in NHRC, Folder 2548. For a biography of Bernard Schriever, see Sheehan, *A Fiery Peace*; for the definitive von Braun biography, see Neufeld, *Von Braun*.
10. "Outline" for April 24 meeting , presumably prepared by Edward Welsh, National Aeronautics and Space Council Files, Box 3, JFKL.
11. Author's interview with Vice Admiral John T. Hayward, September 11, 1967, cited in Logsdon, *Decision*, 115.
12. Author's interview with General Bernard Schriever, November 3, 1967, cited in Logsdon, *Decision*, 114–115.
13. Such a demonstration of the U.S. ability to rendezvous with and capture a satellite in orbit and return it to Earth would certainly have been seen by the Soviet Union as a very provocative act.
14. Lieutenant General Bernard Schriever, Commander, Air Force Systems Command, Memorandum for the Vice President, April 30, 1961, Air Force History Archives, 168-7171-151, Schriever Papers, Roll 35262.
15. Letter from Wernher von Braun to the Vice President of the United States, April 29, 1961, reprinted in Logsdon et al., eds., *Exploring the Unknown,* Volume I, 429–433.
16. Notes from April 24, 1961 meeting, presumably drafted by Edward Welsh, National Aeronautics and Space Council Files, Box 3, JFKL.
17. On the consultation with Dean Rusk, see Logsdon, *Decision*, 118. Letter from Richard N. Gardner to the Vice President, April 24, 1961, NHRC, Folder 12288.
18. Cook's letter is cited in Logsdon, *Decision,* 116. The other two private citizens invited by Vice President Johnson to participate in the space review did not submit written inputs. Frank Stanton in an August 18, 1967, letter to the author wrote that his involvement was "minimal," and George Brown, asked by the NASA History Office for his recollection of the consultations, said that he did not remember them at all. See Logsdon, *Decision,* 114.
19. John F. Kennedy, "Statement by the President upon Signing Bill Amending the Aeronautics and Space Act," April 25, 1961, http://www.presidency.ucsb.edu/ws/index.php?pid=8086&st=&st1=.
20. Lyndon B. Johnson, Memorandum for the President, "Evaluation of Space Program," April 28, 1962, in Logsdon et al., *Exploring the Unknown*, Volume I, 427–429. Oral history interview with Edward Welsh, February 20, 1969, NHRC.
21. Author's interview with Jerome Wiesner, cited in Logsdon, *Decision*, 118.
22. Lambright, *Powering Apollo,* 95.
23. Author's interviews with James Webb and Letter from James E. Webb to Dr. Jerome B. Wiesner, May 2, 1961, cited in Logsdon, *Decision*, 119.
24. Lambright, *Powering Apollo*, 95.

25. Memorandum from Edward Welsh to the Vice President, April 25, 1961, NHRC, Folder 012506 and Memorandum from Welsh to the Vice President, May 1, 1961, Vice Presidential Papers, 1961 Subject Files, Box 117, LBJL.
26. Opening Statement for the Vice President, May 3, 1961, 2:30 p.m. NHRC, Folder 12288.
27. A summary transcript of the meeting is reprinted in Logsdon et al., *Exploring the Unknown*, Volume I, 433–439.
28. Author's telephone interview with James Fulton, September 20, 1967, cited in Logsdon, *Decision*, 121.
29. Letter from Overton Brooks to the Honorable Lyndon B. Johnson, May 5, 1961 with attached memorandum "Recommendations re the National Space Program," dated May 4, 1961, and a second undated memorandum on "The National Space Booster Program," NHRC. The first of the memos is quoted in Logsdon, *Decision,* 121.
30. Lambright, *Powering Apollo*, 96.
31. This letter is quoted in McDougall, *Heavens and the Earth,* 320. I have not been able to find a copy in the NHRC. See also a August 26, 1966, letter from James Webb to the President [Lyndon Johnson], LBJL.
32. Letter to author from James E. Webb, May 28, 1969.
33. Memorandum from James Webb to O.B. Lloyd (who was NASA's public affairs chief), May 5, 1961.
34. Sorensen, *Counselor*, 337–338.
35. Jerome Wiesner, Memorandum for McGeorge Bundy, "Some Aspects of Project Mercury," March 9, 1961, in Logsdon with Launius, *Exploring the Unknown*, Vol. VII, 176.
36. Swenson, Grimwood, and Alexander, *This New Ocean,* 349–350.
37. Donald Hornig, Memorandum for Mr. Sorensen, April 18, 1961, Sorensen Files, Box 38, JFKL.
38. E.C. Welsh, Memorandum for the Vice President, April 26, 1961, NHRC, Folder 012506.
39. Swenson, Grimwood, and Alexander, *This New Ocean*, 349.
40. Author's interview with Jerome Wiesner, cited in Logsdon, *Decision*, 122.
41. Oral History Interview with Dr. Edward C. Welsh, May 16, 1964, JFKL.
42. Swenson, Grimwood, and Alexander, *This New Ocean*, 350.
43. Sorensen, *Counselor*, 338.
44. Evelyn Lincoln, *My Twelve Years with John F. Kennedy* (New York: David McKay Company, 1965), 257–258.
45. Office of Research and Analysis, United States Information Agency, "A Comparison of World Reactions to the U.S. and USSR Space Flights," Report R-23-61, May 25, 1961, in Office of Science and Technology Papers, Box 16, JFKL.
46. John F. Kennedy, "Statement of the President on the Flight of Astronaut Alan B. Shepard, May 5, 1961," http://www.presidency.ucsb.edu/ws/index.php?pid=8112&st=&st1=; and *The New York Times,* May 6, 1961, 12.

### Chapter 7 "A Great New American Enterprise"

1. This account of the May 6–7 meetings is drawn largely from an oral history interview of Robert Seamans, March 27, 1964, JFKL, oral history interviews

of Robert Seamans, November 2, 1987, and of Willis Shapley, July 13, 1994, Glennan-Webb-Seamans Project for Research in Space History, National Air and Space Museum, and the author's interview with John Rubel, August 27, 1968, cited in Logsdon, *Decision,* 123–125. These sources differ on exactly who was at the meeting. Seamans does not mention Dryden or Harold Brown; Shapley does. Shapley does not mention Roswell Gilpatric; Seamans does. See also Seamans, *Aiming at Targets,* 88–90.

2. Central Intelligence Agency, "Soviet Technical Capabilities in Guided Missiles and Space Vehicles," National Intelligence Estimate Number 11-5-61, 25 April 1961, 34, 42, 44.
3. This highly classified annex could not be located in the JFKL or the NHRC.
4. Letter from Jerome Wiesner to David Bell, May 10, 1961, NHRC, Folder 12288. Given the date on the letter, it appears that Wiesner had not yet seen the May 8 report that resulted from the May 6–7 meetings, which had already proposed setting as a new goal the creation of such a system.
5. "Substantive Objectives of U.S. Space Programs," paper likely prepared by Willis Shapley, May 5, 1961, Sorensen Papers, Box 63, JFKL.
6. McDougall, *Heavens and the Earth,* 321.
7. The cover letter is reprinted in John M. Logsdon et al., *Exploring the Unknown*, Volume I, 440–452.
8. John F. Kennedy, "Remarks at the Presentation of NASA's Distinguished Service Medal to Astronaut Alan B. Shepard," May 8, 1961, http://www.presidency.ucsb.edu/ws/index.php?pid=8119&st=&st1=.
9. The account in these two paragraphs is drawn from Neal Thompson, *Light This Candle: The Life & Times of Alan Shepard, America's First Spaceman* (New York: Crown Publishers, 2004), 265, oral history interview with Alan Shepard, June 12, 1964, JFKL, and oral history interview with Robert R. Gilruth, October 2, 1987, Glennan-Webb-Seamans Project for Research in Space History, National Air and Space Museum.
10. The memorandum is reprinted in Logsdon et al., *Exploring the Unknown*, Volume I, 440–452. All quotations are taken from this source.
11. Lyndon B. Johnson, Memorandum for the President, May 8, 1961, Sorensen Papers, Box 38, JFKL.
12. John Finney, "600 Million More Planned to Spur Space Programs," *The New York Times*, May 10, 1961, 1.
13. Author's interview with McGeorge Bundy, cited in Logsdon, *Decision*, 126.
14. Oral History interview with Theodore Sorensen, March 26, 1964, JFKL. Sorensen, "Notes on presidential appearances before Congress," undated, and "Schedule re Second State of the Union Message," Sorensen Subject Files, Box 63, JFKL.
15. Hobart Rowen, *The Free Enterprisers: Kennedy, Johnson, and the Business Establishment* (New York: G.P. Putnam's Sons, 1965), 169–172.
16. Memorandum from David Bell to Mr. Sorensen, "Review of Space Proposals," with attached "Staff Review of Proposed Increases in Space Programs," May 18, 1961, Sorensen Subject Files, Box 63, JFKL. The final "Staff Report on Proposed Increases in Space Programs," dated May 20, 1961, can be found in NHRC, Folder 012457.
17. Bill Lloyd, Memorandum for Mr. Webb, "Anticipated questions from news media," May 21, 1961, 2, NHRC, Folder 012505. Another possible question

was "Might women be used in the lunar program?" The suggested answer was "women demonstrate every day versatility reaching far beyond the confines of the kitchen. I'm confident their day in space will come."

18. Seamans, *Aiming at Targets*, 90–91; Lambright, *Powering Apollo*, 101.
19. Lambright, *Powering Apollo*, 99–100; McDougall, *Heavens and the Earth*, 361.
20. A copy of the 1945 report can be found at http://www.nsf.gov/od/lpa/nsf50/vbush1945.htm
21. Letter from James E. Webb to Vannevar Bush, May 15, 1961, NHRC.
22. Lambright, *Powering Apollo*, 77.
23. James E. Webb, Memorandum for the Vice President, May 23, 1961, Vice President's Security Files, Box 17, LBJL.
24. Lambright, *Powering Apollo*, 100.
25. Ibid,123–124.
26. The text of the speech as delivered can be found at http://www.presidency.ucsb.edu/ws/index.php?pid=8151&st=&st1=.
27. Sorensen, *Counselor*, 336. See also Sorensen, *Kennedy*, 526.
28. A copy of the reading text of the speech, with Kennedy's handwritten insertions, can be found in POF, Box 34, JFKL.
29. Sorensen, *Counselor*, 336 and *Kennedy*, 526. It is interesting to contrast this first-hand account of the reception to Kennedy's address with accounts in other Kennedy biographies. For example, Richard Reeves in *President Kennedy* reports (138) that when Kennedy said that Americans were going to the Moon, "the place went wild." Perret, in his *Jack: A Life Like No Other* says (312) that upon hearing Kennedy's call for the moon mission, "his audience jumped to its feet as one man, applauding wildly, cheering lustily, frantic to beat the Russians in space." Colorful writing is a poor substitute for factual accuracy.
30. Alvin Shuster, "Congress Wary on Cost, But Likes Kennedy Goals," *The New York Times*, May 26, 1961, 13.
31. *The New York Times*, May 26, 1961, 1, 13, 32.
32. Holmes, *Americans on the Moon*, 204–205.
33. Seamans, *Aiming at Targets*, 90–91; Sorensen, *Counselor*, 337.
34. McDougall, *Heavens and the Earth*, 322–324.
35. Sorensen, *Counselor*, 336. In an interview with the author on May 28, 2009, Sorensen suggested that his use of the word "hundreds" was an exaggeration. The author has not been able to locate in the John F. Kennedy Presidential Library archives the "stream of written questions" to which Sorensen refers.
36. Robert Kennedy, interview with journalist John Bartow Martin, in Edwin O. Guthman and Jeffrey Shulman, eds., *Robert Kennedy in His Own Words: The Unpublished Recollections of the Kennedy Years* (New York: Bantam Books, 1988), 340.
37. Author's interview of Willis Shapley, December 14, 1967, copy in NHRC.

## Chapter 8 First Steps on the Way to the Moon

1. David A. Mindell, *Digital Apollo: Human and Machine in Spaceflight* (Cambridge, MA; The MIT Press, 2008), 112.
2. Jane Van Nimmen and Leonard Bruno with Robert Rosholt, *NASA Historical Data Book*, Vol. I, NASA Resources, 1958–1968, NASA SP-4012

(Washington, DC: Government Printing Office, 1988), 137–141, 134, 63–119.

3. Author's interview with Theodore C. Sorensen, May 28, 2009.
4. Ibid.
5. James E. Webb Oral History Interview 1, April 29, 1969, by T.H. Baker, Internet Copy, LBJL.
6. Much of the factual information on the activities of the Space Council is drawn from *The National Aeronautics and Space Council During the Tenure of Lyndon B. Johnson As Vice President and During His Administration as President,* LBJL. This document is one of the histories of all Federal agencies prepared at the end of the Johnson administration.
7. Memorandum from the Vice President to Ed Welsh, July 26, 1961; Memorandum to the Vice President from E.C. Welsh, July 29, 1961; Note to Walter (Jenkins) from LBJ, undated; and Memorandum from GER (George Reedy) to the Vice President, August 14, 1961, all in Vice Presidential Papers, 1961 Subject Files, Box 116, LBJL. Also, Leonard Baker, *The Johnson Eclipse: A President's Vice Presidency* (New York: The MacMillan Company, 1966), 126–137.
8. Merle Miller, *Lyndon: An Oral Biography* (New York: G.P. Putnam's Sons, 1980), 278.
9. For a general discussion of how policy decisions were made during the Kennedy administration, see Theodore C. Sorensen, *Decision-Making in the White House: The Olive Branch or the Arrows* (New York: Columbia University Press, 1963).
10. Memorandum from Frederic G. Dutton (Kennedy's Assistant for Intergovernmental and Interdepartmental Affairs) to the Honorable James Webb, August 15, 1961, WHCF, Box 176, JFKL.
11. For an account of the decision on where to launch the missions to the Moon, see Charles B. Benson and William B. Faherty, *Gateway to the Moon: Building the Kennedy Space Center Launch Complex* (Gainesville, FL: University Press of Florida, 2001), chapter 5. For a discussion of locating Saturn production and testing facilities, see Bilstein, *Stages to Saturn,* chapter 2.
12. Glennan, *Birth of NASA,* 14–15.
13. Henry C. Dethloff, *Suddenly, Tomorrow Came: A History of the Johnson Space Center,* NASA SP-4307 (Washington, DC: Government Printing Office, 1993), 36–37.
14. James E. Webb, Memorandum for the Vice President, May 23, 1961, Vice President's Security Files, Box 17, LBJL.
15. See Robert Sherill, *The Accidental President* (New York: Grossman Publishers, 1967), 240 for a skeptical account of the close relationships among those interested in bringing the new NASA facility to Houston.
16. Memorandum from James Webb to the President, September 14, 1961, with attached "Site Selection Procedure," POF, Box 82, JFKL. This account of the site selection process is taken from this attachment. Also, Letter from Governor John Volpe to the Honorable John F. Kennedy, July 19, 1961, NHRC. Kenneth O'Donnell's views on the site selection process can be found in a telephone interview with Robert Sherrod, May 13, 1971. Sherrod was a veteran journalist who started, but never completed, a detailed history of the events leading up to the Apollo 11 mission; his interview notes can be found in NHRC, Folder 13288.

17. Lambright, *Powering Apollo,* 238.
18. James Webb, in response to a question after a September 12, 1961, address to the National Press Club. A transcript of the speech and the question and answer session is in the NHRC. For the later account, see James E. Webb Oral History Interview 1, April 29, 1969, by T.H. Baker, Internet Version, LBJL.
19. Dethloff, *Suddenly, Tomorrow Came*, 39.
20. Memorandum from James Webb to the President, September 14, 1961, POF, Box 82, JFKL. There are two memos from Webb to the President dated September 14. One communicates the NASA site selection decision. The other, cited earlier, was prepared at Kennedy's request to put on the record the site selection process.
21. Memorandum for Theodore C. Sorensen from Jerome B. Wiesner, November 20, 1961, Sorensen Subject Files, Box 38, JFKL. Unless otherwise noted, Wiesner quotations in this section are from the November 20 memorandum.
22. Memorandum from James E. Webb to Dr. [Hugh] Dryden and Dr. [Robert] Seamans, November 21, 1961, NHRC.
23. Reflecting the controversies surrounding its efforts, the group's report was not issued until almost a year after it had completed its work. Excerpts from the report can be found in John M. Logsdon, Dwayne Day, and Roger Launius, *Exploring the Unknown: Selected Documents in the History of the U.S. Civil Space Program*, Volume II: External Relationships, NASA SP-4407 (Washington, DC: Government Printing Office, 1996), 318–337.
24. For an account of these difficulties and the other decisions needed before NASA decided on how to carry out the lunar landing mission, see John M. Logsdon, "Project Apollo: Americans to the Moon," in Logsdon with Launius, *Exploring the Unknown,* Vol. VII, Murray and Cox, *Apollo: Race to the Moon*, and Seamans, *Project Apollo.*
25. For a discussion of the debate over the number of engines in the Saturn first stage, see Neufeld, *Von Braun,* 364–366, 370–372.
26. Logsdon, "Project Apollo," 398–399 and Neufeld, *Von Braun*, 371.
27. Letter from Robert McNamara to James Webb, November 17, 1961, NHRC.
28. David Bell, Director, Bureau of the Budget, Memorandum for the President, December 14, 1961, Theodore Sorensen Papers, Box 44, JFKL. Bell's quotes in the following paragraphs are from this document.
29. Memorandum from James E. Webb to Dr. [Hugh] Dryden and Dr. [Robert] Seamans, November 21, 1961, NHRC.
30. Memorandum from James E. Webb for Dr. [Hugh] Dryden, Dr. [Robert] Seamans, and Mr. [Abe] Hyatt, May 4, 1962, NHRC, Folder 012518. On Richard Callaghan, see Robert Dallek, *Flawed Giant: Lyndon Johnson and His Times, 1961–1973* (New York: Oxford University Press, 1998), 23.
31. Letter from James E. Webb to the President, June 1, 1962, NHRC, Folder 012518.
32. See Sorensen, *Kennedy,* 412–427 for a discussion of the overall economic situation in 1962.
33. Memorandum from Willis Shapley for Mr. [Roswell] Gilpatric and Mr. [James] Webb, "Special Space Review," August 22, 1962, NSF, Box 308, JFKL. Military Division, Bureau of the Budget; "Draft Staff Report—Special

Space Review," August 1962, I-1, III-17, III-19, II-21, NSF, Box 308, JFKL. Willis Shapley, Oral History Interview, August 2, 1994, Glennan-Webb-Seamans Project for Research in Space History, National Air and Space Museum.

34. Memorandum from John Kennedy for James E. Webb, August 15, 1962 and Letter from James E. Webb to the President, August 18, 1962, NHRC, Folder 012518. Memorandum from John F. Kennedy to David Bell, August 23, 1962, Theodore Sorensen Papers, Box 38, JFKL.
35. See Siddiqi, *Challenge to Apollo*, 292–295 and 356–361 for details on the two Soviet missions.
36. A transcript of the press conference can be found at http://www.presidency.ucsb.edu/ws/index.php?pid=8826&st=&st1=. A draft of a proposed press conference statement on space can be found in Theodore Sorensen Papers, Box 69, JFKL.
37. Memorandum from Chief, Space Division, Office of Scientific Intelligence, Central Intelligence Agency to Carl Kaysen, "Soviet Interplanetary Probes," August 25, 1962, Memorandum from Carl Kaysen to Pierre [Salinger], August 31, 1962, and Memorandum from Lieutenant General Marshall Carter to The President, "Publicity on Failure of Soviet Space Probes," NSF, Box 307, JFKL. John Finney, "The Space Shots: Detection of Soviet Failures Indicates Scope of U.S. Surveillance System," *The New York Times*, September 9, 1962, 1.
38. Much of the following account is based on the discussion of NASA's Electronic Research Center in Thomas P. Murphy, *Science, Geopolitics, and Federal Spending* (Lexington, MA: D.C. Heath and Company, 1971), 225–264; in addition, see Andrew Butrica, "The Electronics Research Center: NASA's Little Known Venture into Aerospace Electronics," AIAA Paper 2002-1138; and House Committee on Science and Astronautics, *Toward the Endless Frontier*. This last study was authored by former Congressman Kenneth Hechler.
39. Memorandum from James Webb to Dr. [Hugh] Dryden and Dr. [Robert] Seamans, May 24, 1963, NHRC, Administrator's Chronological Files, January—July 1963.
40. Ibid and Letter from James Webb to the President, March 21, 1963, NHRC, Folder 012518.
41. Committee on Science and Astronautics, *Toward the Endless Frontier*, 224–225.
42. Ibid, 225–231. The quote is on 229.
43. Murray and Cox, *Apollo: Race to the Moon*, 84. Also, Seamans, *Project Apollo*, 23–24. The assignment of the DX priority was communicated in National Security Action Memorandum No. 144, "Assignment of Highest National Priority to the APOLLO Manned Lunar Landing Program," April 11, 1962, NSF, Box 307, JFKL.

### Chapter 9 "I Am Not That Interested in Space"

1. Bilstein, *Stages to Saturn*, 67. Interview of Jerome Wiesner by W. Henry Lambright, November 15, 1990, NHRC, Folder 7106. Oral history interview of Wernher von Braun, March 31, 1964, JFKL.

2. For accounts of NASA's selection of lunar orbit rendezvous, see Murray and Cox, *Apollo: Race to the Moon*, chapters 8 and 9; Neufeld, *Von Braun*, 372–378; and James R. Hansen, *Enchanted Rendezvous: John C. Houbolt and the Genesis of the Lunar-Orbit Rendezvous Concept,* NASA, Monographs in Aerospace History, No. 4, 1995. Houbolt's letter is reprinted in Logsdon with Launius, *Exploring the Unknown,* Vol. VII, 522–530.
3. Murray and Cox, *Apollo: Race to the Moon*, 141.
4. The following account is drawn from Courtney G. Brooks, James M. Grimwood, and Loyd S. Swenson, Jr., *Chariots for Apollo: A History of Manned Lunar Spacecraft*, NASA SP-4205 (Washington, DC: Government Printing Office, 1979), chapter 4.
5. Letter from Jerome Wiesner to James Webb, July 17, 1962 with attached Memorandum from Donald Hornig to Dr. Jerome B. Wiesner, "Summary of View of Space Vehicle Panel," July 11, 1962, NSF, Box 307, JFKL.
6. Letter from James Webb to Jerome Wiesner, August 20, 1962, NSF, Box 307, JFKL. The "C-5 direct" approach would have reduced the crew size to two rather than three astronauts and sent their spacecraft directly to the lunar surface without any need for a rendezvous.
7. Letter from Jerome Wiesner to James Webb, September 5, 1962, NASA Files, Box 11, JFKL.
8. Murray and Cox, *Apollo: Race to the Moon*, 143.
9. Memorandum from N. E. Golovin to Members, Space Vehicle Panel, November 27, 1962, Office of Science and Technology Files, Box 67, JFKL.
10. Letter from James Webb to Jerome Wiesner, October 24, 1962, with attached "Manned Lunar Mode Comparison," reprinted in Logsdon with Launius, *Exploring the Unknown*, Volume VII, 577–585.
11. Record of telephone conversation between Mr. Webb, NASA and Dr. Wiesner, Monday, October 29, 1962, Wiesner Papers, Box 8, JFKL. The November 2 meeting is mentioned in the "Status Report on Activities of the President's Science Advisory Committee and Its Staff," November 13, 1962, POF, Box 67, JFKL.
12. Memorandum from McG. B. [McGeorge Bundy] to Dr. [Jerome] Wiesner, November 7, 1962, reprinted in Logsdon with Launius, *Exploring the Unknown*, Volume VII, 585.
13. Letter from James Webb to The President, undated but November 1962, reprinted in Ibid, 586–588.
14. Office of the Press Secretary, The White House, "Remarks of the President," Cape Canaveral, FL, September 11, 1962, POF, Box 39, JFKL.
15. For a discussion of the process through which Kennedy's speeches were developed, see Sorensen, *Counselor*, chapters 12 and 18. Agency inputs into the Rice University speech and Sorensen's first draft can be found in Theodore Sorensen Papers, Box 69, JFKL. The Rice University speech as delivered can be accessed at http://www.presidency.ucsb.edu/ws/index.php?pid=8862&st=&st1=. For a report on the speech that mentions the deviations from the prepared text, see E.W. Kenworthy, "Kennedy Asserts Nation Must Lead in Probing Space," *The New York Times*, September 13, 1962, 1.
16. NASA History Office, exit interview with Robert C. Seamans, Jr, May 8, 1968, 41–43, NHRC.
17. Letter from James E. Webb to the President, October 29, 1962, Bureau of the Budget Files, Box 120, JFKL.

18. Seamans, *Project Apollo,* 44.
19. Memorandum from Jerome Wiesner to The President, November 16, 1962, POF, Box 84, JFKL.
20. "In Earthly Trouble," *Time*, November 23, 1962, 15.
21. Memorandum from David Bell for The President, draft, November 13, 1962, POF, Box 84, JFKL. Apparently this memorandum was never put in final form and signed by Bell.
22. See Sorensen, *Counselor,* 318 for a discussion of this taping system, which was unknown to all except John and possibly Robert Kennedy, Kennedy's secretary Evelyn Lincoln, and the Secret Service. See also the discussion of the Kennedy taping system at http://tapes.millercenter.virginia.edu/tapes/kennedy/overview.
23. Background information on the November 21 meeting, the audio record of the meeting, and a transcript of the discussion can be found at http://history.nasa.gov/JFK-Webbconv/pages/backgnd.html.
24. Letter from James E. Webb to The President, November 30, 1962, reprinted in Logsdon et al., *Exploring the Unknown,* Volume I, 461–467.
25. Memorandum from David Bell for the Vice President, "Current budget issues regarding NASA," November 28, 1962, Theodore Sorensen Papers, Box 44, JFKL. Letter from Lyndon B. Johnson to the President, December 4, 1962, Vice-Presidential Security Files, Box 17, LBJL.
26. Memorandum from Jerome Wiesner to The President, "Acceleration of the Lunar Landing Program," January 10, 1963, POF, Box 67, JFKL. Letter from John Kennedy to Lyndon Johnson, January 10, 1963, and letter from Lyndon Johnson to the President, January 18, 1963, White House Famous Names Files, John F. Kennedy, Box 6, LBJL.
27. Seamans, *Project Apollo*, 47.

### Chapter 10 Early Attempts at Space Cooperation

1. See Sorensen's interview, September 19, 1995, in Aleksandr Fursenko and Timothy Naftali, *One Hell of a Gamble: Khrushchev, Castro and Kennedy, 1958–1964* (New York: Norton, 1997), 121.
2. Interview with John F. Kennedy published in *Bulletin of the Atomic Scientists,* November 1960, 347. Even as he discussed cooperation with the USSR, Kennedy also emphasized matching the Soviets in rocket thrust.
3. "Report to the President-Elect of the Ad Hoc Committee on Space," January 10, 1961, Pre-Presidential Papers, Box 1072, JFKL.
4. John F. Kennedy, "Annual Message to the Congress on the State of the Union," January 30, 1961, http://www.presidency.ucsb.edu/ws/index.php?pid=8045&st=&st1=.
5. This point is made in Dodd Harvey and Linda Ciccoritti, *U.S.-Soviet Cooperation in Space* (Washington, DC: Center for International Studies, University of Miami, 1974), 64.
6. For an account of the origins of NASA's cooperative efforts, see Arnold Frutkin, *International Cooperation in Space* (Englewood Cliffs, NJ: Prentice-Hall, 1965). Frutkin was the NASA official in charge of international affairs during the period covered by this study.
7. Ibid, 89–91.

8. General Assembly Resolution 1472 (XIV), December 1959, http://daccess-dds-ny.un.org/doc/RESOLUTION/GEN/NR0/142/95/IMG/NR014295.pdf.
9. Dean Rusk, Memorandum for the President, "United Nations Outer Space Activities," February 2, 1961, Papers of the Office of Science and Technology, Box 16, JFKL.
10. McGeorge Bundy, Memorandum for the Secretary of State, "UN Outer Space Activities," February 28, 1961, Papers of the Office of Science and Technology , Box 16, JFKL.
11. Philip Farley, Special Assistant to the Secretary of State for Atomic Energy and Outer Space, Memorandum for the Secretary, "Informal Task Force on Outer Space Cooperation," with attached "Terms of Reference for Task Force on Possibilities for International Cooperation in Outer Space," February 9, 1961, Papers of the Office of Science and Technology, Box 16, JFKL.
12. The work of the task force is described in Harvey and Ciccoritti, *U.S.-Soviet Cooperation*, 66–74 and Eugene B. Skolnikoff, *Science, Technology, and American Foreign Policy* (Cambridge, MA: The MIT Press, 1967), 32–34.
13. NASA, "Minutes—Meeting of Task Force on International Cooperation," February 20, 1961, National Archives and Record Administration, Record Group 59, Special Assistant for Atomic Energy and Outer Space, Box 252.
14. Department of State, Memorandum of Conversation, "Outer Space," March 8, 1961, National Archives and Record Administration, Record Group 59, Special Assistant for Atomic Energy and Outer Space, Folder 14A21 (Part 1 of 2).
15. "Report of the Panel on International Cooperation in Space Activities," March 20, 1961, 2, 7, Papers of the Office of Science and Technology, Box 16, JFKL; Letter, Jerome B. Wiesner to Bruno Rossi, April 6, 1961, Papers of Jerome Wiesner, Chronological Files, Box 2, JFKL.
16. Memorandum from Philip Farley to Mr. Bohlen, Mr. McSweeney, and Mr. Meeker, "U.S.-Soviet Space Cooperation," April 4, 1961 in National Archives and Record Administration, Record Group 59, Special Assistant for Atomic Energy and Outer Space, Folder 14A21 (Part 1 of 2). Also, "Draft Proposals for US-USSR Space Cooperation" in Logsdon, Day and Launius, *Exploring the Unknown*, Volume II, 143–147.
17. "Draft Proposals for US-USSR Space Cooperation," April 13, 1961, POF, Box 126, JFKL.
18. The Kennedy-Khrushchev exchanges, including the February 1961 one, can be found at http://www.state.gov/www/about_state/history/volume_vi/exchanges.html.
19. Memorandum from James Webb to Hugh Dryden, March 23, 1961, quoted in Harvey and Ciccoritti, *U.S.-Soviet Cooperation*, 74.
20. Schlesinger, *A Thousand Days*, 344–349.
21. Department of State telegram from John F. Kennedy to Chancellor [Konrad Adenauer], May 16, 1961, POF, Box 126, JFKL.
22. Jerome Wiesner, Memorandum for the President, May 18, 1961, with attached Department of State, Memorandum for the President, "Possible U.S.-Soviet Cooperative Space Projects," May 12, 1961, and "Draft Proposals for US-USSR Space Cooperation," Redraft, April 13, 1961. JFKL.

23. Foy D. Kohler, "Foreword," in Harvey and Ciccoritti, *U.S.-Soviet Cooperation*, xxiv. Kohler was a Soviet specialist in the Department of State who in 1962 became the U.S. ambassador to the Soviet Union.
24. The following account of communications between the White House and the Kremlin in advance of the summit meeting is drawn from Fursenko and Naftali, *One Hell of a Gamble*, 101–131.
25. Ibid, 122.
26. For background on the Kennedy-Bolshakov interactions, see also Michael R. Beschloss, *The Crisis Years: Kennedy and Khrushchev, 1960–1963* (New York: HarperCollins, 1991), 152–157.
27. McGeorge Bundy, Memorandum, "President's Meeting with Khrushchev, Vienna, June 3–4, Scientific Cooperation," no date, NSF, Box 234, JFKL.
28. Memorandum from McGeorge Bundy to the President, "Specific Answers to Your Questions of May 29th Related to the USSR," May 29, 1961, POF, Box 126, JFKL.
29. Kennedy is quoted in Sorensen, *Kennedy*, 551. See also Schlesinger, *A Thousand Days*, 361. Beschloss, *The Crisis Years*, 194–236, provides a detailed account of the summit discussions. He quotes (225) President Kennedy as saying about Khrushchev, "he just beat hell out of me."
30. Department of State, Memorandum of Conversation, Vienna Meeting between the President and Chairman Khrushchev, 3:15 p.m., June 4, 1961, POF, Box 126, JFKL. Richard Reeves, in his *President Kennedy*, suggests (171) that Kennedy's final words were even stronger. He quotes the President as saying "Then, Mr. Chairman, there will be war. It will be a cold winter."
31. Department of State, Memorandum of Conversation, Vienna Meeting between the President and Chairman Khrushchev, Luncheons of June 3 and June 4, POF, Box 126, JFKL. The description of Khrushchev's June 3 reaction as "half joking" is in Sorensen, *Kennedy*, 544.
32. Sergey Khrushchev, "The First Earth Satellite: a Retrospective View from the Future," in Roger D. Launius, John M. Logsdon, and Robert W. Smith, eds., *Reconsidering Sputnik: Forty Years Since the Soviet Satellite* (Amsterdam: Harwood Academic Publishers, 2000), 276, 281–282. Khrushchev added (282) "from the perspective of today this was definitely a mistake. A combined lunar project would not only have saved face for us and saved a bundle of money, but it might also have been a turning point in the relations between our two countries." It is worth noting that Sergey Khrushchev is not always a reliable source of specific historical information; for example, it is unlikely that his father consulted Korolev in the twenty-four hours between the two Kennedy-Khrushchev luncheons.
33. Strobe Talbott, translator and editor, *Krushchev Remembers: The Last Testament* (Boston, MA: Little Brown, 1974), 54.
34. Harvey and Ciccoritti, *U.S.-Soviet Cooperation*, 86.
35. Memorandum from Roger Hilsman to The Secretary [of State], "Intelligence Note: Khrushchev Proposes Combined US-Soviet Space Effort," February 21, 1962, National Archives and Record Administration, Record Group 59, Box 231.
36. Harvey and Ciccoritti, *U.S.-Soviet Cooperation*, 86–87.
37. National Security Action Memorandum No. 129, "U.S.-U.S.S.R. Cooperation in the Exploration of Space," February 23, 1962 (Revised February 27, 1962), NSF, Box 334, JFKL. See also Memorandum, McGeorge Bundy to Holders

of National Security Memorandum No, 129, February 27, 1962, NSF, Box 334, JFKL.

38. Memorandum from McGeorge Bundy to the Honorable James Webb, February 23, 1962, NSF, Box 334, JFKL.
39. NASA's philosophy is spelled out in Frutkin, *International Cooperation*, 32–36. For Wiesner's position, see Eugene Skolnikoff, *Science, Technology, and American Diplomacy*, 35–37 and author's interview with Skolnikoff, June 5, 2001.
40. Memorandum from Dean Rusk for the President, "US-USSR Cooperation in the Exploration of Space," March 6, 1962, NSF, Box 334, JFKL. In a handwritten note on the memo, Rusk added "The attached seems to me an excellent start." Rusk also noted that the Department of Defense had been involved in preparing the letter. Such involvement had not been mentioned in NSAM 129.
41. The fact that Kennedy had made a proposal to Khrushchev in Vienna for joint lunar missions was not widely known, either at the time of the summit or in early 1962.
42. The letter is reprinted in Logsdon, Day, and Launius, *Exploring the Unknown*, Volume II, 147–149.
43. Ibid, 149–152.
44. Department of State Policy Directive, "Initial Technical Discussions of U.S.-Soviet Space Cooperation," March 19, 1962, Vice Presidential Papers, 1962 Subject Files, Box 183, LBJL.
45. Memorandum from Philip Farley to Mr. [George] McGhee, "Space Negotiations with Soviets," March 22, 1962, National Archives and Record Administration, Record Group 59, Special Assistant for Atomic Energy and Outer Space, Box 251.
46. Hugh Dryden, "Preliminary Summary Report: U.S.-Soviet Space Cooperation Talks," no date, NSF, Box 334, JFKL.
47. Philip Farley, Memorandum for the Record, "Meeting with Under Secretary McGhee Concerning US-USSR Cooperation in Outer Space Activities," April 24, 1962, and Memorandum from Philip Farley to Mr. [George] McGhee, "Meeting on U.S.-Soviet Space Cooperation," April 18, 1962, both in National Archives and Record Administration, Record Group 59, Special Assistant for Atomic Energy and Outer Space, Box 251.
48. Dean Rusk, Memorandum for the President, "Bilateral Talks Concerning US-USSR Cooperation in Outer Space Activities," May 15, 1962, NSF, Box 334, JFKL.
49. Harvey and Ciccoritti, *U.S.-Soviet Cooperation*, 96.
50. George Ball, Memorandum for the President, "Bilateral Talks Concerning US-USSR Cooperation in Outer Space Activities," July 5, 1962, NSF, Box 337, JFKL.
51. McGeorge Bundy, National Security Action Memorandum No. 172, "Bilateral Talks Concerning US-USSR Cooperation in Outer Space Activities," July 18, 1962, Office of Science and Technology Files, Box 68, JFKL, and Memorandum for the President from McGeorge Bundy, July 13, 1962, reprinted in Logsdon, Day, and Launius, *Exploring the Unknown,* Volume II , 163. For documents associated with the Dryden-Blagonravov talks, see Ibid, 153–162. See also Frutkin, *International* Cooperation, 92–97. The reference to Congressman Olin (Tiger) Teague (D-TX), who was chairman

on the subcommittee in the House of Representatives that oversaw NASA's manned flight program, stems from Teague's communicating to the White House his concern that U.S.-Soviet cooperation could threaten U.S. security interests.

52. Letter from Donald Hornig to Jerome Wiesner, June 28, 1962, NSF, Box 377, JFKL.

### Chapter 11 To the Moon Together: Pursuit of an Illusion?

1. John F. Kennedy, "Address before the 18th General Assembly of the United Nations," September 20, 1963, http://www.presidency.ucsb.edu/ws/index.php?pid=9416&st=&st1=.
2. Schlesinger, *A Thousand Days,* 919.
3. Memorandum for the President from Arthur Schlesinger, Jr., September 16, 1963, Schlesinger White House Papers, Box 45, JFKL.
4. Schlesinger, *A Thousand Days,* 919–920.
5. The phrase "pursuit of an illusion" is the title of the chapter in Harvey and Ciccoritti, *U.S.-Soviet Cooperation in Space* discussing Kennedy's UN proposal. The principal author of this study, Dodd Harvey, and his brother Mose, also at the University of Miami at the time the study was prepared, were close associates of James Webb. In the preface to their book, the authors thank both Webb and Hugh Dryden, and note that Webb "not only helped us greatly in our collection of materials [much of which was not otherwise available to other researchers] but also gave us wise council [*sic*] through the years." (v) It is fair to surmise that the analysis in the book's chapter on the cooperative initiative, and indeed the portions of the study that deal with the 1961–1968 period when Webb was NASA administrator, reflect to a significant degree Webb's views. The quote is on 126 of this book.
6. McDougall, *Heavens and the Earth,* 345, 350.
7. Oral history interview of Theodore C. Sorensen, March 26, 1964, JFKL.
8. See Sorensen, *Kennedy,* 714–746 and Schlesinger, *A Thousand Days,* 889–923 for a discussion of Kennedy's moves towards détente. The quote is the title of chapter 25 in Sorensen's book.
9. John F. Kennedy, "Commencement Address at American University in Washington," June 10, 1963. http://www.presidency.ucsb.edu/ws/index.php?pid=9266&st=&st1=.
10. Harvey and Ciccoritti, *U.S.-Soviet Cooperation,* 112. As suggested previously, this analysis appears to reflect the perspective of James Webb on Kennedy's views at the time.
11. Central Intelligence Agency, Memorandum, "The New Phase of Soviet Policy," August 9, 1963, Summary Page.
12. Oral history interview of Theodore C. Sorensen, March 26, 1964, JFKL.
13. Central Intelligence Agency, National Intelligence Estimate Number 11-1-62, "The Soviet Space Program," December 5, 1962.
14. John F. Kennedy, Memorandum for Director McCone, April 29, 1963, NSF, Box 307, JFKL.
15. Bureau of Scientific Intelligence, Central Intelligence Agency, "A Brief Look at the Soviet Space Program," October 1, 1963, 2–3, NSF, Box 308, JFKL.

16. See Siddiqi, *Challenge to Apollo*, 395–408 and John Logsdon and Alain Dupas, "How Real was the Race to the Moon?" *Scientific American*, June 1994, for discussions of Soviet lunar plans.
17. *The New York Times*, July 17, 1963, 12. The quote is taken from Bernard Lovell, "Soviet Aims in Astronomy and Space Research," *New Scientist*, July 25, 1963, 175. Also, Frutkin, *International Cooperation*, 105.
18. The President's News Conference, July 17, 1963, http://www.presidency.ucsb.edu/ws/index.php?pid=9348&st=&st1=.
19. The letter is quoted in Harvey and Ciccoritti, *U.S.-Soviet Cooperation in Space*, 114–115.
20. Memorandum from Ray Cline, Deputy Director (Intelligence), Central Intelligence Agency, to McGeorge Bundy, with attached memorandum "Soviet Views on Future Space Operations," July 31, 1963, NSF, Box 308, JFKL and Memorandum from Jerome Wiesner to the President with attached article, August 1, 1963, POF, Box 67, JFKL.
21. Harvey and Ciccoritti, *U.S.-Soviet Cooperation*, 114, 117. Letter from Hugh Dryden to M. W. Keldysh, August 21, 1963, NASA Files, Box 8, JFKL.
22. Hugh Dryden, Memorandum for the Record, "Luncheon with Academician Blagonravov in New York, September 11, 1963," Ibid.
23. Frutkin, *International Cooperation*, 113. It is not clear whether Frutkin intended to include President Kennedy as among the "wishful thinkers."
24. Harvey and Ciccoritti, *U.S.-Soviet Cooperation*, 120.
25. Memorandum of Conversation, August 26, 1963, 11:00 a.m., Document 350, Department of State, *Foreign Relations of the United States, 1961–1963*, Volume V, Soviet Union (Washington, DC: Government Printing Office, 1998), 751–752.
26. Telegram from the Embassy in the Soviet Union to the Department of State, Document 354, in Ibid, 762.
27. Memorandum of Conversation, "Soviet-American Relations and Negotiations," Document 355, in Ibid, 763.
28. McGeorge Bundy, Memorandum for the President, "Your 11:00 a.m. Appointment with Jim Webb," September 18, 1963, POF, Box 84, JFKL.
29. James E. Webb Oral History Interview, April 29, 1969 by T. H. Baker, Internet Copy, LBJL. An audio tape of the September 18 meeting between Webb and Kennedy exists, but as of November 2010 it has not yet been released by the Kennedy Library.
30. Skolnikoff, *Science, Technology, and American Diplomacy*, 37, Harvey and Ciccoritti, *U.S.-Soviet Cooperation*, 122–123, and John W. Finney, "U.S. Aide Rebuffs Soviet's Moon Bid," *The New York Times*, September 18, 1963, 11. It was presumably Finney's story, which may have been given to President Kennedy before he met with Webb that morning, that prompted Kennedy's comment about not being undercut by NASA, although Kennedy was also likely to have been aware of a general skepticism about the wisdom of the proposal among NASA's leaders. Memorandum from E. C. Welsh to the Vice President, "Space Council Meeting," September 23, 1963, Vice Presidential Papers, 1963 Subject Files, Box 238, LBJL. Welsh in this memorandum suggested a Space Council meeting "to ascertain what steps need to be taken if the Russians show a willingness to explore this proposal." Such a meeting was not called.

31. Thomas Hamilton, "Soviet Proposes Summit Meeting on Arms in 1964," *The New York Times*, September 20, 1963, 1. The draft of Kennedy's UN speech can be found in Theodore C. Sorensen Papers, Box 77, JFKL. Schlesinger, *A Thousand Days*, 920 and Harvey and Ciccoritti, quoting Webb, in *U.S.-Soviet Cooperation*, 122.
32. Sorensen, *Kennedy*, 520–521 and Harvey and Ciccoritti, *U.S.-Soviet Cooperation*, 123.
33. "Let Us Explore the Stars —II," *The New York Times*, September 21, 1963, 20; John Finney, "Washington is Surprised by the President's Proposal," *The New York Times*, September 21, 1963, 1; "Europe's Press Praises Kennedy," *The New York Times*, September 22, 1963, 34; Richard Witkin, "Joint Moon Trip Held Inefficient," *The New York Times*, October 7, 1963, 23; William J. Coughlin, "The Reasons Why," *Missiles and Rockets*, September 30, 1963, 10.
34. Drew Pearson, "JFK Vetoed Experts on Space Bid,' *Washington Post*, September 26, 1963, F11. In the same column, Pearson persists in calling NASA administrator James Webb "Jimmy."
35. Letter to President Kennedy from Albert Thomas, September 21, 1963, NSF, Box 308, JFKL.
36. Letter from John F. Kennedy to the Honorable Albert Thomas, September 23, 1963, Sorensen Speech Files, Box 38, JFKL. The language of the letter bears the unmistakable eloquence of Sorensen's writing style.
37. Robert C. Toth, "House Opposes Joint Moon Trip; Votes NASA Funds," *The New York Times*, October 11, 1963, 1.
38. Quoted in Harvey and Ciccoritti, *U.S.-Soviet Cooperation*, 130. A similar restriction was added in each of the subsequent three years.
39. Both press mentions are from Ibid, 123.
40. *The Washington Post*, October 8, 1963, 1. Sedov was one of the few Soviet scientists (together with Keldysh and Blagonravov) connected to the Soviet space program whose identities were publicly known. The real "father of Sputnik," Sergei Korolev, was publicly identified only as the "chief designer." Wiesner's comments are in Jerome Wiesner, Memorandum for the President, October 8, 1963, Wiesner Files, Box 11, JFKL.
41. Khrushchev's remarks were reported in the government newspaper *Izvestia* and are quoted in Harvey and Ciccoritti, *U.S.-Soviet Cooperation*, 124.
42. Office of Current Intelligence, Central Intelligence Agency, Current Intelligence Memorandum, "Khrushchev's Press Conference on 25 October 1963," OCI No. 2377/63; Memorandum from Ray S. Cline, Deputy Director (Intelligence), CIA, to the Special Assistant for National Security, October 29, 1963, NSF, Box 308, JFKL. Memorandum from Thomas Hughes to The Secretary, "Khrushchev's Obscure and Noncommittal Statements about Moon Shots," November 5, 1963, NSF, Box 308, JFKL. It should be noted that this memorandum was discussing both the October 25 and November 1 statements by Khrushchev. Memorandum to Jerome Wiesner from S.R. Rivkin, "Some Thoughts on Khrushchev's Statement on Space," October 28, 1963, Office of Science and Technology Files, Box 67, JFKL. For a discussion of the conflict within the CIA over Soviet intent and indeed of the CIA's monitoring of Soviet space efforts, see the two-part article by Dwayne A. Day and Asif Siddiqi, "The Moon in the Crosshairs: CIA Intelligence

on the Soviet Manned Lunar Program," *Spaceflight*, November 2003 and March 2004.

43. James E. Webb, "Policy Guidance for NASA Staff," Draft, September 23, 1963, with handwritten note "agreed by phone," initialed MgB [McGeorge Bundy] and dated 9/23, NSF, Box 308, JFKL. Frutkin, *International Cooperation*, 116–117.
44. "Space Chief Backs Cooperation with Soviets on Moon Programs," *The New York Times*, September 26, 1963, 8.
45. Letter from U. Alexis Johnson to James Webb, October 14, 1963, National Archives and Record Administration, Record Group 59, Box 1483.
46. NASA response to letter of U. Alexis Johnson to James Webb, undated, Office of Science and Technology Files, Box 67, JFKL.
47. Memorandum for Mr. [McGeorge] Bundy from Charles E. Johnson, October 28, 1963 with attached National Intelligence Brief, "A Brief Look at the Soviet Space Program," October 1, 1963, NSF, Box 308, JFKL.
48. Memorandum for the President from Jerome Wiesner, "The Proposal for a Joint US-USSR Lunar Program," October 29, 1963, NSF, Box 377, JFKL.
49. The memorandum is in POF, Box 84, JFKL.
50. The President's News Conference, October 31, 1963, http://www.presidency.ucsb.edu/ws/index.php?pid=9507&st=&st1=.
51. Memorandum from E.C. Welsh, National Aeronautics and Space Council, to Andrew Hatcher, The White House, November 1, 1963, WHCF, Box 114, JFKL.
52. Harvey and Ciccoritti, *U.S.-Soviet Cooperation*, 125, quoting a *Pravda* report. The Welsh memorandum quoted in the prior footnote translated Khrushchev as saying "What could be better than to send a Russian and an American together, or, better still, a Russian man and an American woman."
53. Arthur Schlesinger, Jr., in a November 7, 1963, memorandum to McGeorge Bundy characterized State Department cable #1542 from the U.S. embassy in Moscow as discussing "Khrushchev's apparent acceptance in principle of the idea of a joint moon shot," NSF, Box 308, JFKL. The shift in Khrushchev's views, as reported by his son, is described in Siddiqi, *Challenge to Apollo*, 400.
54. Sergey Khrushchev, in Launius, Logsdon, and Smith, eds., *Reconsidering Sputnik*, 282.
55. Arthur Schlesinger, Jr., Memorandum for McGeorge Bundy, November 7, 1963, NSF, Box 308, JFKL.
56. Schlesinger, *A Thousand Days*, 1018.
57. Memorandum to the Vice President from E.C. Welsh, November 1, 1963, 1963 Subject Files, Box 238, LBJL.
58. Charles E. Johnson, Memorandum for Mr. [McGeorge] Bundy, November 8, 1963, NSF, Box 342, JFKL.
59. John F. Kennedy, National Security Action Memorandum 271, "Cooperation with the USSR on Outer Space Matters," reprinted in Logsdon, Day, and Launius, *Exploring the Unknown*, Volume II, 166–167.
60. Memorandum for McGeorge Bundy from Arthur Schlesinger, Jr., November 9, 1963, NSF, Box 308, JFKL.
61. Memorandum from Harlan Cleveland to the Acting Secretary, "In Outer Space, Too, It Takes Two to Tango," November 21, 1963, National Archives and Record Administration, Record Group 59, Box 4183.

62. Harlan Cleveland, Memorandum for the Secretary, "Presidential Approval of Outer Space Statement," November 23, 1963, National Archives and Record Administration, Record Group 59, Box 4183 and Memorandum for the President from Dean Rusk, November 24, 1963, 1963 Subject Files, LJBL.
63. The text of the speech can be found in the *Department of State Bulletin*, December 30, 1963, 1011. Memorandum from Charles Johnson to Mr. [McGeorge] Bundy, November 24, 1963, 1963 Subject Files, LBJL.
64. Harvey and Ciccoritti, *U.S.-Soviet Cooperation*, 135.
65. Webb's transmittal letter and the report, "US-USSR Cooperation in Space Research Programs," January 31, 1964 are reprinted in Logsdon, Day and Launius, *Exploring the Unknown*, Volume II, 168–182. The quoted material is on 168–170.
66. Harvey and Ciccoritti, *U.S.-Soviet Cooperation*, 138.

## Chapter 12 Apollo under Pressure

1. The President's News Conference of July 17, 1963, http://www.presidency.ucsb.edu/ws/index.php?pid=9348&st=&st1= and The President's News Conference of October 31, 1963, http://www.presidency.ucsb.edu/ws/index.php?pid=9507&st=&st1=.
2. *The New York Times* editorial is quoted in NASA, *Astronautics and Aeronautics,1963: Chronology on Science, Technology, and Policy,* NASA SP-4004 (Washington, DC: Government Printing Office, 1964), 48–49. A copy of this chronology can be found at http://history.nasa.gov/AAchronologies/1963.pdf. Kennedy's press conference remarks can be found at http://www.presidency.ucsb.edu/ws/index.php?pid=9124&st=&st1=.
3. The Eisenhower letter is quoted in NASA, *Astronautics and Aeronautics, 1963,* 112. For Kennedy's press conference remarks, see http://www.presidency.ucsb.edu/ws/index.php?pid=9139&st=&st1=.
4. James Reston, "The Man on the Moon and the Men on the Dole," *The New York Times*, April 5, 1963, 35 and James Reston, "What Government Officials Do You Believe?" *The New York Times*, April 24, 1963, 26.
5. P.H.A. (Philip H. Abelson), "Manned Lunar Landing," *Science*, April 19, 1963, 267.
6. John F. Kennedy, "Remarks and Question and Answer Period before the American Society of Newspaper Editors," April 19, 1963, http://www.presidency.ucsb.edu/ws/index.php?pid=9154&st=&st1=.
7. Letter from McGeorge Bundy to the Honorable James E. Webb, April 5, 1963, NSF, Box 307, JFKL.
8. Memorandum from John F. Kennedy to the Vice President, April 9, 1963, POF, Box 84, JFKL. The content and language of this memorandum suggests that it was drafted by the staff of the Space Council, who may have been unhappy that the Council had been bypassed by Kennedy's earlier request to James Webb for information to help defend Apollo.
9. Letter from Arnold Fritsch, Technical Assistant to the Chairman, AEC to Edward Welsh, April 24, 1963, and Memorandum from Robert Packard, Office of International Scientific Affairs, Department of State to Executive Secretary of the National Aeronautics and Space Council, April 24, 1963, POF, Box 84, JFKL.

10. Letter from Jerome Wiesner to the Vice-President, April 29, 1963, Wiesner Files, Box 11, JFKL. Memorandum from Robert S. McNamara to the Vice President, "National Space Program," May 3, 1963, POF, Box 84, JFKL.
11. Letter from James E. Webb to the Vice President, May 3, 1963, NHRC, Folder 012518, and Memorandum from James E. Webb to the Vice President, May 10, 1963, POF, Box 84, JFKL.
12. National Aeronautics and Space Council, "Summary of Meeting on President's Memorandum of April 9, 1963," National Aeronautics and Space Council Files, Box 2, JFKL and memorandum from Walter (Jenkins) to Dr. Welsh, April 16, 1963, White House Famous Names Files, John F. Kennedy, Box 6, LBJL.
13. Letter from Lyndon B. Johnson to the President with attached report, May 13, 1963, POF, Box 84, JFKL.
14. Wiesner's conversation with the President is summarized in a memorandum prepared by BOB official Willis Shapley and sent from the head of the BOB Military Division to the Director of the Bureau of the Budget, "Space Council Reply to the President's Questions of April 9, 1963," May 16, 1963, NSF, Box 307, JFKL.
15. Ibid.
16. Bundy's question to Charles Johnson is handwritten on a copy of a Memorandum from E. C. Welsh to the Honorable McGeorge Bundy, May 17, 1963. Johnson's reply is in a Memorandum from Charles E. Johnson to Mr. Bundy, "Space Council's Reply to Questions of April 9, 1963," May 21, 1963. Both documents are in NSF, Box 307, JFKL.
17. The report is quoted in Robert C. Toth, "G.O.P. Questions Urgency on Space," *The New York Times*, May 11, 1963, 1, 6. Eisenhower's comment is in John W. Finney, "NASA Loses Chief of Moon Project," *The New York Times*, June 13, 1963, 1, 10.
18. Webb and Johnson's remarks and the *Life* and *Aviation Week* editorials are quoted in NASA, *Astronautics and Aeronautics, 1963*, 180, 192, 200, 209–210, 215–216, 226. Also, Walter Sullivan, "Manned Moon Flight Supported in 8 Scientists' Retort to Critics," *The New York Times*, May 27, 1963, 1–2.
19. Senate Committee on Aeronautical and Space Sciences, *Scientists' Testimony on Space Goals*, Hearings, 88/1, June 10–11, 1963. Excerpts from the hearings can be found in NASA, *Astronautics and Aeronautics, 1963*, 237–238.
20. Francis Vivian Drake, "We're Running the Wrong Race with Russia!" *Reader's Digest*, August 1963, 49–55.
21. Memorandum from JFK to Secretary McNamara and Jim Webb, July 22, 1963, NSF, Box 308, JFKL, and John F. Kennedy, Memorandum to the Vice President, July 29, 1963, National Aeronautics and Space Council Files, Box 2, JFKL.
22. Memorandum from James Webb for the Vice President, July 30, 1963, and Memorandum from Roswell Gilpatric for the President, "Reader's Digest Article on Space," July 31, 1963, both in NSF, Box 308, JFKL.
23. National Aeronautics and Space Council, "Summary—Meeting Re the Lunar Program," July 31, 1963, and Letter from Lyndon B. Johnson to the President, July 31, 1963, National Aeronautics and Space Council Files, Box 2, JFKL.

24. Memorandum from James E. Webb for the President, "*Reader's Digest* Article on Space," August 9, 1963, NHRC, Folder 012518; James E. Webb Oral History Interview I, April 29, 1969, by T. H. Baker, Internet Copy, LBJL.
25. Memorandum of Conversation, August 26, 1963, 11:00 a.m, Document 350, Department of State, *Foreign Relations of the United States, 1961–1963,* Volume V, Soviet Union (Washington, DC: Government Printing Office, 1998 ), 751–752.
26. *Aviation Week and Space Technology*, March 18, 1963, 30.
27. Memorandum from Jerome Wiesner for the President, "The Unmanned Program in Support of NASA," August 2, 1963, Wiesner Files, Box 11, JFKL.
28. This account of the appropriations process is based on information in NASA, *Astronautics and Aeronautics, 1963*, 316–440.
29. Interview of Robert Seamans by NASA Historian Eugene Emme, May 8, 1968, NHRC, Folder 3622 and Oral History Interview of Robert Seamans, March 27, 1964, JFKL.
30. Memorandum from James Webb for the President, June 11, 1963, POF, Box 84, JFKL.
31. John Finney, "Manned Test Flight Lags 9 Months in Moon Project," *The New York Times*, September 1, 1963, 1; "Off Target," *Newsweek*, September 23, 1963, 73.
32. John Disher and Del Tischler, "Apollo Cost and Schedule Evaluation" in Logsdon with Launius, *Exploring the Unknown, Vol.* VII, 609. On the importance of Mueller's "all up" approach, see Murray and Cox, *Apollo: The Race to the Moon*, 155–162.
33. John Finney, "Space Debate Grows Sharper," *The New York Times*, October 6, 1963, 191.

### Chapter 13 Were Changes in the Wind?

1. Letter from Olin E. Teague to the Honorable John F. Kennedy, September 23, 1963, WHCF, Box 652, JFKL.
2. Letter from McGeorge Bundy to The Honorable Olin Teague, October 4, 1963, NSF, Box 404A, JFKL.
3. Meeting between President Kennedy and Secretary of state Dean Rusk, October 3, 1963, Tape 114/A49, Presidential Recordings, JFKL; James E. Webb Oral History Interview 1, April 29, 1969, by T. H. Baker, Internet Copy, LBJL.
4. Harold Brown (Director of Defense Research and Engineering), Memorandum for the Record, October 8, 1963, NSF, Box 308, JFKL.
5. Memorandum from W. F. Boone to Dr. [Robert] Seamans, "Joint NASA-DOD Review of National Space Program," January 7, 1964, NHRC, Folder 3802.
6. John Finney, "Soviet Bars Race with U.S. to Land Men on the Moon," *The New York Times*, October 27, 1963, 1 and John Finney, "U.S. Taking a Fresh Look at Its Lunar Program," *The New York Times*, November 3, 1963, E5.
7. McGeorge Bundy, Memorandum for the President, "Your 11:00 a.m. Appointment with Jim Webb," September 18, 1963, POF, Box 84, JFKL.
8. Memorandum from Charles Johnson to Mr. [McGeorge] Bundy, "Moon Race," October 29, 1963 with attached Memorandum for the Record

prepared by Julian Scheer, October 29, 1963. Scheer's memorandum contained the October 26 statement as Attachment A and the fuller statement as Attachment B. It is not clear that the fuller statement was ever issued by NASA. The document is in NSF, Box 308, JFKL.

9. Dryden by this time had been diagnosed with what turned out to be a fatal cancer and was often away from NASA undergoing medical treatment.
10. Memorandum for the Record from W.H. Shapley, "NASA's suggestion to Mr. Bundy on position on manned lunar landing program at President's news conference," October 30, 1963, NHRC, Folder 012457.
11. Memorandum from Charles Johnson for Mr. Bundy, October 30, 1963, NSF, Box 308, JFKL.
12. The President's News Conference, October 31, 1963, http://www.presidency.ucsb.edu/ws/index.php?pid=9507&st=&st1=
13. The description of President Kennedy's visit to Cape Canaveral is drawn from an oral history interview with Robert Seamans, March 27, 1964, JFKL and Marjorie Hunter, "President, Touring Canaveral, Sees a Polaris Fired," *The New York Times*, November 17, 1963, 1.
14. Letter from Olin E. Teague to NASA Historian Eugene Emme, January 24, 1979, NHRC, Folder 012499.
15. Roger D. Launius, "What Are Turning Points in History, and What Were They for the Space Age?" in Steven J. Dick and Roger D. Launius, eds. *Societal Impact of Spaceflight*, NASA SP2007-4801 (Washington, DC: Government Printing Office, 2007), 35; "Text of Kennedy's Address in Fort Worth and of His Undelivered Dallas Speech," *The New York Times*, November 24, 1963, 12.
16 Memorandum from Associate Administrator [Robert Seamans] to the Deputy Associate Administrator for Manned Space Flight, "Earth-orbit Program in Support of Manned Lunar Landing," October 15, 1963 and Memorandum for Dr. [Robert] Seamans from W. H. Boone, "National Manned Space Flight Program," November 21, 1963, both in NHRC, Folder 3802.
17. Memorandum from W. H. Shapley to Dr. [Robert] Seamans with revised draft of the introduction and manned lunar landing section of the space report, November 13, 1963, NHRC, Folder 10668.
18. Memorandum from W.H. Shapley to Dr. [Robert] Seamans, November 22, 1963, with attached November 20 draft of Special Space Review, NHRC, Folder 10668.
19. Bureau of the Budget, "Special Space Review—Draft Report," November 29, 1963, NSF, Box 307, JFKL.
20. For a summary of the November 30 meeting, see Associate Administrator [Robert Seamans], "Memorandum for the File—National Space Program, December 2, 1963, NHRC, Folder 10668, and W.H. Shapley, "Memorandum for the Record, " December 2, 1963, NHRC, Folder 10668.
21. John F. Kennedy, "Remarks in San Antonio at the Dedication of the Aerospace Medical Health Center," November 21, 1963, http://www.presidency.ucsb.edu/ws/index.php?pid=9534&st=&st1=.

## Chapter 14 John F. Kennedy and the Race to the Moon

1. The renaming of the geographical area in Florida was reversed in 1973 to return to the historical designation Cape Canaveral. Johnson's hastily drafted

November 29, 1963, Executive Order 11129 applied to both the NASA and the Air Force facilities at the Cape; NASA administrator on December 20, 1963, clarified that only the former NASA Launch Operations Center would be called NASA's John F. Kennedy Space Center.

2. Schlesinger is quoted in *Congressional Record*, 94th Congress, 2nd Session, September 30, 1976, p. H11946.
3. For a comprehensive and thoughtful attempt to understand the historical significance of Apollo, see the new study by Roger D. Launius, *After Apollo: The Legacy of the American Moon Landings* (New York: Oxford University Press, 2011).
4. Burton I. Kaufman, "John F. Kennedy as World Leader: A Perspective on the Literature," *Diplomatic History*, Vol. 17, No. 3 (Summer 1993), 469.
5. Quoted material here and in the rest of this chapter is drawn from the previous chapters, unless otherwise noted. I will not repeat here the specific references to such quoted material.
6. I have drawn heavily from my analysis in *The Decision to Go to the Moon* in preparing the current chapter. See also Roger D. Launius, "Interpreting the Moon Landings: Project Apollo and the Historians," *History and Technology*, Vol. 22 (September 2006), 228 for his assessment of my point of view.
7. Schlesinger, *A Thousand Days*, 276.
8. Hugh Sidey, writing in *Time*, November 14, 1983, 69.
9. Memorandum from McG. B. (McGeorge Bundy) to Mr. Sorensen, "Decisions in the White House," March 8, 1963, NSF, Box 327, JFKL.
10. James N. Giglio, "John F. Kennedy and the Nation" in Giglio and Rabe, *Debating the Kennedy Presidency*, 134.
11. Roger D. Launius, "Kennedy's Space Policy Reconsidered: A Post-Cold War Perspective," *Air Power History*, Winter 2003, 18–19 and Roger D. Launius, "Interpreting the Moon Landings: Project Apollo and the Historians," *History and Technology*, Vol. 22 (September 2006), 227.
12. See Linda Krug, *Presidential Perspectives on Space Exploration: Guiding Metaphors from Eisenhower to Bush* (New York: Praeger, 1991) for a discussion of Kennedy's language as he justified Apollo.
13. Van Dyke, *Pride and Power*, 137 and Chapter 9. On soft power, see Joseph S. Nye, Jr., *The Paradox of American Power: Why the World's Only Superpower Can't Go It Alone* (New York: Oxford University Press, 2002), 4–9.
14. John F. Kennedy, "Address at Rice University in Houston on the Nation's Space Effort," http://www.presidency.ucsb.edu/ws/index.php?pid=8862&st=&st1=.
15. Michael R. Beschloss, "Kennedy and the Decision to Go to the Moon," in Launius and McCurdy, *Spaceflight and the Myth of Presidential Leadership*, 56, 60–61, 63.
16. John F. Kennedy, "Commencement Address at American University," June 10, 1963 at http://www.presidency.ucsb.edu/ws/index.php?pid=9266&st=&st1=
17. McDougall, *Heavens and Earth*, 305–306.
18. Eggers and O'Leary, *If We Can Put a Man on the Moon*, xi.
19. Murray and Cox, *Project Apollo*, provide the best account of the many NASA managers and employees who made Apollo successful.
20. Logsdon, *Decision*, 181, 178.
21. F. S. Oliver, *The Endless Adventure*, Volume II, quoted in Giles Alston, *International Prestige and the American Space Program*, unpublished doctoral

dissertation, St. Antony's College, Oxford University, 1989, 8–9. A copy of Alston's thesis, which is an excellent analysis of the links between space achievement and national prestige, can be found in the NASA Headquarters Library in Washington, DC.

22. U. Alexis Johnson as quoted in Alston, *International Prestige and the American Space Program,* 258–259.
23. Stephen Rabe, "John F. Kennedy and the World" in Giglio and Rabe, *Debating the Kennedy Presidency,* 7.
24. See Sidiqqi, *Challenge to Apollo* and John Logsdon and Alain Dupas, "How Real was the Race to the Moon?" *Scientific American*, June 1994, for a discussion of the Soviet lunar program. See also *Dwayne* A. *Day* and Asif A. Siddiqi, "The *Moon in the Crosshairs*: CIA Intelligence on the Soviet Manned Lunar Programme," Spaceflight 45 (2003), 466–475.
25. Walter A. McDougall, "Apollo and Technocracy" in Space Policy Institute, The George Washington University, Washington, DC, *Apollo in Its Historical Context,* April 1990, 12. Charles Krauthammer, "The Moon We Left Behind," *The Washington Post,* July 17, 2009, available at http://www.washingtonpost.com/wp-dyn/content/linkset/2005/03/24/LI2005032401690_1.html.
26. Richard M. Nixon, "Statement about the Future of the United States Space Program," March 7, 1970, http://www.presidency.ucsb.edu/ws/index.php?pid=2903&st=&st1=
27. Richard S. Lewis, "The Kennedy Effect," *Bulletin of the Atomic Scientists,* March 1968, 2.
28. Columbia Accident Investigation Board, "Report," Volume 1, August 2003, 209. The report can be accessed at http://caib.nasa.gov/.
29. Garry D. Brewer, "Perfect Places: NASA as an Idealized Institution" in Radford Byerly, Jr., ed., *Space Policy Reconsidered* (Boulder, CO: Westview Press, 1989), 157–173.
30. Daniel Boorstin, "The Rise of Public Discovery" in Space Policy Institute, *Apollo in Its Historical Context,* 21. The global televising of Neil Armstrong's first steps on the Moon was possible because the International Telecommunications Satellite Consortium had completed its global satellite network just a few weeks before the lunar landing.
31. McLeish's poem appeared in *The New York Times,* December 25, 1968, 1.

# Selected Bibliography


Baker, Leonard. *The Johnson Eclipse: A President's Vice Presidency*. New York: The MacMillan Company, 1966.

Benson, Charles, and William B. Faherty. *Gateway to the Moon: Building the Kennedy Space Center Launch Complex* (Gainesville: University Press of Florida), 2001.

Beschloss, Michael R. *The Crisis Years: Kennedy and Khrushchev, 1960–1963*. New York: HarperCollins, 1991.

Bilstein, Roger E. *Stages to Saturn: A Technological History of the Apollo/Saturn Launch Vehicles*, NASA SP-4206. Washington, DC: Government Printing Office, 1980.

Bizony, Piers. *The Man Who Ran the Moon: James E. Webb, NASA, and the Secret History of Project Apollo*. New York: Thunder's Mouth Press, 2006.

Brooks, Courtney G., James M. Grimwood, and Loyd S. Swenson, Jr., *Chariots for Apollo: A History of Manned Lunar Spacecraft*, NASA SP-4205. Washington, DC: Government Printing Office, 1979.

Bzdek, Vincent. *The Kennedy Legacy: Jack, Bobby and Ted and a Family Dream Fulfilled*. New York: Palgrave Macmillan, 2009

Dallek, Robert. *An Unfinished Life: John F. Kennedy, 1917–1963*. Boston: Little Brown and Company, 2003.

Day, Dwayne A., John M. Logsdon, and Brian Latell, eds. *Eye in the Sky: The Story of the CORONA Spy Satellites*. Washington, DC: Smithsonian Institution Press, 1998

Dethloff, Henry C. *Suddenly, Tomorrow Came: A History of the Johnson Space Center*, NASA SP-4307. Washington, DC: Government Printing Office, 1993.

Diamond, Edwin. *The Rise and Fall of the Space Age*. Garden City, NY: Doubleday & Company, 1964.

Dick, Steven J. and Roger D. Launius, eds. *Societal Impact of Spaceflight*. NASA SP2007-4801, Washington: Government Printing Office, 2007.

Divine, Robert A. *The Sputnik Challenge: Eisenhower's Response to the Soviet Satellite*. New York: Oxford University Press, 1993.

Evans, Rowland, and Robert Novak, *Lyndon B. Johnson: The Exercise of Power*. New York: New American Library, 1966.

Frutkin, Arnold. *International Cooperation in Space*. Englewood Cliffs, NJ: Prentice-Hall, 1965.

Fursenko, Aleksandr and Timothy Naftali, *One Hell of a Gamble: Khrushchev, Castro and Kennedy, 1958–1964*. New York: Norton, 1997.

Giglio, James N. and Stephen G. Rabe. *Debating the Kennedy Presidency*. Lanham, MD: Rowman & Littlefield Publishers, Inc., 2003.

Glennan, T. Keith. *The Birth of NASA: The Diary of T. Keith Glennan.* NASA SP-4105, Washington, DC: Government Printing Office, 1993.

Griffith, Alison. *The National Aeronautics and Space Act: A Study of the Development of Public Policy.* Washington, DC: Public Affairs Press, 1962.

Harvey, Dodd, and Linda Ciccoritti. *U.S.-Soviet Cooperation in Space.* Washington, DC: Center for International Studies, University of Miami, 1974.

Hilsman, Roger. *To Move a Nation: The Politics of Foreign Policy in the Administration of John F. Kennedy.* Garden City, NY: Doubleday & Company, 1967.

Holmes, Jay. *Americans on the Moon: The Enterprise of the 60's.* Philadelphia: J.B. Lippincott Company, 1962

Killian, James R. Jr. *A Scientist in the White House.* Cambridge, MA: The MIT Press, 1977.

Kistiakowsky, George B. *A Scientist at the White House: The Private Diary of President Eisenhower's Special Assistant for Science and Technology.* Cambridge, MA: Harvard University Press, 1976.

Krug, Linda. *Presidential Perspectives on Space Exploration: Guiding Metaphors from Eisenhower to Bush.* New York: Praeger, 1991.

Lambright, W. Henry. *Powering Apollo: James E. Webb of NASA.* Baltimore, MD: Johns Hopkins University Press, 1995.

Launius, Roger D., and Howard E. McCurdy, eds. *Spaceflight and the Myth of Presidential Leadership.* Urbana, IL: University of Illinois Press, 1997.

———, John M. Logsdon, and Robert W. Smith, eds. *Reconsidering Sputnik: Forty Years Since the Soviet Satellite.* Amsterdam: Harwood Academic Publishers, 2000.

———. *After Apollo: The Moon Landings in American Memory.* New York: Oxford University Press, 2011.

Lincoln, Evelyn. *My Twelve Years with John F. Kennedy.* New York: David McKay Company, 1965.

Logsdon, John M. *The Decision to Go to the Moon: Project Apollo and the National Interest.* Cambridge, MA: The MIT Press, 1970.

———, et al., eds. *Exploring the Unknown: Selected Documents in the History of the U.S. Civil Space Program.* Volume I: Organizing for Exploration, NASA SP-4407, Washington, DC: Government Printing Office, 1995.

———, Dwayne Day, and Roger Launius, eds. *Exploring the Unknown: Selected Documents in the History of the U.S. Civil Space Program.* Volume II: External Relationships, NASA SP-4407, Washington, DC: Government Printing Office, 1996.

———, ed., with Roger D. Launius. *Exploring the Unknown: Selected Documents in the History of the U.S. Civil Space Program.* Volume VII: Human Spaceflight: Mercury, Gemini, and Apollo, NASA SP-2008-4407, Washington, DC: Government Printing Office, 2008.

McDougall, Walter A. *. . . the Heavens and the Earth: A Political History of the Space Age.* New York: Basic Books, 1985.

Miller, Merle. *Lyndon: An Oral Biography.* New York: G.P. Putnam's Sons, 1980.

Mindell, David A. *Digital Apollo: Human and Machine in Spaceflight.* Cambridge, MA: The MIT Press, 2008.

Murphy, Thomas P. *Science, Geopolitics, and Federal Spending.* Lexington, MA: D.C. Heath and Company, 1971.

Murray, Charles and Catherine Bly Cox. *Apollo: Race to the Moon.* New York: Simon & Schuster, 1989.

Neufeld, Michael. *Von Braun: Dreamer of Space, Engineer of War.* New York: Alfred P. Knopf, 2007.
Neustadt, Richard E. *Presidential Power: The Politics of Leadership.* New York: Wiley, 1960.
Perret, Geoffrey. *Jack: A Life Like No Other.* New York: Random House, 2001.
Reeves, Richard. *President Kennedy: Profile of Power.* New York: Simon & Schuster, 1993.
Reeves, Thomas C. *A Question of Character: A Life of John F. Kennedy.* New York: The Free Press, 1991.
Rosholt, Robert L. *An Administrative History of NASA, 1958–1963.* Washington, DC: Government Printing Office, 1966.
Salinger, Pierre. *With Kennedy.* Garden City, NY: Doubleday & Company, 1966.
Schlesinger, Arthur M. Jr. *A Thousand Days: John F. Kennedy in the White House.* Boston, MA: Houghton Mifflin, 1965.
Seaborg, Glenn T., with the assistance of Benjamin S. Loeb. *Kennedy, Khrushchev, and the Test Ban.* Berkeley, CA: University of California Press, 1981.
Seamans, Robert C. Jr. *Aiming at Targets: The Autobiography of Robert C. Seamans, Jr.* NASA SP-4106, Washington, DC: Government Printing Office, 1996.
———. *Project Apollo: The Tough Decisions.* NASA Monographs in Aerospace History No. 31, SP-2005-4537, Washington, DC: Government Printing Office, 2005.
Sheehan, Neil. *A Fiery Peace in the Cold War: Bernard Schriever and the Ultimate Weapon.* New York: Random House, 2009.
Sidey, Hugh. *John F. Kennedy, President.* New York: Atheneum, 1964.
Siddiqi, Asif A. *Challenge to Apollo: the Soviet Union and the Space Race, 1945–1974.* NASA SP-2000-4408, Washington, DC: Government Printing Office, 2000.
Skolnikoff, Eugene B. *Science, Technology, and American Foreign Policy.* Cambridge, MA: The MIT Press, 1967.
Sorensen, Theodore C. *Decision-Making in the White House: The Olive Branch or the Arrows.* New York: Columbia University Press, 1963.
———. *Kennedy.* New York: Harper & Row, 1965.
———. *Counselor: A Life at the Edge of History.* New York: HarperCollins, 2008.
Spires, David N. *Beyond Horizons: A Half Century of Air Force Space Leadership,* Revised Edition, Colorado Springs, CO: Air University Press, 1998.
Strober, Deborah Hart, and Gerald S. Strober. *The Kennedy Presidency: An Oral History of an Era.* Washington, DC: Brassey's, Inc., 1993.
Swenson, Loyd Jr., James M. Grimwood, and Charles C. Alexander. *This New Ocean: A History of Project Mercury.* NASA SP-4201, Washington, DC: Government Printing Office, 1966.
Thompson, Neal. *Light This Candle: The Life & Times of Alan Shepard, America's First Spaceman.* New York: Crown Publishers, 2004.
U.S. House of Representatives, Committee on Science and Astronautics. *Toward the Endless Frontier: History of the Committee on Science and Technology, 1959–1979.* Washington, DC: Government Printing Office, 1980.
Van Dyke, Vernon. *Pride and Power: The Rationale of the Space Program.* Urbana, IL: University of Illinois Press, 1964.
White, Theodore H. *The Making of the President, 1960.* New York: Atheneum Publishers, 1961.
Woods, Randall B. *LBJ: Architect of American Ambition.* New York: Simon and Schuster, 2006.

# Index